Lecture Notes in Computer Science 4169

Commenced Publication in 1973
Founding and Former Series Editors:
Gerhard Goos, Juris Hartmanis, and Jan van Leeuwen

Hans L. Bodlaender   Michael A. Langston (Eds.)

Parameterized and Exact Computation

Second International Workshop, IWPEC 2006
Zürich, Switzerland, September 13-15, 2006
Proceedings

 Springer

Volume Editors

Hans L. Bodlaender
Utrecht University
Department of Information and Computing Sciences
P.O. Box 80089, 3508 TB Utrecht, The Netherlands
E-mail: hansb@cs.uu.nl

Michael A. Langston
University of Tennessee
Department of Computer Science
203 Claxton Complex, 1122 Volunteer Boulevard, Knoxville, TN 37996-3450, USA
E-mail: langston@cs.utk.edu

Library of Congress Control Number: 2006931895

CR Subject Classification (1998): F.2, F.1, E.1, I.3.5, G.2

LNCS Sublibrary: SL 1 – Theoretical Computer Science and General Issues

ISSN	0302-9743
ISBN-10	3-540-39098-7 Springer Berlin Heidelberg New York
ISBN-13	978-3-540-39098-5 Springer Berlin Heidelberg New York

Springer is a part of Springer Science+Business Media

springer.com

© Springer-Verlag Berlin Heidelberg 2006
Printed in Germany

Typesetting: Camera-ready by author, data conversion by Scientific Publishing Services, Chennai, India
Printed on acid-free paper   SPIN: 11847250   06/3142   5 4 3 2 1 0

Preface

The Second International Workshop on Parameterized and Exact Computation (IWPEC) was held in Zürich, Switzerland, during September 13–15, 2006. It was organized as a component of ALGO 2006, which also hosted the 14th Annual European Symposium on Algorithms, the 6th Workshop on Algorithms in Bioinformatics, the 4th Workshop on Approximation and Online Algorithms, and the 6th Workshop on Algorithmic Methods and Models for Optimization of Railways.

This meeting was the second in the IWPEC series, with the first having been held in Bergen, Norway, during September 14–16, 2004. The field continues to experience rapid growth, in part due to its appeal as an alternative to traditional complexity theory, and in part due to the powerful practical applications it has spawned. IWPEC events are intended to cover research in all aspects of parameterized and exact computation and complexity, including but not limited to new techniques for the design and analysis of parameterized and exact algorithms, parameterized complexity theory, relationships between parameterized complexity and traditional complexity, applications of parameterized and exact computation, implementation issues and high-performance computing. A major goal is to disseminate the latest research results, including significant work-in-progress, and to identify, define and explore directions for future study.

The papers accepted for presentation and printed in these proceedings represent a diverse spectrum of the latest developments on parameterized and exact algorithm design, analysis, application and implementation. We hope that you will read them, and that you find the time spent a rewarding experience. Each submission was thoroughly reviewed by at least three members of the IWPEC 2006 Program Committee. We are certain that many of them will find their way to archival journal publication in more complete and polished form. We wish to thank all authors for contributing their work for review. Many more meritable papers were submitted than can be accommodated in the schedule. In addition, three invited lectures were given by leading experts in the field of parameterized and exact computation: Frank Dehne, Uwe Schöning, and Michael Fellows.

It has been a privilege to serve as Program Committee Co-chairs. Assembling this slate of first-rate papers would not have been possible without the tireless and professional efforts of the remainder of the IWPEC 2006 Program Committee, who are:

Jianer Chen (USA)
Frank Dehne (Canada)
Erik D. Demaine (USA)
Rodney G. Downey (New Zealand)
Michael R. Fellows (Australia)

Henning Fernau (UK)
Jörg Flum (Germany)
Fedor V. Fomin (Norway)
Martin Grohe (Germany)
Edward A. Hirsch (Russia)
Kazuo Iwama (Japan)
Dániel Marx (Germany)
Catherine McCartin (New Zealand)
Naomi Nishimura (Canada)
Venkatesh Raman (India)
Peter Rossmanith (Germany)
Uwe Schöning (Germany)
Ulrike Stege (Canada)
Jan Arne Telle (Norway)
Dimitrios M. Thilikos (Spain)
Sue Whitesides (Canada)
Gerhard J. Woeginger (The Netherlands)

We also wish to acknowledge the assistance of the numerous external reviewers who have been an immense technical help during committee deliberations.

Zürich, Switzerland   Hans Bodlaender and
September 2006   Michael A. Langston

Organization

Referees

Peter Brass
Yijia Chen
Frederic Dorn
John D. Eblen
Serge Gaspers
Magdalene Grantson
Magdalena Grüber
Magnus Halldorsson
Michael Kaufmann
Joachim Kneis
Stephan Kreutzer
Alexander Kulikov
Yury Lifshits
Songjian Lu
Daniel Mölle
Catherine Mccartin
Jie Meng
Moritz Müller
Andy D. Perkins
Konstantin Pervyshev
Artem Pyatkin
Mark Ragan
Stefan Richter
Saket Saurabh
Somnath Sikdar
Yngve Villanger
Mark Weyer
Fenghui Zhang
Yun Zhang

Table of Contents

Applying Modular Decomposition to Parameterized Bicluster Editing

Fábio Protti[1], Maise Dantas da Silva[2],
and Jayme Luiz Szwarcfiter[3]

[1] Instituto de Matemática and Núcleo de Computação Eletrônica
Universidade Federal do Rio de Janeiro
Caixa Postal 2324, 20001-970, Rio de Janeiro, RJ, Brasil
fabiop@nce.ufrj.br
[2] COPPE-Sistemas
Universidade Federal do Rio de Janeiro
Caixa Postal 68511, 21945-970, Rio de Janeiro, RJ, Brasil
maiseds@cos.ufrj.br
[3] Instituto de Matemática, Núcleo de Computação Eletrônica and COPPE-Sistemas
Universidade Federal do Rio de Janeiro
Caixa Postal 68511, 21945-970, Rio de Janeiro, RJ, Brasil
jayme@nce.ufrj.br

Abstract. A graph G is said to be a *cluster graph* if G is a disjoint union of cliques (complete subgraphs), and a *bicluster graph* if G is a disjoint union of bicliques (complete bipartite subgraphs). In this work, we study the parameterized version of the NP-hard BICLUSTER GRAPH EDITING problem, which consists of obtaining a bicluster graph by making the minimum number of modifications in the edge set of an input bipartite graph. When at most k modifications are allowed in the edge set of any input graph (BICLUSTER(k) GRAPH EDITING problem), this problem is FPT, solvable in $O(4^k m)$ time by applying a search tree algorithm. It is shown an algorithm with $O(4^k + n + m)$ time, which uses a new strategy based on modular decomposition techniques. Furthermore, the same techniques lead to a new form of obtaining a problem kernel with $O(k^2)$ vertices for the CLUSTER(k) GRAPH EDITING problem, in $O(n+m)$ time. This problem consists of obtaining a cluster graph by modifying at most k edges in an input graph. A previous FPT algorithm for this problem was presented by Gramm *et al.* [11]. In their solution, a problem kernel with $O(k^2)$ vertices and $O(k^3)$ edges is built in $O(n^3)$ time.

Keywords: NP-complete problems, fixed-parameter tractability, edge modification problems, cluster graphs, bicluster graphs.

1 Introduction

Many NP-hard problems can be formulated with a parameter k, so that polynomial-time algorithms can be designed for them when k is fixed. The parameterized complexity theory was developed by Downey and Fellows [6,7], as an alternative to deal with such problems. They defined the class of *fixed-parameter*

H.L. Bodlaender and M.A. Langston (Eds.): IWPEC 2006, LNCS 4169, pp. 1–12, 2006.

tractable (FPT) problems, which admit algorithms of complexity $O(f(k)n^\alpha)$, where f is an arbitrary function and α is a constant independent of both n and k. They also defined a hierarchy of parameterized decision problem classes, $FPT \subseteq W[1] \subseteq W[2] \subseteq \cdots$, with appropriate reducibility and completeness notions, and conjectured that each of the containments in this hierarchy is proper. More details about this theory can be found in [6,7,8,9].

Let $u, v \in V(G)$. An *edge modification* or *edge edition* with respect to u, v is either the deletion of (u, v) if $(u, v) \in E(G)$, or the addition of (u, v) if $(u, v) \notin E(G)$.

In this paper, we study the parameterized version of the BICLUSTER GRAPH EDITING problem. In the optimization version, this problem consists of editing the minimum number of edges in a bipartite graph so that it becomes a vertex-disjoint union of bicliques (complete bipartite subgraphs), called *bicluster graph*. The NP-hardness of this problem was proved by Amit [1]. The parameterized version of this problem is the BICLUSTER(k) GRAPH EDITING problem, whose goal is to obtain a bicluster graph by editing at most k edges from any input graph (not necessarily bipartite). This problem is FPT, solvable in $O(4^k m)$ time by applying a search tree algorithm. We propose an $O(4^k + n + m)$ time algorithm, which works in two stages: firstly, a problem kernel is built, using a new strategy based on modular decomposition. Following, a bounded seach tree is applied.

By using this strategy based on modular decomposition, we propose a new form of obtaining a problem kernel for the CLUSTER(k) GRAPH EDITING problem, whose optimization version consists of editing the minimum number of edges from a graph so that it becomes a vertex-disjoint union of cliques (complete subgraphs), called *cluster graph*. This problem were studied by Shamir *et al.* [19] and proved to be NP-hard. The tractability of the parameterized version proceeds directly from Cai's result [4]. From the more general view of graph modification problems, Cai proved the fixed-parameter tractability of deciding whether an input graph can be transformed into a graph with a specified hereditary property by deleting vertices and/or edges, and adding edges, when the hereditary property can be characterized by a finite set of forbidden induced subgraphs. His result provides an $O(3^k n^4)$ time algorithm for CLUSTER(k) GRAPH EDITING. In [11], Gramm *et al.* present an $O(2.27^k + n^3)$ time algorithm, which builds a problem kernel with $O(k^2)$ vertices and $O(k^3)$ edges in $O(n^3)$ time, and then applies a bounded search tree in $O(2.27^k)$ time. In [12], the time complexity of the search tree algorithm was improved to $O(1.92^k)$ time. We propose an algorithm that builds a problem kernel with $O(k^2)$ vertices in $O(n + m)$ time. For a more detailed study on edge modification problems, see [17].

This paper is organized as follows. Section 2 contains the definitions and notation used in the paper. In Section 3, we provide an overview about modular decomposition of graphs and propose some definitions used along this paper. In Section 4, the new FPT solution for BICLUSTER(k) GRAPH EDITING is explained. Finally, Section 5 presents how to obtain a problem kernel for CLUSTER(k) GRAPH EDITING in $O(n + m)$ time.

2 Basic Definitions and Notation

In this work, G denotes a finite graph, without loops nor multiple edges. The vertex set and the edge set of G are denoted by $V(G)$ and $E(G)$, respectively. Assume $|V(G)| = n$ and $|E(G)| = m$.

A *clique* is a complete subgraph. A *cluster graph* is a vertex-disjoint union of cliques. It is easy to see that G is a cluster graph if and only if it contains no P_3 as an induced subgraph.

A *biclique* is a complete bipartite subgraph. A *bicluster graph* is a vertex-disjoint union of bicliques. This class of graphs is the intersection of two other classes, bipartite graphs and cographs (graphs containing no P_4 as an induced subgraph), inheriting therefore their forbidden induced subgraphs. Thus, G is a bicluster graph if and only if it contains no P_4 nor C_{2k+1} as induced subgraphs.

An *edge modification set* F is a set of pairs of vertices, where each pair has a mark $+$ or $-$ such that:

$+(a, b)$ represents the addition of the edge (a, b),
$-(a, b)$ represents the deletion of the edge (a, b).

$G + F$ represents the addition to G of the edges (a, b) marked by $+(a, b)$ in F (assuming that they are not in $E(G)$) and the deletion of the edges (a, b) marked by $-(a, b)$ in F (assuming that they belong to $E(G)$). Similarly, $G - F$ represents the addition of the edges marked by $-(a, b)$ in F and the deletion of the edges marked by $+(a, b)$ in F. Clearly, $G' = G + F$ if and only if $G = G' - F$.

In the remainder of this work, F denotes an edge modification set for G, and G' denotes the graph $G + F$.

3 Modular Decompositions of Graphs

Important references for this section are [2,5,10,13,14,15,16].

The subset $M \subseteq V(G)$ is a *module* in G if for all $u, v \in M$ and $w \in V(G) \backslash M$, $(u, w) \in E(G)$ if and only if $(v, w) \in E(G)$.

Theorem 1. *If X and Y are disjoint modules of a graph, then either every element of X is adjacent to every element of Y ("adjacent modules"), or no element of X is adjacent to an element of Y ("nonadjacent modules").*

If $M = V(G)$ or $|M| = 1$, then M is a *trivial* module. All graphs have trivial modules. If G has no nontrivial modules, then G is called *prime*. A module M is *strong* if, for every module M', either $M \cap M' = \emptyset$ or one module is included into the other.

There exist three types of modules: *parallel*, *series* and *neighbourhood*. A module is *parallel* when the subgraph induced by its vertices is disconnected, *series* when the complement of the subgraph induced by its vertices is disconnected, and *neighbourhood* when both the subgraph induced by its vertices and its complement are connected.

The process of decomposing a graph into modules is called *modular decomposition*. The modular decomposition of G is represented by a *modular decompositon tree T_G*. The nodes of T_G correspond to strong modules of G. The root corresponds to $V(G)$, and the leaves correspond of all vertices of G. Each internal node of T_G is labeled P (parallel), S (series) or N (neighbourhood), according to the type of the module. The children of every internal node M of T_G are the maximal submodules of M. The modular decomposition tree of a graph is unique up to isomorphism and can be obtained in linear time [13].

As an important special case, modular decomposition trees containing only series and parallel internal nodes correspond precisely to the class of cographs. It is easy to see that P_4 is the smallest nontrivial prime graph.

3.1   P-Quotient and s-Quotient Graphs

Let Π be a partition of $V(G)$ such that each member of Π is a module. Then Π is said to be a *congruence partition*. The graph whose vertices are the members of Π and whose edges correspond to the adjacency relationships involving members of Π is called *quotient graph G/Π*.

In this section we define special types of auxiliary quotient graphs, namely the *p-quotient* and *s-quotient graphs*. Those graphs are obtained from specific congruence partitions, defined as follows.

Definition 2. *Let Π be a congruence partition of $V(G)$.*

*1. Π is the **p-partition** of $V(G)$ if, for every internal node M of T_G:*
- if M is labeled S or N then every leaf child of M is a unitary member of Π;
- if M is labeled P, then all leaf children of M form a member of Π.

*2. Π is the **s-partition** of $V(G)$ if, for every internal node M of T_G:*
- if M is labeled P or N, then every leaf child of M is a unitary member of Π;
- if M is labeled S then all leaf children of M form a member of Π.

Clearly, every member of the p-partition (s-partition) is a strong module in G. Since the modular decomposition tree of a graph is unique, the p-partition and the s-partition are also unique.

Definition 3. *Let Π be a congruence partition of $V(G)$.*

*1. If Π is the p-partition of $V(G)$, then G/Π is the **p-quotient** graph of G, denoted by G_p.*

*2. If Π is the s-partition of $V(G)$, then G/Π is the **s-quotient** graph of G, denoted by G_s.*

A vertex of G_p (resp. G_s) corresponding to a member of Π with size larger than one is called *p-vertex* (resp. *s-vertex*), whereas a vertex corresponding to a unitary member $\{v\}$ of Π is called *u-vertex*.

For simplicity, if a p-vertex (resp. s-vertex) corresponds to a module $M \subseteq V(G)$, then we write M to stand for both the module and the p-vertex (resp.

s-vertex); and if a u-vertex corresponds to a member $\{v\}$ of Π then we write v to stand for the u-vertex. We also say that a vertex $v \in V(G)$ *belongs* to a p-vertex $M \in V(G_p)$ (resp. s-vertex $M \in V(G_s)$) when $v \in M$.

If H is a p-quotient (resp. s-quotient) graph, denote by $V_p(H)$ (resp. $V_s(H)$) the set of p-vertices (resp. s-vertices) of H, and by $V_u(H)$ the set of u-vertices of H.

Figure 1 depicts a graph G, its modular decomposition tree T_G and the graphs G_p and G_s, where p-vertices are graphically represented by the symbol ⓟ , and s-vertices by ⓢ .

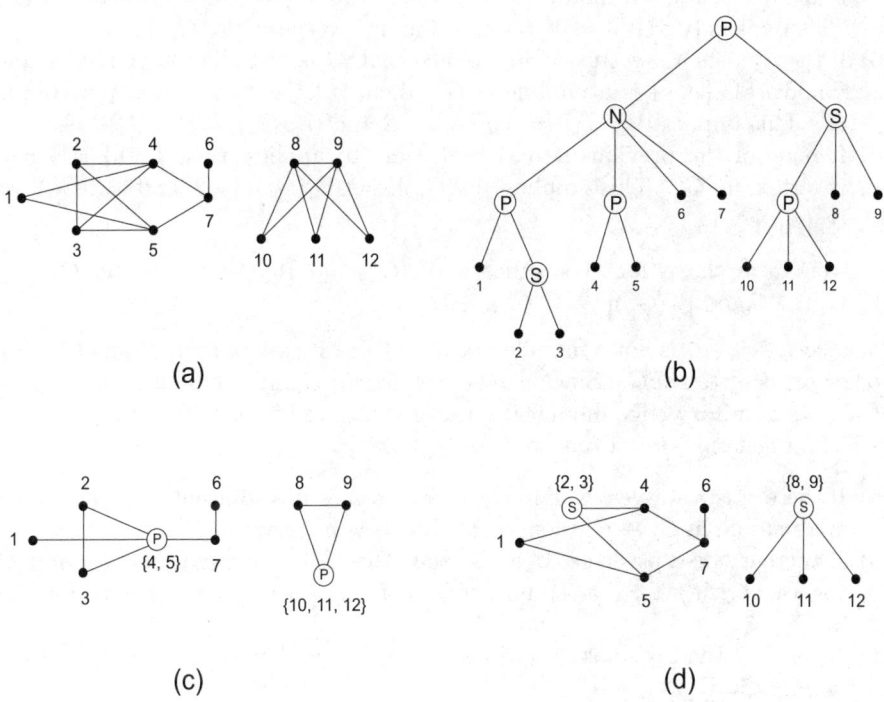

(a)   (b)   (c)   (d)

Fig. 1. (a) A graph G (b) The modular decomposition tree T_G (c) G_p (d) G_s

The next lemma presents useful bounds on the sizes of $V_u(G'_p)$, $V_p(G'_p)$ and $V(G'_p)$ for the case of one edge modification in G.

Lemma 4. *If $|F| = 1$ then the following inequalities hold:*
(1) $|V_u(G'_p)| \leq |V_u(G_p)| + 4$
(2) $|V_p(G'_p)| \leq |V_p(G_p)| + 2$
(3) $|V(G'_p)| \leq |V(G_p)| + 2$

Proof. Let (u, v) be the modified edge. The proof is based on the analysis of the local modifications made in G_p in order to obtain G'_p, by considering the new adjacency relations in G'. There are four cases, described below. We present in detail Case 1, the remaining ones are only sketched.

Case 1: u and v are u-vertices in G_p. This case is divided into the following subcases:

Subcase 1.1: $\{u, v\}$ is a module in G_p. Then u and v are adjacent in G_p, for otherwise u and v would be contained in the same p-vertex of G_p. This implies $F = \{-(u, v)\}$.

a) If there exists a u-vertex w in G_p such that w is nonadjacent to both u and v and $\{u, v, w\}$ is a module in G_p, then $\{u, v, w\}$ is a new p-vertex in G'_p. This implies $|V_u(G'_p)| = |V_u(G_p)| - 3$ and $|V_p(G'_p)| = |V_p(G_p)| + 1$.
b) If there exists a p-vertex M in G_p such that M is nonadjacent to both u and v and $M \cup \{u, v\}$ is a module in G_p, then $M \cup \{u, v\}$ is a new p-vertex in G'_p. This implies $|V_u(G'_p)| = |V_u(G_p)| - 2$ and $|V_p(G'_p)| = |V_p(G_p)|$.
c) If none of the previous situations (a) or (b) applies, then $\{u, v\}$ is a new p-vertex in G'_p. This implies $|V_u(G'_p)| = |V_u(G_p)| - 2$ and $|V_p(G'_p)| = |V_p(G_p)| + 1$.

Overall, we have for this subcase $|V_u(G'_p)| \leq |V_u(G_p)| - 2$, $|V_p(G'_p)| \leq |V_p(G_p)| + 1$ and $|V(G'_p)| \leq |V(G_p)| - 1$.

Subcase 1.2: $\{u, v\}$ is not a module in G_p. In this subcase, u and v cannot belong to a same p-vertex of G'_p. Since u, v are vertices in G_p, it will be useful to regard F also as a unitary edge modification set for G_p, and look at the graph $G_p + F$ (which in general is *not* isomorphic to G'_p).

a) If there exists a u-vertex w in G_p such that w is nonadjacent to u and $\{u, w\}$ is a module in $G_p + F$, then $\{u, w\}$ is a new p-vertex in G'_p.
b) If there exists a p-vertex M in G_p such that M is nonadjacent to u and M together with u form a module in $G_p + F$, then $M \cup \{u\}$ is a new p-vertex in G'_p.
c) If none of the previous situations (a) or (b) applies to u, then u is still a u-vertex in G'_p.

The same possibilities (a)-(c) are applicable to v. Overall, we have for this subcase $|V_u(G'_p)| \leq |V_u(G_p)|$, $|V_p(G'_p)| \leq |V_p(G_p)| + 2$ and $|V(G'_p)| \leq |V(G_p)|$.

Case 2: u is a u-vertex and v belongs to a p-vertex M in G_p. In this case u is also a u-vertex in G'_p. Write $M = \{v, v_1, \ldots, v_\ell\}$. If $\ell = 1$ then v_1 is a new u-vertex in G'_p; with respect to v, there are three possibilities: v can become a new u-vertex, v can form a new p-vertex with some u-vertex w of G_p ($w \neq u$), or v can be added to a pre-existing p-vertex M' of G_p. If $\ell > 1$ then $M \setminus \{v\}$ is a p-vertex in G'_p, and the possibilities for v are the same as in the previous situation. Overall, we have for this case $|V_u(G'_p)| \leq |V_u(G_p)| + 2$, $|V_p(G'_p)| \leq |V_p(G_p)| + 1$ and $|V(G'_p)| \leq |V(G_p)| + 1$.

Case 3: u and v belong to distinct p-vertices M and M' in G_p, respectively. Write $M = \{u, u_1, \ldots, u_\ell\}$ and $M' = \{v, v_1, \ldots, v_r\}$. Then u and v are two new u-vertices in G'_p. If $\ell = 1$ and $r = 1$, u_1 and v_1 are also two new u-vertices in G'_p. If $\ell = 1$ and $r > 1$, u_1 is a new u-vertex and $M'\backslash\{v\}$ is a p-vertex in G'_p. The situation $\ell > 1$ and $r = 1$ is similar to the previous one. Finally, if $\ell, r > 1$ then $M\backslash\{u\}$ and $M'\backslash\{v\}$ are p-vertices in G'_p. Overall, we have for this case $|V_u(G'_p)| \le |V_u(G_p)| + 4$, $|V_p(G'_p)| \le |V_p(G_p)|$ and $|V(G'_p)| = |V(G_p)| + 2$.

Case 4: u and v belong to the same p-vertex M in G_p. Write $M = \{u, v\} \cup W$. Then (u, v) is an added edge, and u, v are new u-vertices in G'_p. If $W = \{w_1\}$, then w_1 is a new u-vertex in G'_p. Otherwise, if $|W| > 1$, we have that W is a p-vertex in G'_p. Overall, we have for this case $|V_u(G'_p)| \le |V_u(G_p)| + 3$, $|V_p(G'_p)| \le |V_p(G_p)|$ and $|V(G'_p)| = |V(G_p)| + 2$.

By considering all the cases, the lemma follows.   □

Lemma 5. *If $|F| = 1$ then the following inequalities hold:*
(1') $|V_u(G'_s)| \le |V_u(G_s)| + 4$
(2') $|V_p(G'_s)| \le |V_p(G_s)| + 2$
(3') $|V(G'_s)| \le |V(G_s)| + 2$

Proof. Similar to the previous lemma.   □

4   Bicluster(k) Graph Editing

In this section, we will show a new FPT algorithm for Bicluster(k) Graph Editing. Our algorithm solve this problem in $O(4^k + n + m)$ time, and works in two stages. Firstly, a problem kernel is built, using the concepts given in the previous section. Following, a bounded search tree is applied on the problem kernel.

4.1   Building the Problem Kernel

Clearly, if a vertex v lies in a component of G which is already a biclique, then no edge modifications involving v need to be made in order to transform G into a bicluster graph. We can assume therefore that no component of the input graph G is a biclique. In this subsection, we will show that if G' is a bicluster graph, $|F| \le k$, then a problem kernel G_k with at most $4k^2 + 6k$ vertices can be built.

Bounding the number of vertices of G_p. If G' is a bicluster graph, note then that it contains at most $2k$ biclique components. In G'_p, each one can have one of the graphical representations illustrated in Figure 2. G'_p has $O(k)$ vertices, and by applying Lemma 4 k times, we get an $O(k)$ bound for the number of vertices of G_p. Lemma 6 and Theorem 7 give a more precise bounds.

Lemma 6 presents bounds on the sizes of $V_u(G_p)$, $V_p(G_p)$ and $V(G_p)$ when $|F| = 1$ and G' is a bicluster graph.

Lemma 6. *Assume $|F| = 1$. If G' is a bicluster graph then $|V_u(G_p)| \le 6$, $|V_p(G_p)| \le 4$ and $|V(G_p)| \le 6$.*

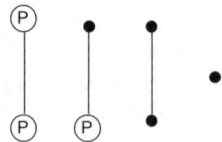

Fig. 2. Possible graphical representations of a biclique component in G'_p

Proof. Observe that $G = G' - F$. Then we can apply the inverse modification to G' and obtain G. The bounds in Lemma 4 can be used by interchanging G and G'.

G' has at most two biclique components. If G' consists of only one biclique component, the proof is immediate by Lemma 4. Otherwise, there are eight possibilities for G'_p, shown in Figure 3.

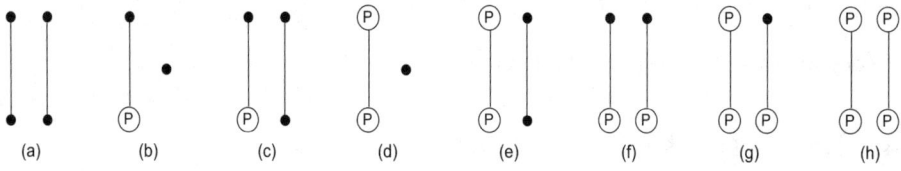

Fig. 3. Possible graphical representations of G'_p with two biclique components

Since no component of G is a biclique, it is easy to see that the vertices u and v of the modified edge lie in different bicliques of G'.

For possibility (a) in Figure 3, Subcase 1.2(c) in the proof of Lemma 4 applies for both u and v. Thus, $|V_u(G_p)| = |V_u(G'_p)| = 4$, $|V_p(G_p)| = |V_p(G'_p)| = 0$ and $|V(G_p)| = |V(G'_p)| = 4$.

For possibilities (b) to (f) in Figure 3, the proof is immediate by applying Inequalities (1) to (3) in Lemma 4.

For possibility (g) in Figure 3, we have two cases:

- u is the u-vertex and v belongs to a p-vertex in G'_p. Then, by examining Case 2 in the proof of Lemma 4, it follows that $|V_u(G_p)| \leq 3$, $|V_p(G_p)| \leq 4$ and $|V(G_p)| \leq 5$.
- u and v belong to distinct p-vertices. Then, by examining Case 3 in the proof of Lemma 4, it follows that $|V_u(G_p)| \leq 5$, $|V_p(G_p)| \leq 3$ and $|V(G_p)| = 6$.

For possibility (h) in Figure 3, by examining again Case 3 in the proof of Lemma 4 we will have $|V_u(G_p)| \leq 4$, $|V_p(G_p)| \leq 4$ and $|V(G_p)| \leq 6$.   □

The next theorem generalizes the previous lemma.

Theorem 7. *Assume $|F| = k$. If G' is a bicluster graph then $|V_u(G_p)| \leq 6k$, $|V_p(G_p)| \leq 4k$ and $|V(G_p)| \leq 6k$.*

Proof. Observe that $G = G' - F$. Then we can apply the inverse modifications to G' and obtain G.

The proof is done by induction on k. The basis of the induction is given by Lemma 6.

Let F^- be a subset of F such that $|F^-| = |F| - 1$, and let $G^- = G' - F^-$. By the induction hypothesis, the result is valid for F^-. Hence the subgraph of $(G^-)_p$ induced by components which are not bicliques contains at most $6(k-1)$ vertices, among which at most $4(k-1)$ are p-vertices. Since G' can contain $2k$ biclique components, $(G^-)_p$ can possibly contain some other components which are bicliques. See Figure 4.

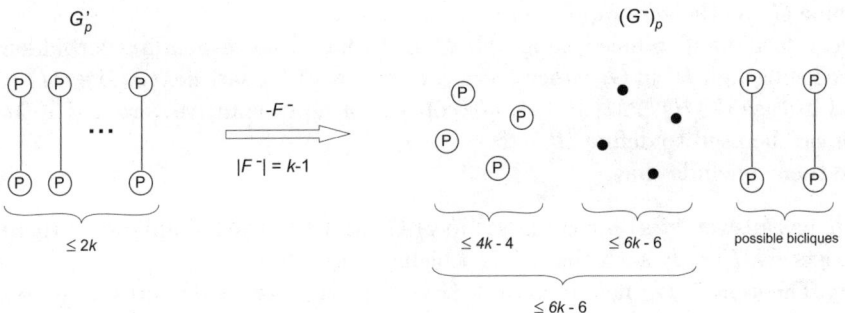

Fig. 4. $(G^-)_p = (G' - F^-)_p$

Let α be the edge modification such that $F = F^- \cup \alpha$. Then, $G = G^- - \alpha$. Let x, y be the vertices of α. We have 3 possibilities:

a) x and y lie in components which are not bicliques in G^-. Then G^- contains no biclique components. By Lemma 4, G_p contains in this case at most $6k-4$ vertices, among which at most $4k-2$ are p-vertices.

b) x lies in a biclique component B, and y in a component which is not a biclique in G^-. Then, B is the only biclique component in G^-. Considering the possible p-partitions of B (see Figure 2), B_p contains at most 2 vertices. By Lemma 4, G_p contains in this case at most $6(k-1) + 2 + 2 = 6k-2$ vertices, among which at most $4(k-1) + 2 + 2 = 4k$ are p-vertices.

c) x and y lie in different biclique components in G^-. Then G^- contains exactly two biclique components B_1 and B_2, and $-\alpha$ creates from B_1 and B_2 a unique component which is not a biclique in G. Other components in G^- remain unchanged in G. By Lemma 6, the p-quotient graph of the subgraph of G induced by $V(B_1) \cup V(B_2)$ has at most 6 vertices, among which at most 4 are p-vertices. Hence, the graph G_p contains in this case at most $6k$ vertices, among which at most $4k$ are p-vertices. $\square$

Finding the kernel's size. Assume that there exists a minimal edge modification set F, $|F| \leq k$, such that G' is a bicluster graph. This implies that if F contains a pair (a, v_a) such that a belongs to a p-vertex M of G_p, then (a, v_a) is an edge or a non-edge of some forbidden induced subgraph H of G ($H = P_4$ or $H = C_{2k+1}$). Also, F must contain a pair (b, v_b) for every other vertex $b \in M$, since these pairs also correspond to edges or non-edges of forbidden induced subgraphs. Therefore, if a p-vertex M of G_p satisfies $|M| \geq k+1$ then F contains no pair (a, v_a) with $a \in M$.

The above argument suggests a form of obtaining a problem kernel G_k by restricting the size of the p-vertices of G_p to $k+1$.

Let $V_k \subseteq V(G)$ constructed as follows. Initially, set $V_k := \emptyset$. Next, for every u-vertex $v \in V(G_p)$, set $V_k := V_k \cup \{v\}$. Finally, for every p-vertex $M \in V(G_p)$, if $|M| \leq k+1$ then set $V_k := V_k \cup M$, otherwise select $M' \subseteq M$ such that $|M'| = k+1$ and set $V_k := V_k \cup M'$.

Define G_k as the subgraph of G induced by V_k.

Every forbidden induced subgraph H in G has a corresponding forbidden induced subgraph H' in G_k, since every p-vertex M of G_p satisfies $|V(H) \cap M| \leq 1$, and in case $|V(H) \cap M| = \{v\}$, G_k contains a representative vertex $v' \in M$ which can be used to define $H' = H - v + v'$.

We then conclude that:

(i) G_k has answer "yes" for BICLUSTER(k) GRAPH EDITING if and only if there exists F, $|F| \leq k$, such that G' is a bicluster graph.
(ii) By Theorem 7, G_k has at most $4k(k+1) + 2k = 4k^2 + 6k$ vertices (corresponding to $6k$ vertices, where $4k$ are p-vertices and $2k$ are u-vertices).

By (i) and (ii), G_k is a problem kernel for BICLUSTER(k) GRAPH EDITING with input G.

Theorem 8. *A problem kernel with $O(k^2)$ vertices can be built for* BICLUSTER(k) GRAPH EDITING.

4.2   Applying a Bounded Search Tree

The second stage of the algorithm consists of applying a bounded search tree T on G_k, which corresponds to the root. In each node X of the tree, a search for a forbidden subgraph (P_4 or C_{2k+1}) is made in the graph G_X corresponding to X. As odd cycles of size at least 5 contain P_4 as an induced subgraph, it is sufficient to consider P_4 and K_3 as forbidden subgraphs. One child of X is created for each possible way of "destroying" the forbidden subgraph of G_X, without creating another forbidden subgraphs. As an edge modification is applied to obtain each child, the depth of the tree is bounded by k. A node will be a leaf of T either when its corresponding graph is a bicluster graph or its level is k (the root has level 0). If the first case occurs, the edge modification set applied along the root-leaf path constitutes a solution for the problem.

4.3   Running Time

First, each biclique component is removed from G. This takes linear time. Next, the modular decomposition tree T_G is built in $O(n+m)$ time, with $O(n)$ nodes.

For the construction of G_k, all nodes of T_G are visited. When visiting a node labeled P, we check whether the number of leaf children of this node is larger than $k+1$, and surplus vertices are eliminated. At the end, every node of T_G is visited once. Therefore, G_k can be built in $O(n+m)$ time.

The running time of the second stage of the algorithm is mainly determined by the size of the search tree T. The maximum number of children a node of T can have is 4 (this corresponds to the total number of different ways of destroying a P_4 by adding or removing one edge, whithout creating a C_3).

Theorem 9. BICLUSTER(k) GRAPH EDITING *can be solved in* $O(4^k + n + m)$ *time.*

Proof. The search tree T applied on G_k has size bounded by $O(4^k)$. In each node X of T, the corresponding graph G_X has at most $4k^2 + 6k$ vertices. This implies an $O((4k^2 + 6k)^3)$ local work, since the search for a P_4 can be made in linear time [3] and the search for a K_3 can be made exhaustively in cubic time. This yields an $O(4^k(4k^2 + 6k)^3 + (n+m)) = O(4^k + n + m)$ time complexity, due to the interleaving technique by Niedermeier and Rossmanith [18].   □

5   Building a Problem Kernel for CLUSTER(k) GRAPH EDITING

In this section, we will show a new form of obtaining a problem kernel with $O(k^2)$ vertices for CLUSTER(k) GRAPH EDITING in $O(n+m)$ time.

Similarly to the previous problem, F denotes an edge modification set for G, and G' denotes the graph $G+F$. Assume that G contains no clique components.

Lemma 10. *Assume* $|F| = 1$. *If* G' *is a cluster graph then* $|V_u(G_s)| \leq 4$, $|V_s(G_s)| \leq 2$ *and* $|V(G_s)| \leq 4$.

Proof. By using Lemma 5, the proof structure is similar to that of Lemma 6. □

Theorem 11. *Assume* $|F| = k$. *If* G' *is a cluster graph then* $|V_u(G_s)| \leq 4k$, $|V_s(G_s)| \leq 2k$ *and* $|V(G_s)| \leq 4k$.

Proof. By using Lemma 10, the proof structure is similar to that of Theorem 7.   □

As in previous problem, the size of all s-vertices of G_s can be bounded by $k+1$. By removing from G surplus vertices, we obtain a problem kernel G_k for CLUSTER(k) GRAPH EDITING in $O(n+m)$ time. By Theorem 11, G_k has at most $2k(k+1) + 2k = 2k^2 + 4k$ vertices.

Theorem 12. *A problem kernel with* $O(k^2)$ *vertices can be built for* CLUSTER(k) GRAPH EDITING *in* $O(n+m)$ *time.*

6   Conclusion

Future research is to apply modular decomposition to l-clique editing. (A l-clique is a complete l-partite graph.)

References

1. Amit, N. *The Bicluster Graph Editing Problem*, M.Sc. Thesis, Tel Aviv University, 2004.
2. Bauer, H. and Möhring, R. H. A fast algorithm for the decomposition of graphs and posets. *Mathematics of Operations Research* 8 (1983) 170–184.
3. Bretscher, A., Corneil, D., Habib, M., Paul, C. A simple linear time lexBFS cograph recognition algorithm, *WG 2003, Lecture Notes in Computer Science* 2880 (2003) 119–130.
4. Cai, L. Fixed-parameter tractability of graph modification problems for hereditary properties. *Information Processing Letters* 58 (1996) 171–176.
5. Dahlhaus, E., Gustedt, J., McConnel, R. M. Efficient and practical algorithms for sequential modular decomposition. *Journal of Algorithms* 41 (2001) 360–387.
6. Downey, R. G. and Fellows, M. R. Fixed-parameter tractability and completeness I: Basic results. *SIAM Journal on Computing* 24,4 (1995) 873–921.
7. Downey, R. G. and Fellows, M. R. Fixed-parameter tractability and completeness II: On completeness for W[1]. *Theoretical Computer Science* 141 (1995) 109–131.
8. Downey, R. G. and Fellows, M. R. *Parameterized Complexity*. Springer-Verlag.
9. Fernau, H. *Parameterized Algorithms: A Graph-Theoretic Approach*. University of Newcastle, 2005.
10. Gallai, T. Transitiv orientierbare graphen. *Acta Math. Acad. Sci. Hungar.* 18 (1967) 26–66.
11. Gramm, J., Guo, J., Hüffner, F., Niedermeier, R. Graph-modeled data clustering: Fixed-parameter algorithms for clique generation. *Theory of Computing Systems* 38,4 (2005) 373–392.
12. Gramm, J., Guo, J., Hüffner, F., Niedermeier, R. Automated generation of search tree algorithms for hard graph modification problems. *Algorithmica* 39,4 (2004) 321–347.
13. Habib, M., Montgolfier, F., Paul, C. A simple linear-time modular decomposition algorithm for graphs, using order extension. *9th Scandinavian Workshop on Algorithm Theory (SWAT 2004), Lecture Notes in Computer Science* 3111 (2004).
14. Möhring, R. H. and Radermacher, F. J. Substitution decomposition and connections with combinatorial optimization. *Ann. Discrete Math.* 19 (1984) 257–356.
15. McConnell, R. M. and Spinrad, J. P. Linear-time modular decomposition and efficient transitive orientation of comparability graphs. *Proceedings of the Fifth Annual ACM-SIAM Symposium on Discrete Algorithms* 5 (1994) 536–545.
16. McConnell, R. M. and Spinrad, J. P. Ordered vertex partitioning. *Discrete Mathematics and Theoretical Computer Science* 4 (2000) 45–60.
17. Natanzon, A., Shamir, R., Sharan, R. Complexity classification of some edge modification problems, *Discrete Applied Mathematics* 113 (1999) 109–128.
18. Niedermeier, R. and Rossmanith, P. A general method to speed up fixed-parameter-tractable algorithms. *Information Processing Letters* 73 (2000) 125—129.
19. Shamir, R., Sharan, R., Tsur, D. Cluster graph modification problems. *Discrete Applied Mathematics* 144 (2004) 173–182.

The Cluster Editing Problem: Implementations and Experiments[*]

Frank Dehne[1], Michael A. Langston[2], Xuemei Luo[1],
Sylvain Pitre[1], Peter Shaw[3], and Yun Zhang[2]

[1] School of Computer Science, Carleton University, Ottawa, Canada
[2] Department of Computer Science, University of Tennessee,
Knoxville, TN, USA
[3] Department of Computer Science, University of Newcastle, Newcastle, Australia

Abstract. In this paper, we study the cluster editing problem which is fixed parameter tractable. We present the first practical implementation of a FPT based method for cluster editing, using the approach in [6,7], and compare our implementation with the straightforward greedy method and a solution based on linear programming [3]. Our experiments show that the best results are obtained by using the refined branching method in [7] together with interleaving (re-kernelization). We also observe an interesting lack of monotonicity in the running times for "yes" instances with increasing values of k.

1 Introduction

The *CLUSTER EDITING* problem is defined as follows. *Input*: An undirected graph $G = (V, E)$, and a non-negative integer k. *Question*: Can we transform G, by inserting and deleting at most k edges, into a graph that consists of a disjoint union of cliques? The *CLUSTER EDIT DISTANCE* for a graph G is the smallest k for which cluster editing is possible.

Our main target applications are centered around computational biology. We are in particular interested in the analysis of putative gene co-regulation (transcriptomics via microarray analysis), putative gene product co-occurrence (proteomics via mass spec or MALDI), and pathway/network elucidation in data from synthetic genetic arrays (double knockout arrays). In all these applications, the underlying biological data is very expensive and, in some cases, requires years to produce. For example, RI strains take more than 8 years to make isogenetically pure, pathway/network elucidation for yeast requires 4,500 synthetic genetic arrays, and Affy U133 arrays contain more than 30k probesets. Because of the high value of the underlying biological data, saving computation time for the analysis by using approximation is often not acceptable. Hence, we turn to FPT based approaches for solving such problems. The cluster editing problem is an important such problem in the context outlined above.

In this paper, we present the first practical implementation of a FPT based method for cluster editing, using the approach in [6,7]. In order to evaluate the effectiveness

[*] This research has been supported in part by the Natural Sciences and Engineering Research Council of Canada and by the U.S. National Institutes of Health under grants 1-R01-MH-074460-01, 1-P01-DA-015027-01 and U01-AA13512.

H.L. Bodlaender and M.A. Langston (Eds.): IWPEC 2006, LNCS 4169, pp. 13–24, 2006.

of the FPT approach, we also implemented a well known previous method based on linear programming [3] as well as a greedy approach for cluster editing. The total programming effort was approx. 120 person hours, producing approx. 2500 lines of code.

Our experiments show that the best results for cluster editing are obtained by using the refined branching method in [7] together with interleaving (re-kernelization). Our experiments show that the refined branching method in [7] is vastly superior to the basic branching method, which is not obvious because the refined branching method is considerably more complicated and incurs larger constant factors. We also demonstrate that, in practice, branching with interleaving is indeed decidedly faster than branching without interleaving.

A surprising observation from our experiments comes with respect to the problem of determining the optimum edit value, k. In practice, we do of course not know k and the general approach would be to determine k via binary search. Things turns out to be quite different with cluster editing. If we happen to be advancing from below (that is, solving a "no" instance), then as e.g. with vertex cover run times predictably increase with rising parameter values. On the other hand, if we are advancing from above (facing a "yes" instance), then run times may decrease, increase or even stay the same with sinking parameter values. This is very different from the standard behavior e.g. for vertex cover. For cluster editing it is not the case that run times are highest around the optimum value of k. We conclude that a binary search may not be the best way to implement an FPT based approach for clustering editing, and that one may be better off steering the parameter as much as possible from below.

The remainder of this paper is organized as follows. Section 2 outlines our experimental setup used throughout the paper. Section 3 outlines our implementation of LP based and greedy based methods for cluster editing which serve as a baseline for evaluating the FPT based approach. In Section 4 we present a first practical implementation of an FPT based approach for cluster editing. Section 5 presents our experimental results for the FPT based approach and Section 6 concludes the paper.

2   Experimental Setup

In the remainder of this paper, we compare the performance of various algorithms under various criteria. Experiments were performed on a Dell OptiPlex GX280 using a 3.2GHz Pentium 4 dual processor, with 1.0 gigabytes (GB) of SDRAM, and running a Linux 2.6.8-2-686-smp kernel. Algorithms were implemented in C and compiled using gcc version 3.4.4. The various costs of implementation are listed in Table 1.

Table 1. Implementation costs for the three cluster editing algorithms studied in this paper

Algorithm	Development Time	Lines of Code	Library Used
LP-based method	45 hours	250	lp_solve version 5.5
Greedy method	15 hours	250	None
FPT approach	60 hours	2000	None

Unless otherwise stated, we used as input synthetic graphs for which we know the optimum edit distances. For this, we built a graph generator which operates as follows. Our graph generator takes as input parameters the desired number of vertices (n), clusters (c) and the required edit distance (d). First, a random size (number of vertices) in the range $[0.5(n/c), 1.5(n/c)]$ is assigned to each cluster. We produce a clique graph G' containing c fully connected cliques (clusters) of the sizes determined above, with no edges between those cliques. Then, we execute d random edits on G' resulting in an output graph G. An edit consists of randomly inserting or deleting an edge. More precisely, for each edit we randomly decide whether to insert or delete an edge. For an insert operation, we randomly chose two vertices i and j that are not connected by an edge and then add an edge (i, j). For a delete operation, we randomly select an edge in the graph and remove that edge. Once an edge is inserted in G it cannot be deleted by a future edit. Similarly once an edge has been deleted from G it cannot be re-inserted by another edit. For random number generation we used the Mitchell-Moore algorithm as described in [9]. Note that, the above method creates in most cases, but not always, a graph G with edit distance d. In some cases, as observed in our test runs, the edit distance of G is smaller than d because a different set of clusters than those used by our generator can be created with fewer edits.

3   A Baseline for Comparisons

In order to evaluate the effectiveness of a fixed parameter tractability approach for cluster editing, we need to establish a baseline to which we can compare our implementation. One well known previous method [3] is based on linear programming. Another alternative is to use a greedy approach for cluster editing. Needless to say, both methods only provide an approximation of the edit distance.

We implemented the LP based cluster editing method described in [3]. Given a graph, we first build a LP model by setting the objective function and constraints for every pair of vertices and every triple of vertices. For each pair of vertices i and j, a partitioning into clusters can be represented with a binary variable x_{ij}, where $x_{ij} = 0$ if i and j are in the same cluster, and $x_{ij} = 1$ if they are in different clusters. The integer constraints can be relaxed to allow real values for x_{ij}, that is, $0 \leq x_{ij} \leq 1$. For each triple of vertices i, j and k, the triangle inequality $x_{ik} \leq x_{ij} + x_{jk}$ holds because if $x_{ij} = 0$ and $x_{jk} = 0$ then $x_{ik} = 0$. The objective is to minimize the number of edge edits: the number of edges $(i, j) \in E$ for which $x_{ij} = 1$ and the number of pairs of vertices that are not adjacent $(i, j) \notin E$ for which $x_{ij} = 0$. After the LP model is built, a linear programming C library lp_solve version 5.5 is used to solve the LP model and get the values for every variable x_{ij}. Finally, the graph is partitioned into clusters based on the variables x_{ij} in the following way: two vertices i and j are put into the same cluster if $x_{ij} \leq 0.5$. The edit distance is calculated as the summation of the number of edges that are needed to be added, which do not exist but the two endpoints are in same cluster, and the number of edges that are needed to be deleted, which exist but the two endpoints are in different clusters.

We also implemented the following greedy method for cluster editing. Consider a graph G. For an edge insertion consider all possible edges, for an edge deletion consider only the edges in G. Calculate a cost for each insertion/deletion as follows. For an edge $e = (i, j)$ define the common neighborhood as the set of common neighbors of i and j, and define the non-common neighborhood as the set of non-common neighbors of i and j. For an insertion of an edge $e = (i, j)$, define the cost as the number of edge insertions required to transform the common and non-common neighborhood of e into a clique. For a deletion of an edge $e = (i, j)$, define the cost as the number of edge deletions required to disconnect the common neighborhood of e. Select the edit operation with smallest cost and iterate until a graph of disjoint cliques is obtained. To implement this greedy method, we initially mark every pair of vertices as unmarked. For each unmarked pair of vertices i and j, the smaller of the cost of having an edge between them and the cost of not having an edge between them is chosen as the cost of i and j. We select the pair with least cost, perform the edits associated, and mark the pair. This is repeated until all pairs are marked, which will give a set of connected components. The edit distance is calculated as the number of edge editions to get the set of connected components plus the number of edge editions to transform those connected components into cliques.

The results of our experimental evaluation of the LP and greedy methods are shown in Figures 1 and 2. Our experiments show that the LP based cluster editing method is consistently better than the greedy method with respect to both, the computation time and the value for k obtained. Hence, for the remainder of this paper, we will compare our implementation of a fixed-parameter tractability approach only to the LP based cluster editing method.

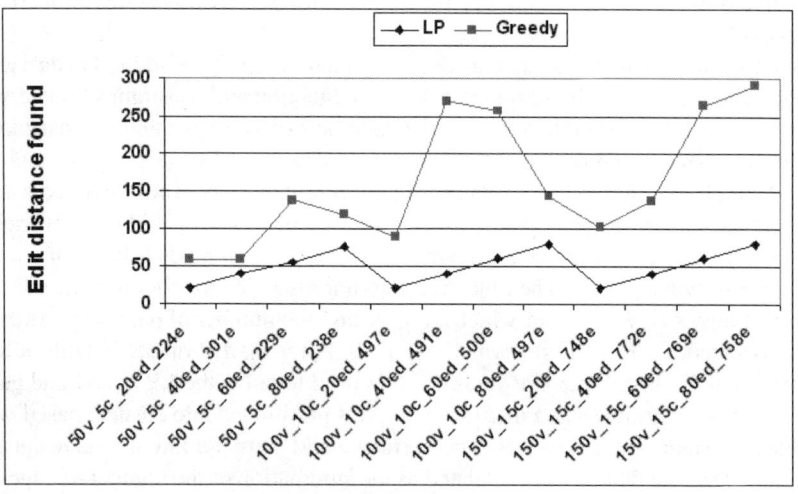

Fig. 1. A comparison of edit distances computed by LP versus the greedy method

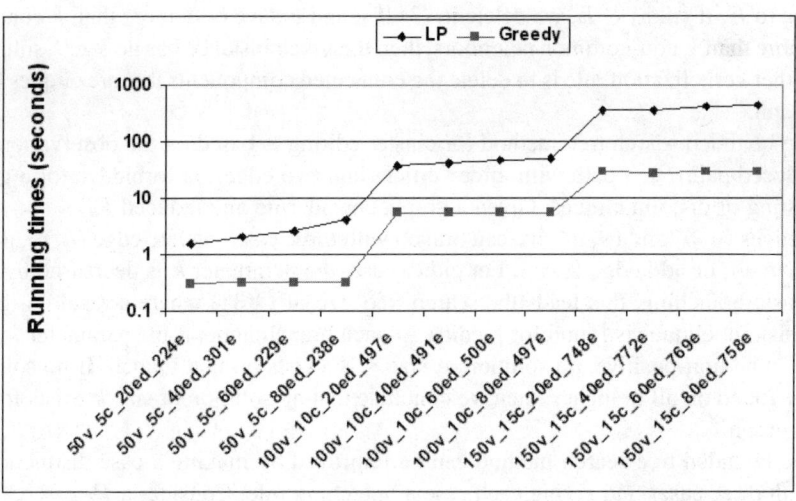

Fig. 2. A comparison of run times required by LP versus the greedy method

4 An FPT-Based Approach

In this section we present an outline of our adaptation of the algorithm in [7,6] that we used to obtain the first practical implementation for exact cluster edit distance computation.

Algorithm 1. Solving the Cluster Edit Problem Via a Fixed-Parameter Tractability Approach

(1) Extract highly connected (e.g., 2- and 3-connected) components. Our motivation is to eliminate sparse parts of the input. (Note: level of connectivity depends on the application.)
(2) Bound the search space for k:
    (a) Let k_{LP} be the edit distance determined by the linear programming method [3].
    (b) The search interval for the true edit distance k is $[k_{LP}/4, k_{LP}]$ (see [3]).
(3) For increasing k, starting with $k_{LP}/4$:
    (a) Execute the kernelization method described in [7], Section 7.2 (see also [6]).
    (b) Execute either the basic or the refined bounded tree search method described in [7], Section 10.1 or Section 10.2, respectively (see also [6]).
    (c) Use "interleaving": at each branch node in the bounded tree search, execute again the kernelization method from Step 3a.
— End of Algorithm —

Two kernelization rules for the cluster editing problem have been described. The first rule is based on the neighborhood of every pair of vertices $u, v \in V$. (1) If u and v have more than k common neighbors, then (u, v) has to belong to E; if $(u, v) \notin E$, we add it to E. (2) If u and v have more than k non-common neighbors, then (u, v) cannot belong to E; if $(u, v) \in E$, we delete it. (3) If u and v have both more than k common and more than k non-common neighbors, then the given instance has no size k solution. The other kernelization rule is to delete the connected components that are cliques from the graph.

The bounded search tree method for cluster editing is based on the observation that an induced path P_3, a path with three vertices and two edges, is forbidden for a graph consisting of disjoint cliques. Given a graph, considering any induced $P_3 = \{u, v, w\}$ with edges (u, v) and (u, w), we can branch with three cases: delete edge (u, v), delete edge (u, w), or add edge (v, w). For either case, the parameter k is decreased by one. For basic branching, this leads the search tree size of $O(3^k)$ where a resulting graph with disjoint cliques is found for k edits. At each branch node, if the parameter k goes down to be non-positive, no solution of size $\leq k$ exists on that branch. If no solution can be found on all branches, then we conclude that no solution of size k exists for the given graph.

The bounded tree search method can be improved by making a case distinction of P_3 with three cases and giving each case a branching rule. Consider a $P_3 = \{u, v, w\}$ with edges (u, v) and (u, w). There are three cases based on the neighborhood of u, v and w: (1) v and w do not share a common neighbor other than u; (2) v and w have a common neighbor x other than u, and x is adjacent to u; (3) v and w have a common neighbor x other than u, but x is not adjacent to u. For each pair of vertices, an annotation mapping is employed to facilitate the branching rules. Each vertex pair u and v is assigned one of the following annotations: "permanent" meaning $(u, v) \in E$ and (u, v) cannot be deleted, "forbidden" meaning $(u, v) \notin E$ and (u, v) cannot be inserted, or "none" meaning no information available and it can be edited. For every three vertices $u, v, w \in V$, if (u, v) and (u, w) are permanent, (v, w) has to be permanent, and if (u, v) is permanent and (u, w) is forbidden, (v, w) has to be forbidden.

Algorithm 2. Given a graph $G = (V, E)$ and parameter k, consider a $P_3 = \{u, v, w\}$ with edges (u, v) and (u, w). The refined branching strategy using the above annotation mapping works as follows.

(1) If v and w do not share a common neighbor other than u, then branch with
   (a) $(G \setminus \{(u, v)\}, k - 1)$, and
   (b) $(G \setminus \{(u, w)\}, k - 1)$.
(2) If v and w have a common neighbor $x \neq u$ and $(u, x) \in E$, then branch with five subcases:
   (c) $(G \cup \{(v, w)\}, k - 1)$;
   (d) Set (v, w) to forbidden, and branch with $(G \setminus \{(u, v), (v, x)\}, k - 2)$;
   (e) Set (v, w) to forbidden, (v, x) to permanent, and branch with $(G \setminus \{(u, v), (u, x), (w, x)\}, k - 3)$;
   (f) Set (v, w) to forbidden, and branch with $(G \setminus \{(u, w), (w, x)\}, k - 2)$;
   (g) Set (v, w) to forbidden, (w, x) to permanent, and branch with $(G \setminus \{(u, w), (u, x), (v, x)\}, k - 3)$.
(3) If v and w have a common neighbor $x \neq u$ and $(u, x) \notin E$, then branch with five subcases:
   (h) $(G \setminus \{(u, v)\}, k - 1)$;

(i) Set (u, v) to permanent, (v, w) to forbidden, and branch with $(G \setminus \{(u, w), (v, x)\}, k - 2)$;

(j) Set (u, v) to permanent, (v, w) to forbidden, (v, x) to permanent, and branch with $(G \cup \{(u, x)\} \setminus \{(u, w), (w, x)\}, k - 3)$;

(k) Set (u, v) and (u, w) to permanent, and branch with $(G \cup \{(v, w)\} \setminus \{(w, x), (v, x)\}, k - 3)$;

(l) Set (u, v) and (u, w) to permanent, and branch with $(G \cup \{(v, w), (u, x)\}, k - 2)$.

— End of Algorithm —

Initially, all vertex pairs are set to "none". When an edge is added it is set to "permanent", and when an edge is deleted it is set to "forbidden". The algorithm also stops when the parameter k reaches 0 or below or when the graph G contains no induced P_3. The search tree size for the refined branching strategy is $O(2.27^k)$.

For both basic and refined bounded tree search methods, we applied the kernelization method at each branch node. Both methods are implemented as recursive functions. For future improvement, we plan to implement them as iterative functions to achieve better performance.

5   Experimental Results

To gauge the practical merit of an approach based on fixed-parameter tractability, we tested the methods just described against each other and against our LP implementation on a variety of both synthetic and real graphs. We have already described the process by which we generate synthetic graphs. By "real" graphs, we mean those that naturally arise in application domains, in the present case, from protein domain sequence similarity. In all cases we report branching times only because the time needed for initial preprocessing and kernelization is insignificant compared to that required during branching.

It seems that refined branching is vastly superior to basic branching. The run times reported in Figure 3, where the edit distance is set to 20, are typical of those we observe. This was not obvious in advance. It is simply not always the case that asymptotically faster methods in the worst case translate into better algorithms in the average case. Unless the data is contrived, additional overhead and complexities incurred by ever more sophisticated branching strategies can ofttimes negate any real gains in efficiency.

It also seems that branching with interleaving is decidedly faster than branching without interleaving. The run times reported on larger instances in Figure 4, where the edit distance is set to 40, are typical. Again, this makes sense, but is neither obvious nor necessarily the case in general. [8]

Of course we do not know the optimum edit value in advance, and so generally determine it by performing a binary search. One might expect run times to be rather predictable. With vertex cover, for example, we have long observed [1,2] that the most difficult computations are centered around the point at which a "no" instance becomes a "yes" instance, with the "no" the harder of the two (with "no," our algorithms cannot find a solution and halt early). Because we are interested in clique, the situation is

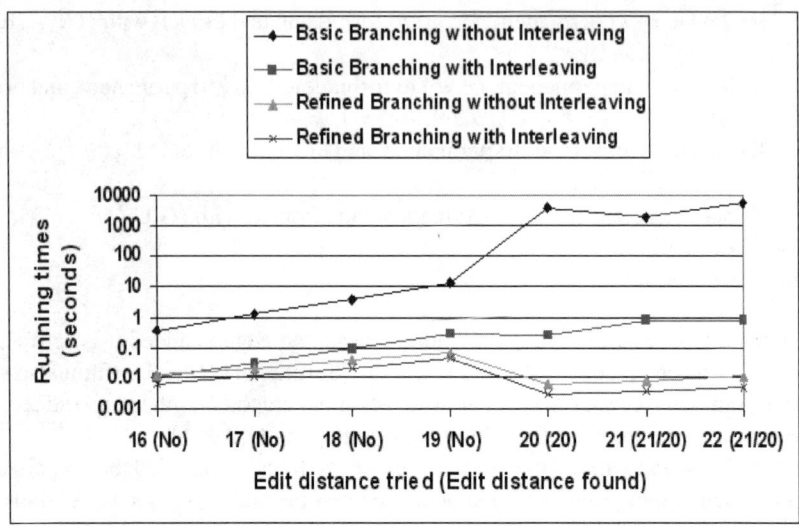

Fig. 3. FPT run times on a graph with 50 vertices, 5 clusters and edit distance 20

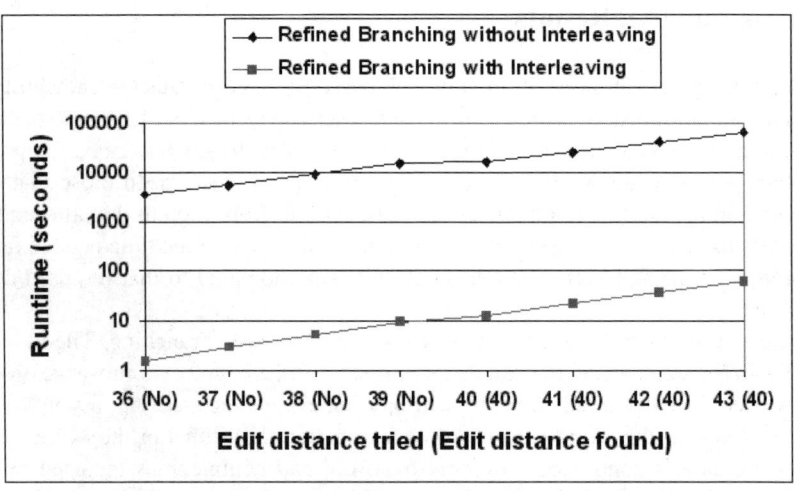

Fig. 4. FPT run times on a graph with 100 vertices, 10 clusters and edit distance 40

reversed but still monotonic above and below the optimum value. A standard example is illustrated in Figure 5.

Things turn out to be quite different with cluster editing. If we happen to be advancing from below (that is, solving a "no" instance), then as with vertex cover run times predictably increase with rising parameter values. On the other hand, if we are advancing from above (facing a "yes" instance), then run times may decrease, increase or even stay the same with sinking parameter values. See Figures 6 and 7.

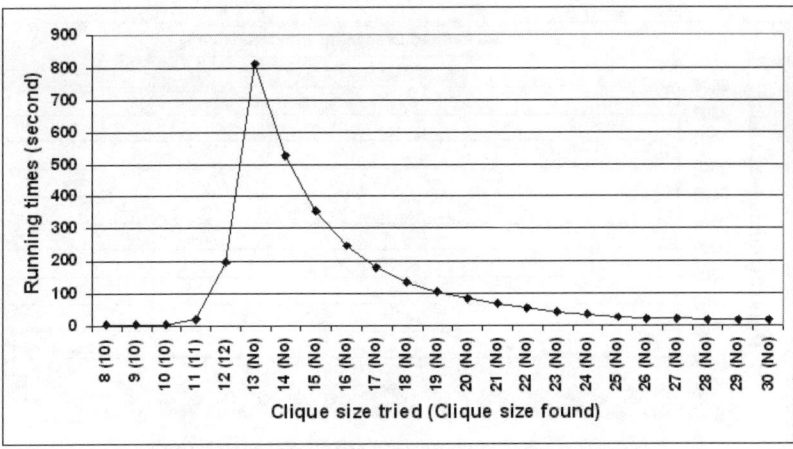

Fig. 5. The monotonicity of FPT parameter effects as seen when solving clique with vertex cover

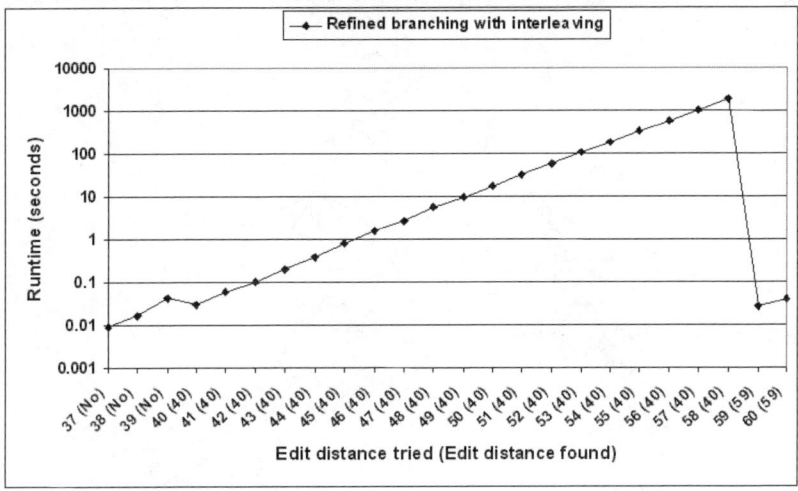

Fig. 6. FPT parameter effects on a graph with 100 vertices, 5 clusters and edit distance 40

In retrospect, we find the answer to this conundrum lies in the way solutions are distributed and the way branching works with parameter values above the optimum. With vertex cover, for example, a cover of size i ensures a cover of size $i + 1$ (as long as $i < |V|$). With cluster editing, however, it is conceivable that there is a solution with edit distance i yet no solution with distance $i + 1$ (or $i + 2$ and so forth). Thus a higher parameter value may mean only that a cluster editing algorithm has to do more work. See Figure 8, which depicts search tree traversals on the graph used to report run times in Figure 7. From this we conclude that a binary search may not be the best way to implement an FPT-based approach for clustering editing, and that one may be better off steering the parameter as much as possible from below.

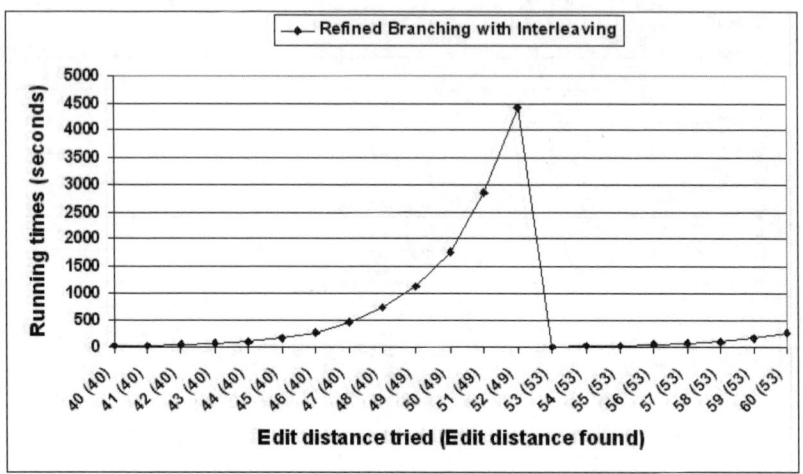

Fig. 7. FPT parameter effects on a graph with 100 vertices, 10 clusters and edit distance 40

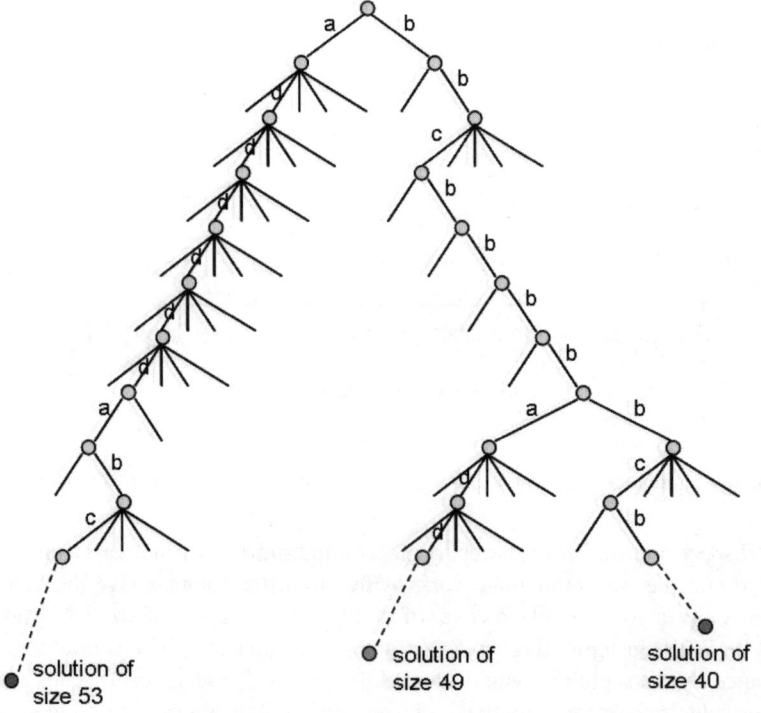

Fig. 8. Search trees depend on parameter values

6   Concluding Remarks

The issue of scalability deserves scrutiny, especially if we are to scale to genome-sized problem instances. Even for covers and cliques, supercomputers and monolithic memory may be heavily taxed [10]. In this respect, it is noteworthy that well-known problems such as vertex cover require searching a parameter space of size $|V|$, while cluster editing possesses a search space of size $|E|$. At some problem size, of course, approximation should better optimization. The exact size probably depends on many factors, including graph density, relative efficiency of implementations, and even machine architecture. Initial experiments have produced interesting comparisons. See Figure 9.

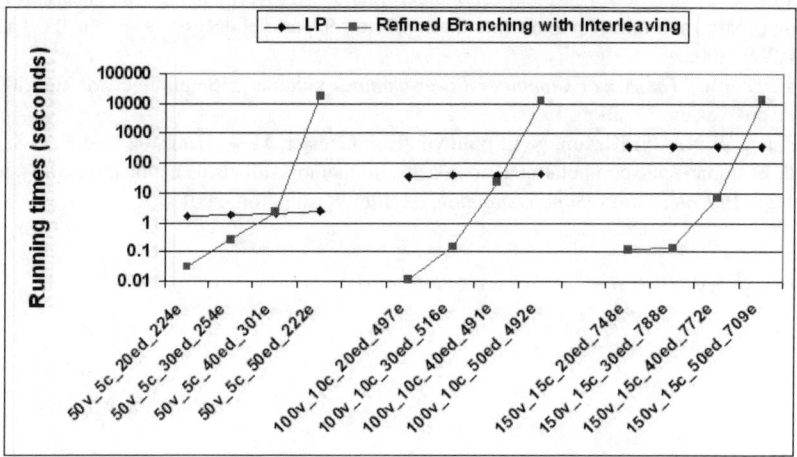

Fig. 9. The scalability of approximation via LP versus optimization via FPT

Related questions abound. For example, we are interested in the difficulty of enumerating solutions within some fixed number of edge additions and deletions [4]. Such an enumeration may prove useful in making decisions between multiple and possibly ambiguous solutions. We are also interested in relaxing the requirement that cluster editing cliques be disjoint [5]. Such a relaxation makes particular sense in applications for which vertices represent genes or gene products, because these are often pleiotropic and thus may rightfully belong in overlapping cliques.

References

1. F. N. Abu-Khzam, R. L. Collins, M. R. Fellows, M. A. Langston, W. H. Suters, and C. T. Symons. Kernelization algorithms for the vertex cover problem: Theory and experiments. In *Proceedings, Workshop on Algorithm Engineering and Experiments,* New Orleans, Louisiana, 2004.
2. F. N. Abu-Khzam, M. A. Langston, P. Shanbhag, and C. T. Symons. Scalable parallel algorithms for FPT problems. *Algorithmica,* 2006, accepted for publication.

3. M. Charikar, V. Guruswami, and A. Wirth. Clustering with qualitative information. *Journal of Computer and System Sciences*, 71:360–383, 2005.

4. P. Damaschke. On the fixed-parameter enumerability of cluster editing. In *Proceedings, International Workshop on Graph-Theoretic Concepts in Computer Science*, pages 283–294, 2005.

5. P. Damaschke. Fixed-parameter tractable generalizations of cluster editing. In *Proceedings, International Conference on Algorithms and Complexity*, 2006.

6. J. Gramm, J. Guo, F. Hueffner, and R. Niedermeier. Graph-modeled data clustering: Fixed-parameter algorithms for clique generation. *Theory of Computing Systems*, 38(4):373 – 392, 2005.

7. J. Guo. *Algorithm design techniques for parameterized graph modification problems*. PhD thesis, Univ. Jena, 2005.

8. Edith Hemaspaandra, Jörg Rothe, and Holger Spakowski. Recognizing when heuristics can approximate minimum vertex covers is complete for parallel access to np. In *WG*, pages 258–269, 2002.

9. D. E. Knuth. *The Art of Computer Programming*, volume 2: Seminumerical algorithms. Addison-Wesley, 3 edition, 1997.

10. Y. Zhang, F. N. Abu-Khzam, N. E. Baldwin, E. J. Chesler, M. A. Langston, and N. F. Samatova. Genome-scale computational approaches to memory-intensive applications in systems biology. In *Proceedings, Supercomputing*, Seattle, Washington, 2005.

The Parameterized Complexity of Maximality and Minimality Problems

Yijia Chen[1] and Jörg Flum[2]

[1] BASICS, Department of Computer Science, Shanghai Jiaotong University,
Shanghai 200030, China
yijia.chen@cs.sjtu.edu.cn
[2] Abteilung für Mathematische Logik, Universität Freiburg, Eckerstr. 1,
79104 Freiburg, Germany
joerg.flum@math.uni-freiburg.de

Abstract. Many parameterized problems (as the clique problem or the dominating set problem) ask, given an instance and a natural number k as parameter, whether there is a solution of size k. We analyze the relationship between the complexity of such a problem and the corresponding maximality (minimality) problem asking for a solution of size k maximal (minimal) with respect to set inclusion. As our results show maximality problems may increase the parameterized complexity, while "in terms of the W-hierarchy" minimality problems do not increase the complexity.

1 Introduction

Suppose we know (or at least have an upper bound on) the complexity of deciding whether a problem has a solution of size k. What can we say about the complexity of the existence of maximal (or minimal) solutions with respect to set inclusion of size k? This paper addresses this type of questions. By complexity we always mean the parameterized complexity, the parameter being the size of the solution.

We start by mentioning three results, the first and the second one are well-known (cf. [7]) and the third one will be derived in Section 3:

(a) The problem p-VERTEX-COVER ("Does a graph have a vertex cover of size k") is fixed-parameter tractable and so is the problem p-MINIMAL-VERTEX-COVER ("Does a graph have a minimal vertex cover of size k").

(b) The problem p-INDEPENDENT-SET ("Does a graph have an independent set of size k") is W[1]-complete and the problem p-MAXIMAL-INDEPENDENT-SET is W[2]-complete.

(c) The problem p-DOMINATING-SET ("Does a graph have a dominating set of size k") is W[2]-complete and so is p-MINIMAL-DOMINATING-SET.

As we show these are not isolated results but special cases of general phenomena: Maximality problems (may) increase the complexity, while minimality problems do not.

Let us first introduce a framework appropriate to discuss this type of questions. A set S of vertices of a graph $\mathcal{G}$ is a vertex cover if in $\mathcal{G}$ it satisfies the formula $vc(Z)$ of first-order logic with the set variable Z, where

$$vc(Z) := \forall x \forall y (\neg Exy \vee Zx \vee Zy)$$

H.L. Bodlaender and M.A. Langston (Eds.): IWPEC 2006, LNCS 4169, pp. 25–37, 2006.

(here the quantifiers range over the vertices, Exy means that there is an edge between x and y, and Zx means that x is an element of Z). We say that $vc(Z)$ Fagin-defines the problem p-VERTEX-COVER (on the class of graphs). Similarly the problems p-INDEPENDENT-SET and p-DOMINATING-SET are Fagin-defined by

$$indep(Z) := \forall x \forall y (\neg Exy \vee \neg Zx \vee \neg Zy) \text{ and } ds(Z) := \forall y \exists x (Zx \wedge (x = y \vee Exy)),$$

respectively. Note that the formulas $vc(Z)$ and $ds(Z)$ are positive in Z (no occurrence of Z is in the scope of a negation symbol) and the formula $indep(Z)$ is negative in Z (every occurrence of Z is in the scope of exactly one negation symbol).

If $\varphi(Z)$ is an arbitrary formula, we denote by p-WD$_\varphi$ the problem Fagin-defined by $\varphi(Z)$ (see Section 2.1 for the precise definition). It should be clear what we mean by p-MAXIMAL-WD$_\varphi$ and by p-MINIMAL-WD$_\varphi$. The problem p-MAXIMAL-DOMINA-TING-SET is trivial, since the set of all vertices is the only maximal dominating set in a given graph. Similarly, p-MINIMAL-INDEPENDENT-SET is trivial. More generally, one easily verifies (cf. Section 4) that the problem p-MAXIMAL-WD$_\varphi$ is trivial for $\varphi(Z)$ positive in Z and so is the problem p-MINIMAL-WD$_\varphi$ for $\varphi(Z)$ negative in Z.

We collect the main known results concerning the W-hierarchy and Fagin-defined problems (cf. [5,11]). We use the following notation: If C is a class of parameterized problems, [C]^{fpt} is the class of problems fpt-reducible to some problem in C.

Theorem 1. *Let $t \geq 1$.*
(1) $W[t] = \left[\{p\text{-WD}_\varphi \mid \varphi(Z) \text{ a } \Pi_t\text{-formula}\} \right]^{\text{fpt}}$.
(2) If t is odd, then

$$W[t] = \left[\{p\text{-WD}_\varphi \mid \varphi(Z) \text{ a } \Pi_t\text{-formula negative in } Z\} \right]^{\text{fpt}}$$
$$= \left[\{p\text{-WD}_\varphi \mid \varphi(Z) \text{ a } \Pi_{t+1}\text{-formula negative in } Z\} \right]^{\text{fpt}}.$$

(3) If t is even, then

$$W[t] = \left[\{p\text{-WD}_\varphi \mid \varphi(Z) \text{ a } \Pi_t\text{-formula positive in } Z\} \right]^{\text{fpt}}$$
$$= \left[\{p\text{-WD}_\varphi \mid \varphi(Z) \text{ a } \Pi_{t+1}\text{-formula positive in } Z\} \right]^{\text{fpt}}.$$

The second equalities in (2) and (3) are formulations in terms of Fagin-definable problems of the Antimonotone Collapse Theorem and the Monotone Collapse Theorem, respectively.

In this paper we first analyze maximality problems. We start by observing that p-MAXIMAL-WD$_\varphi$ can be considerably harder than p-WD$_\varphi$. In fact, there is a Π_1-formula φ such that p-WD$_\varphi$ is W[P]-hard. In more conventional terms, we show that the maximal weighted satisfiability problem for formulas in 4-CNF is W[P]-hard. We then turn to formulas $\varphi(Z)$ negative in Z and derive the following. A comparison with part (2) of Theorem 1 shows that the transition from p-WD$_\varphi$ to p-MAXIMAL-WD$_\varphi$ increases the complexity one level in the W-hierarchy. Moreover, note that the first equality generalizes the result for the independent set problem mentioned in (b).

Theorem 2. *If $t \geq 1$ is odd, then*

$$W[t+1] = \left[\{p\text{-MAXIMAL-WD}_\varphi \mid \varphi(Z) \text{ } a \text{ } \Pi_t\text{-formula negative in } Z\}\right]^{\text{fpt}}$$
$$= \left[\{p\text{-MAXIMAL-WD}_\varphi \mid \varphi(Z) \text{ } a \text{ } \Pi_{t+1}\text{-formula negative in } Z\}\right]^{\text{fpt}}.$$

The proof of this result implies, for example, that the maximal weighted satisfiability problem for formulas in 2-CNF with only negative literals is W[2]-complete.

We then turn to minimality problems. The following theorem generalizes the results for the vertex cover problem and the dominating set problem mentioned in (a) and (c), respectively. Moreover, a comparison of (1) and (2) with (1) and (3) in Theorem 1, respectively, show that for minimality problems we do not have an increase of complexity.

Theorem 3. *(1) If $t \geq 2$, then*

$$W[t] = \left[\{p\text{-MINIMAL-WD}_\varphi \mid \varphi(Z) \text{ } a \text{ } \Pi_t\text{-formula}\}\right]^{\text{fpt}}.$$

(2) If $t \geq 2$ is even, then

$$W[t] = \left[\{p\text{-MINIMAL-WD}_\varphi \mid \varphi(Z) \text{ } a \text{ } \Pi_t\text{-formula positive in } Z\}\right]^{\text{fpt}}$$
$$= \left[\{p\text{-MINIMAL-WD}_\varphi \mid \varphi(Z) \text{ } a \text{ } \Pi_{t+1}\text{-formula positive in } Z\}\right]^{\text{fpt}}.$$

(3) $\text{FPT} = \left[\{p\text{-MINIMAL-WD}_\varphi \mid \varphi(Z) \text{ } a \text{ } \Pi_1\text{-formula}\}\right]^{\text{fpt}}.$

For the weighted satisfiability problem of propositional formulas in $\Gamma_{t,4}$, $\Gamma_{t,4}^+$, and $\Gamma_{t,4}^-$ (these sets are defined in Section 2.2) we exemplify our results on maximality and minimality problems in the following table.

	maximality problem	minimality problem
$\Gamma_{t,4}$	W[P]-hard	$t = 1$: FPT $t > 1$: W[t]-complete
$\Gamma_{t,4}^+$	FPT	t even: W[t]-complete t odd: W[t − 1]-complete
$\Gamma_{t,4}^-$	t even: W[t]-complete t odd: W[t + 1]-complete	FPT

Of course, the existence of maximal and minimal solutions has been studied for various problems from the classical ("unparameterized") point of view; e.g., TRANSVERSAL HYPERGRAPH is the problem of generating all minimal satisfying assignments for a positive formula in CNF. We refer to [6].

Since due to space limitations we have to defer the main part of the proofs of Theorem 2 and Theorem 3 to the full version of the paper, we finish this introduction with some remarks concerning the proof methods. Based on [5], in [10,11] the relationship between weighted satisfiability problems for fragments of propositional logic, model-checking problem for fragments of first-order logic, and Fagin-definable problems has been analyzed systematically and corresponding "translation procedures" were developed. Partly, our proofs build on these procedures. We should mention that Theorem 2 and Theorem 3 remain true if Z is replaced by a relation symbol of arbitrary arity.

2   Preliminaries

The set of natural numbers isdenoted by $\mathbb{N}$. For a natural number n let $[n] := \{1, \ldots, n\}$. We assume that the reader is familiar with the basic notions of parameterized complexity theory (cf. [4,11]). We denote by FPT the class of all fixed-parameter tractable problems.

2.1   First-Order Logic

A *vocabulary* τ is a finite set of relation symbols. Each relation symbol has an *arity*. A τ-*structure* $\mathcal{A}$ consists of a set A called the *universe*, which we assume to be finite, and an interpretation $R^{\mathcal{A}} \subseteq A^r$ of each r-ary relation symbol $R \in \tau$. For example, we view a *graph* as a structure $\mathcal{G} = (G, E^{\mathcal{G}})$, where E is a binary relation symbol and $E^{\mathcal{G}}$ is an irreflexive and symmetric binary relation on the set of vertices G. Nevertheless, often we denote the vertex set of a graph $\mathcal{G}$ by V and the edge set by E (instead of G and $E^{\mathcal{G}}$) and use the set notation $\{v, w\}$ for undirected edges.

Formulas of first-order logic are built up from atomic formulas using the boolean connectives $\neg$, $\wedge$, $\vee$ and existential and universal quantification. (The connectives $\rightarrow$ and $\leftrightarrow$ are understood as abbreviations.) For $t \geq 1$, let Σ_t denote the class of all first-order formulas of the form

$$\exists x_{11} \ldots \exists x_{1k_1} \forall x_{21} \ldots \forall x_{2k_2} \ldots Q x_{t1} \ldots Q x_{tk_t} \, \psi,$$

where $Q = \forall$ if t is even and $Q = \exists$ otherwise, and where ψ is quantifier-free. Π_t-formulas are defined analogously starting with a block of universal quantifiers. Let $t, u \geq 1$. A formula φ is $\Sigma_{t,u}$, if it is Σ_t and all quantifier blocks after the leading existential block have length $\leq u$.

For a class Φ of first-order formulas we consider the *parameterized model-checking problem*:

$p\text{-MC}(\Phi)$
> *Input:* A structure $\mathcal{A}$ and a sentence $\varphi \in \Phi$.
>> *Parameter:* $|\varphi|$.
>> *Question:* Does $\mathcal{A} \models \varphi$ hold, that is, is $\mathcal{A}$ a model of φ?

Let Z be a fixed set variable (that is, unary relation variable). We consider first-order formulas that may contain atomic subformulas of the form Zx. For $t, d \geq 1$ we denote by $\Pi_{t/d}$ the set of Π_t-formulas $\varphi(Z)$ with at most d occurrences of Z.

A first-order formula $\varphi = \varphi(Z)$ *Fagin-defines* the problem:

$p\text{-WD}_\varphi$
> *Input:* A structure $\mathcal{A}$ and $k \in \mathbb{N}$.
>> *Parameter:* k.
>> *Question:* Is there a subset $S \subseteq A$ of size k such that $\mathcal{A} \models \varphi(S)$?

2.2 Propositional Logic

Formulas of propositional logic are built up from propositional variables by taking conjunctions, disjunctions, and negations. We distinguish between *small conjunctions*, denoted by $\wedge$, which are just conjunctions of two formulas, and *big conjunctions*, denoted by $\bigwedge$, which are conjunctions of nonempty finite sets of formulas. Analogously, we distinguish between *small disjunctions*, $\vee$, and *big disjunctions*, $\bigvee$.

For $t \geq 0$ and $d \geq 1$ we inductively define the following classes $\Gamma_{t,d}$ and $\Delta_{t,d}$ of formulas:

$$\Gamma_{0,d} := \{\lambda_1 \wedge \ldots \wedge \lambda_c \mid c \in [d], \lambda_1, \ldots, \lambda_c \text{ literals}\},$$
$$\Delta_{0,d} := \{\lambda_1 \vee \ldots \vee \lambda_c \mid c \in [d], \lambda_1, \ldots, \lambda_c \text{ literals}\},$$

$$\Gamma_{t+1,d} := \left\{\bigwedge_{i \in I} \delta_i \mid I \text{ finite nonempty index set, and } \delta_i \in \Delta_{t,d} \text{ for all } i \in I\right\},$$

$$\Delta_{t+1,d} := \left\{\bigvee_{i \in I} \gamma_i \mid I \text{ finite nonempty index set, and } \gamma_i \in \Gamma_{t,d} \text{ for all } i \in I\right\}.$$

If in the definition of $\Gamma_{0,d}$ and $\Delta_{0,d}$ we require that all literals are positive (negative) we obtain the sets denote by $\Gamma_{t,d}^+$ and $\Delta_{t,d}^+$ ($\Gamma_{t,d}^-$ and $\Delta_{t,d}^-$), respectively.

We denote by $\text{Var}(\alpha)$ the set of propositional variables of a propositional formula α. Let V be a set of propositional variables. We identify an assignment $S : V \to \{\text{TRUE}, \text{FALSE}\}$ with the set $\{X \in V \mid S(X) = \text{TRUE}\}$. The *weight* of an assignment S is $|S|$, the number of variables set to TRUE. A propositional formula α is *k-satisfiable* (where $k \in \mathbb{N}$), if there is an assignment for the set of variables of α of weight k satisfying α. For a set Γ of propositional formulas, the *parameterized weighted satisfiability problem p-WSAT(Γ) for formulas in Γ* is the following problem:

---

p-WSAT(Γ)
>   *Input:* A propositional formula $\alpha \in \Gamma$ and $k \in \mathbb{N}$.
>   *Parameter:* k.
>   *Question:* Is α k-satisfiable?

2.3 The W-Hierarchy

The following theorem contains well-known characterizations (or definitions) of the W-hierarchy in terms of model-checking problems and weighted satisfiability problems. Characterizations in terms of Fagin-definable problems were given in Theorem 1.

Theorem 4. *(1) Let $t, u \geq 1$. Then p-MC($\Sigma_{t,u}$) is W[t]-complete under fpt-reductions.*
(2) Let $t, d \geq 1$ and $t + d \geq 3$. Then p-WSAT($\Gamma_{t,d}$) is W[t]-complete under fpt-reductions.
(3) Let $t, d \geq 1$ and $t + d \geq 3$. If t is even (odd), then p-WSAT($\Gamma_{t,d}^+$) (p-WSAT($\Gamma_{t,d}^-$)) is W[t]-complete under fpt-reductions.

3   Maximality and Minimality Problems

First we derive the result mentioned in (b) at the beginning of the Introduction:

Theorem 5. p-MINIMAL-DOMINATING-SET *is* W[2]-*complete.*

Proof: We first show that p-MINIMAL-DOMINATING-SET $\in$ W[2] by reducing it to p-MC($\Sigma_{2,1}$). For $k \in \mathbb{N}$ we have

$$(\mathcal{G}, k) \in p\text{-MINIMAL-DOMINATING-SET} \iff \mathcal{G} \models \varphi_k,$$

where $(x_1, \ldots, x_k$ correspond to the elements of a minimal dominating set and $z_1, \ldots, z_k$ witness the minimality)

$$\varphi_k := \exists x_1 \ldots \exists x_k \exists z_1 \ldots \exists z_k \left(\bigwedge_{1 \leq i < j \leq k} x_i \neq x_j \wedge \forall y \bigvee_{i \in [k]} (x_i = y \vee Ex_i y) \wedge \right.$$
$$\left. \bigwedge_{j \in [k]} \bigwedge_{i \in [k], i \neq j} (x_i \neq z_j \wedge \neg Ex_i z_j) \right).$$

Since φ_k is (logical equivalent to) a $\Sigma_{2,1}$-sentence, this gives the desired reduction.

To show the W[2]-hardness of p-MINIMAL-DOMINATING-SET we present an fpt-reduction from p-DOMINATING-SET to it. Let $\mathcal{G} = (V, E)$ be a graph and $k \leq |V|$. We construct the graph $\mathcal{G}' = (V', E')$ as follows:

$$V' := \left(V \times [k] \right) \dot\cup V \dot\cup [k]$$

$$E' := \bigcup_{\ell \in [k]} \{\{(u, \ell), (v, \ell)\} \mid u, v \in V, u \neq v\}$$

$$\cup \bigcup_{\ell \in [k]} \{\{(u, \ell), v\} \mid u, v \in V \text{ and } (u = v \text{ or } \{u, v\} \in E)\}$$

$$\cup \{\{(u, \ell), \ell\} \mid u \in V \text{ and } \ell \in [k]\}.$$

One easily verifies that

$\mathcal{G}$ has a dominating set of size k $\iff$ $\mathcal{G}'$ has a minimal dominating set of size k. $\square$

For p-WSAT(Γ), where Γ is a class of propositional formulas, it should be clear what we mean by p-MINIMAL-WSAT(Γ) or by p-MAXIMAL-WSAT(Γ), as it was for p-DOMINATING-SET. But in many cases it is not clear what a maximal or minimal solution should be. For example, what is a maximal solution of the parameterized halting problem?

---

p-SHORT-NSTM-HALT
  Input: A nondeterministic single-tape Turing machine $\mathbb{M}$
     and $k \in \mathbb{N}$.
  Parameter: k.
  Question: Does $\mathbb{M}$ accepts the empty string in at most k steps?

The situation is different for Fagin-defined problems.

Definition 6. Let $\varphi(Z)$ be a first-order formula of vocabulary τ. Let $\mathcal{A}$ be a τ-structure and $S \subseteq A$.
(1) S is a *solution (of $\varphi(Z)$ in $\mathcal{A}$)* if $\mathcal{A} \models \varphi(S)$.
(2) S is a *minimal solution (of $\varphi(Z)$ in $\mathcal{A}$)* if S is a solution and no subset $S' \subset S$ is a solution.
(3) S is a *maximal solution (of $\varphi(Z)$ in $\mathcal{A}$)* if S is a solution and no superset $S' \supset S$ is a solution.

For the parameterized problem $p\text{-WD}_\varphi$ Fagin-defined by φ (see Section 2.1) we define the *maximality problem Fagin-defined by φ* as

---

p-MAXIMAL-WD$_\varphi$
    Input: A structure $\mathcal{A}$ and $k \in \mathbb{N}$.
Parameter: k.
    Question: Does there exist a maximal solution of $\varphi(Z)$ in $\mathcal{A}$
        of size k?

---

and the *minimality problem Fagin-defined by φ* as

---

p-MINIMAL-WD$_\varphi$
    Input: A structure $\mathcal{A}$ and $k \in \mathbb{N}$.
Parameter: k.
    Question: Does there exist a minimal solution of $\varphi(Z)$ in $\mathcal{A}$
        of size k?

In particular, on the class of graphs, the problem $p\text{-WD}_{indep}$ (recall that $indep(Z) = \forall x \forall y(\neg Exy \vee \neg Zx \vee \neg Zy)$) coincides with p-INDEPENDENT-SET and p-MAXIMAL-WD$_{indep}$ with p-MAXIMAL-INDEPENDENT-SET.

In general, the problem p-MAXIMAL-WD$_\varphi$ can be considerably harder than $p\text{-WD}_\varphi$ as shown by the following result. When reading this result recall that $p\text{-WD}_\varphi \in \text{W}[1]$ for all Π_1-formulas.

Theorem 7. *There exists a Π_1-formula $\varphi(Z)$ such that p-MAXIMAL-WD$_\varphi$ is W[P]-hard.*

Proof: We give a reduction from the weighted Boolean circuit satisfiability problem p-WSAT(CIRCUIT), a W[P]-complete problem, to p-MAXIMAL-WD$_\varphi$ for an appropriate $\varphi(Z) \in \Pi_1$.

We view circuits as τ_{circ}-structures for $\tau_{\text{circ}} := \{E, IN, OUT, AND, OR, NEG\}$ with unary relation symbols IN, OUT, AND, OR, NEG and binary E. Hence, a circuit has the form $\mathcal{C} = (C, E^{\mathcal{C}}, IN^{\mathcal{C}}, OUT^{\mathcal{C}}, AND^{\mathcal{C}}, OR^{\mathcal{C}}, NEG^{\mathcal{C}})$, where $(C, E^{\mathcal{C}})$ is the directed acyclic graph underlying the circuit, $IN^{\mathcal{C}}$ is the set of all input nodes, $OUT^{\mathcal{C}}$ just contains the output node, and $AND^{\mathcal{C}}, OR^{\mathcal{C}}$, and $NEG^{\mathcal{C}}$ are the sets of and-gates, or-gates, and negation-gates, respectively. Input nodes have in-degree 0, negation-gates in-degree 1, and-gates and or-gates in-degree 2; the output node has out-degree 0.

It is routine to verify that for every circuit $\mathcal{D}$ and every subset S_0 of the set $IN^{\mathcal{D}}$ of input nodes:

$$S_0 \text{ does not satisfy } \mathcal{D} \iff \mathcal{D} \models circunsat(S) \text{ for some } S \text{ with } S \cap IN^{\mathcal{D}} = S_0, \quad (1)$$

where $circunsat(Z) := \forall x \forall y \forall z \psi(x, y, z, Z)$ with

$$
\begin{aligned}
\psi := \ & ((NEG\,x \wedge Eyx) \rightarrow (Zx \leftrightarrow \neg Zy)) \\
& \wedge ((AND\,x \wedge Eyx \wedge Ezx \wedge y \neq z) \rightarrow (Zx \leftrightarrow (Zy \wedge Zz))) \\
& \wedge ((OR\,x \wedge Eyx \wedge Ezx \wedge y \neq z) \rightarrow (Zx \leftrightarrow (Zy \vee Zz))) \\
& \wedge (OUT\,x \rightarrow \neg Zx).
\end{aligned}
$$

Let $(\mathcal{C}, k)$ be an instance of p-WSAT(CIRCUIT) and let $X_1, \ldots X_n$ be the input nodes of $\mathcal{C}$. We turn $\mathcal{C}$ into a new circuit with input nodes $\{X_{i,j} \mid i \in [k] \text{ and } j \in [n]\}$ as follows:

We replace each X_j by a circuit equivalent to $\bigvee_{i \in [k]} X_{i,j}$. We add $2k + 2$ consecutive negation-gates on top of the output node of $\mathcal{C}$. The highest negation-gate is the new output node.

We view the circuit $\mathcal{C}^*$ thus obtained as a $\tau^* := \tau_{\text{circ}} \cup \{DIFF\}$-structure, where $DIFF$ is a binary relation symbol and

$$DIFF^{\mathcal{C}^*} := \{(X_{i,j}, X_{i',j'}) \mid i \neq i' \text{ and } j \neq j'\}.$$

Obviously for $j_1, \ldots, j_k \in [n]$,

$$\{X_{j_1}, \ldots, X_{j_k}\} \text{ satisfies } \mathcal{C} \iff \{X_{1,j_1}, \ldots, X_{k,j_k}\} \text{ satisfies } \mathcal{C}^*. \quad (2)$$

Let

$$\varphi(Z) := \left(circunsat(Z) \vee \forall x(Zx \rightarrow IN\,x) \right) \wedge assign(Z)$$

with $assign(Z) := \forall x \forall y((Zx \wedge Zy \wedge IN\,x \wedge IN\,y \wedge x \neq y) \rightarrow DIFF\,xy)$. Intuitively $\varphi(Z)$ expresses:

(a) either $Z \cap IN$ is not a satisfying assignment and Z is the set of nodes of the circuit set to TRUE by this assignment or Z only contains input gates; and
(b) $Z \cap IN = \{X_{i_1,j_1}, \ldots X_{i_\ell,j_\ell}\}$ for some $\ell \leq k$, pairwise distinct $i_1, \ldots, i_\ell$, and pairwise distinct $j_1, \ldots, j_\ell$.

We claim that

$$\mathcal{C} \text{ is } k\text{-satisfiable} \iff (\mathcal{C}^*, k) \in p\text{-MAXIMAL-WD}_\varphi,$$

which shows the W[P]-hardness of p-MAXIMAL-WD$_\varphi$.

First assume that $\mathcal{C}$ is satisfied by the assignment $S_0 = \{X_{j_1}, \ldots X_{j_k}\}$ of weight k. Let $S_0^* := \{X_{1,j_1}, \ldots X_{k,j_k}\}$. Clearly $\mathcal{C}^* \models \forall x(Zx \rightarrow IN\,x)(S_0^*)$ and $\mathcal{C}^* \models assign(S_0^*)$. Hence $\mathcal{C}^* \models \varphi(S_0^*)$. We show that S_0^* is a maximal solution of $\varphi(Z)$ in $\mathcal{C}^*$. Let $S \supsetneq S_0^*$ and assume that $\mathcal{C}^* \models \varphi(S)$. Since $\mathcal{C}^* \models assign(S)$, we see that $S_0^* = S \cap IN^{\mathcal{C}^*}$, and thus

$S \setminus IN^{\mathcal{C}^*} \neq \emptyset$. Therefore $\mathcal{C}^* \not\models \forall x(Zx \to IN\,x)(S)$. Hence, $\mathcal{C}^* \models circunsat(S)$. Thus, by (1), we know that $S \cap IN^{\mathcal{C}^*}$, that is, the set S_0^*, does not satisfy $\mathcal{C}^*$. Therefore, by (2), S_0 does not satisfy $\mathcal{C}$, a contradiction.

Assume now that $(\mathcal{C}^*, k) \in p\text{-MAXIMAL-WD}_\varphi$ and let S^* be a maximal solution of size k. Then $\mathcal{C}^* \not\models circunsat(S^*)$, as otherwise S^* must contain $k+1$ many of the negation nodes added to $\mathcal{C}$ and hence $|S^*| \geq k+1$. Thus $\mathcal{C}^* \models \big(\forall x(Zx \to IN\,x) \wedge assign\big)(S^*)$. Therefore $S^* = \{X_{1,j_1}, \ldots X_{k,j_k}\}$ for some pairwise distinct $j_1, \ldots, j_k \in [n]$. We set $S_0 := \{X_{j_1}, \ldots, X_{j_k}\}$. Hence S_0 is an assignment of $\mathcal{C}$ of weight k. We claim that S_0 satisfies $\mathcal{C}$. Otherwise by (2), S^* does not satisfy $\mathcal{C}^*$. Then by (1), we have some $S \supset S^*$ with $S^* = S \cap IN^{\mathcal{C}^*}$ such that $\mathcal{C}^* \models circunsat(S)$. It is clear that $\mathcal{C}^* \models assign(S)$. Hence $\mathcal{C}^* \models \varphi(S)$ contradicting the maximality of S^*. $\qquad\square$

Let $circsat(Z)$ be the formula obtained from $circunsat(Z)$ by replacing the conjunct $(OUT\,x \to \neg Zx)$ by $(OUT\,x \to Zx)$. For

$$\varphi(Z) := \forall x(Zx \to OUT\,x) \vee circsat(Z)$$

it is not hard to show that

$$\mathcal{C} \text{ is not satisfiable} \iff (\mathcal{C}, 1) \in p\text{-MAXIMAL-WD}_\varphi$$

for every circuit $\mathcal{C}$ with at least two nodes. From this one easily gets (formulation (2) addresses those readers familiar with the corresponding concepts):

Theorem 8. *(1) There exists a Π_1-formula $\varphi(Z)$ such that $p\text{-MAXIMAL-WD}_\varphi$ is not in W[P] unless P = NP.*
(2) There exists a Π_1-formula $\varphi(Z)$ such that $p\text{-MAXIMAL-WD}_\varphi$ is para-co-NP-hard under fpt-reductions.

4   Maximality Problems for Negative Formulas

Formulas $\varphi(Z)$ negative in Z are antimonotone, that is

$$\mathcal{A} \models \varphi(S) \text{ and } S' \subseteq S \text{ imply } \mathcal{A} \models \varphi(S').$$

For such $\varphi(Z)$ we see that only the empty set can be a minimal solution of $\varphi(Z)$ in $\mathcal{A}$. And the empty set is a minimal solution if and only if $\mathcal{A} \models \varphi(\emptyset)$. Since $\mathcal{A} \models \varphi(\emptyset)$ can be checked in time polynomial in the size of $\mathcal{A}$, the problem $p\text{-MINIMAL-WD}_\varphi$ is fixed-parameter tractable for $\varphi(Z)$ negative in Z.

The antimonotonicity of formulas negative in Z allows to bound the complexity of $p\text{-MAXIMALWD}_\varphi$ as claimed in Theorem 2. By the way it is well-known (cf. [2]) that the following converse of the previous fact is true: If $\varphi(Z)$ is antimonotone in Z in all *finite and infinite* structures, then $\varphi(Z)$ is logically equivalent to a formula $\psi(Z)$ negative in Z. Nevertheless, we do not know whether, in Theorem 2, we can replace "$\varphi(Z)$ a Π_t-formula negative in Z" by "$\varphi(Z)$ a Π_t-formula antimonotone in Z;" in fact, for $t \geq 3$, it is not known if every Π_t-formula antimonotone in Z is equivalent to a formula $\psi(Z)$ in Π_t negative in Z. For $t = 1, 2$ this is true (cf. [2] for $t = 1$ and [14] for $t = 2$).

For a formula $\varphi(Z)$ the following problem asks for solutions that are maximal with respect to extensions by a single element:

---

p-1-MAXIMAL-WD$_\varphi$

   Input: A structure $\mathcal{A}$ and $k \in \mathbb{N}$.

  Parameter: k.

  Question: Is there a solution S of $\varphi(Z)$ in $\mathcal{A}$ of size k such that
     for all $a \in A \setminus S$ the set $S \cup \{a\}$ is not a solution?

For $\varphi(Z)$ negative in Z we get by antimonotonicity:

Lemma 9. *Let* $\varphi(Z)$ *be negative in* Z*. Then* p-MAXIMAL-WD$_\varphi$ = p-1-MAXIMAL-WD$_\varphi$.

For an arbitrary formula $\varphi(Z)$ let

$$1\text{-}max\text{-}\varphi(Z) := \varphi(Z) \wedge \forall y \big(Zy \vee \neg\varphi(Z \cup \{y\}) \big),$$

where $\varphi(Z \cup \{y\})$ denotes the formula obtained from φ by replacing atomic formulas Zx by $(Zx \vee x = y)$. The proof of the following lemma is straightforward.

Lemma 10. *(1)* p-1-MAXIMAL-WD$_\varphi$ = p-WD$_{1\text{-}max\text{-}\varphi}$.
(2) Let $t \geq 1$*. If* $\varphi(Z)$ *is a* Π_t*-formula, then* $1\text{-}max\text{-}\varphi$ *is (equivalent to) a* Π_{t+1}*-formula.*

Using these observations we already get part of Theorem 2:

Lemma 11. *If* $t \geq 1$*, then*

$$\big[\{p\text{-MAXIMAL-WD}_\varphi \mid \varphi(Z) \text{ a } \Pi_t\text{-formula negative in } Z\} \big]^{\text{fpt}} \subseteq W[t+1].$$

Proof: Let $t \geq 1$ and $\varphi(Z)$ be a Π_t-formula negative in Z. Then

$$
\begin{aligned}
p\text{-MAXIMAL-WD}_\varphi &= p\text{-1-MAXIMAL-WD}_\varphi &&\text{(by Lemma 9)}\\
&= p\text{-WD}_{1\text{-}max\text{-}\varphi} &&\text{(by Lemma 10).}
\end{aligned}
$$

Since $1\text{-}max\text{-}\varphi$ is a Π_{t+1}-formula, the problem p-WD$_{1\text{-}max\text{-}\varphi}$ is in $W[t+1]$ by Theorem 1(1). $\qquad\square$

The proof of the remaining parts of Theorem 2 is much more involved and will be presented in the full version of the paper.

In view of Theorem 2 one might conjecture that p-MAXIMAL-WD$_\varphi \in W[1]$ for every Π_1-formula $\varphi(Z)$ negative in Z with p-WD$_\varphi \in$ FPT. In the full version we will disprove this conjecture (unless $W[1] = W[2]$) and show:

Theorem 12. *The problem*

---

p-CLIQUE-OR-INDEPENDENT-SET

   Input: A graph G and $k \in \mathbb{N}$.

  Parameter: k.

  Question: Does G have a clique of size k or an independent set of
     size k?

---

is fixed-parameter tractable and p-MAXIMAL-CLIQUE-OR-INDEPENDENT-SET *is* W[2]-*complete. Moreover, it coincides with p*-WD$_\varphi$ *for some* Π_1-*formula* $\varphi(Z)$.

The fixed-parameter tractability of p-CLIQUE-OR-INDEPENDENT-SET is from [13]. In [1] it is shown that the corresponding counting problem p-#CLIQUE-OR-INDEPEN-DENT-SET is not in FPT (unless FPT = W[1]). We want to remark that most of our results can be extended to counting problems. We do not pursue this topic here (we will address counting problems in the full version of the paper), but only mention the following result which we state for the reader familiar with [9].

Theorem 13. *p*-#MAXIMAL-INDEPENDENT-SET *and* *p*-#MAXIMAL-CLIQUE-OR-INDEPENDENTSET *are* #W[2]-*complete under parsimonious reductions.*

5 Maximal Weighted Satisfiability Problems

We determine the complexity of maximal weighted satisfiability problems for some of the standard classes of propositional formulas considered in parameterized complexity. We obtain our results applying the well-known correspondence between weighted satisfiability problems and Fagin-definable problems stated in the following two lemmas, the first one translates Fagin-definable problems into weighted satisfiability problems and the second one contains a translation in the other direction. For proofs we refer the reader to [12,8].

Lemma 14. *Let* $t, d \geq 1$ *and* $\varphi(Z)$ *be a* $\Pi_{t/d}$-*formula of vocabulary* τ. *Then there is a polynomial time algorithm associating with every* τ-*structure* $\mathcal{A}$ *a propositional formula* $\alpha \in \Gamma_{t,d}$ *such that* Var$(\alpha) \subseteq \{X_a \mid a \in A\}$ *and for all* $S \subseteq A$:

$$\mathcal{A} \models \varphi(S) \iff \{X_b \mid b \in S\} \text{ satisfies } \alpha. \tag{3}$$

If $\varphi(Z)$ *is negative in* Z, *then* α *can be chosen in* $\alpha \in \Gamma_{t,d}^-$.

Lemma 15. *Let* $t, d \geq 1$. *There is a* $\Pi_{t/2d}$-*formula* $\varphi(Z)$ *and a polynomial time algorithm associating with every propositional formula* $\alpha \in \Gamma_{t,d}$ *a structure* $\mathcal{A}$ *in a vocabulary* τ *containing a unary relation symbol VAR with VAR$^{\mathcal{A}}$ = Var(α) and such that for all* $S \subseteq$ Var(α)

$$\mathcal{A} \models \varphi(S) \iff S \text{ satisfies } \alpha. \tag{4}$$

If t *is odd and we only consider formulas* α *in* $\Gamma_{t,d}^-$, *then we can choose* $\varphi(Z)$ *as* $\Pi_{t/d}$-*formula negative in* Z.

From these lemmas one easily obtains:

Corollary 16. *Let* $t, d \geq 1$. *Then:*
(1) $\left[p\text{-MAXIMAL-WSAT}(\Gamma_{t,d}^-) \right]^{\text{fpt}} = \left[\{p\text{-MAXIMAL-WD}_\varphi \mid \varphi(Z) \in \Pi_{t/d} \text{ negative in } Z\} \right]^{\text{fpt}}$.

(2) $\left[\{p\text{-MAXIMAL-WD}_\varphi \mid \varphi(Z) \in \Pi_{t/d'}, \ d' \geq 1\} \right]^{\text{fpt}} = \left[\{p\text{-MAXIMAL-WSAT}(\Gamma_{t,d'}) \mid d' \geq 1\} \right]^{\text{fpt}}$.

Proof: (1) Let $\varphi(Z) \in \Pi_{t/d}$ be negative in Z and $(\mathcal{A}, k)$ an instance of p-MAXIMAL-WD$_\varphi$. Choose $\alpha \in \Gamma_{t,d}^-$ according to Lemma 14. Let S_0 be the set of $a \in A$ such that the variable X_a does not occur in α and let $\ell_0 := |S_0|$. Then, by (3), for all $S \subseteq A$

$$\mathcal{A} \models \varphi(S) \iff \mathcal{A} \models \varphi(S \cup S_0). \tag{5}$$

The equivalences (3) and (5) show that

$$(\mathcal{A}, k) \in p\text{-MAXIMAL-WD}_{\cdot\varphi} \iff (\alpha, k - \ell_0) \in p\text{-MAXIMAL-WSAT}(\Gamma_{t,d}^-),$$

which yields a reduction from p-MAXIMAL-WD$_\varphi$ to p-MAXIMAL-WSAT$(\Gamma_{t,d}^-)$.

Let $t \geq 1$ be odd. Using Lemma 15 one gets a reduction from p-MAXIMAL-WSAT$(\Gamma_{t,d}^-)$ to p-MAXIMAL-WD$_\psi$ for some $\Pi_{t/d}$-formula $\psi(Z)$ negative in Z. In fact, we can choose as $\psi(Z)$ the formula $\varphi(Z) \wedge \forall x(Zx \rightarrow \text{VAR } x)$, where $\varphi(Z)$ negative in Z is as in Lemma 15. For odd one applies the Antimonotone Collapse Theorem.

The proof of (2) is similar.   □

Theorem 17. *(1) Let $t \geq 1$ and $d \geq 2$. Then p-MAXIMAL-WSAT$(\Gamma_{t,d}^-)$ is W$[t+1]$-complete for odd t and W$[t]$-complete for even t.*
(2) p-MAXIMAL-WSAT$(\Gamma_{1,4})$ is W$[P]$-hard.

Proof: Part (1) follows from Theorem 2 and its proof using Corollary 16. For (2) we invoke Theorem 7, thereby noting that when translating the formula $\varphi(Z)$ of the proof of Theorem 7 into a propositional formula according to Lemma 14, one obtains a formula in $\Gamma_{1,d}$ for $d = 4$.   □

We close this section by stating the results for minimality weighted satisfiability problems:

Theorem 18. *(1) p-MINIMAL-WSAT$(\Gamma_{1,d})$ is fixed-parameter tractable for all $d \geq 1$.*
(2) p-MINIMAL-WSAT$(\Gamma_{t,d})$ is W$[t]$-complete for all $t \geq 2$ and $d \geq 1$.
(3) p-MINIMAL-WSAT$(\Gamma_{t,d}^+)$ and p-MINIMAL-WSAT$(\Gamma_{t+1,d}^+)$ are W$[t]$-complete for all even $t \geq 2$ and $d \geq 1$.

6 Maximal Solutions with Respect to Extensions by a Single Element

As seen in Theorem 2, for odd t, the "maximality operation" applied to problems in W$[t]$ yields (up to fpt-reductions) the class W$[t+1]$. In the full version of the paper we will show that the "1-maximality operation" has this property for *all* t, that is:

Theorem 19. *For every $t \geq 1$, $\text{W}[t+1] = \left[\{p\text{-}1\text{-MAXIMAL-WD}_\varphi \mid \varphi(Z) \in \Pi_t\}\right]^{\text{fpt}}$.*

7 Conclusions

Using the notion of Fagin-definability we have analyzed the parameterized complexity of maximality and minimality problems. We believe that our results show that Fagin-definability yields a very appropriate framework for the study of these type of questions. In particular, as already mentioned, in this framework we will address the corresponding counting problems in the full version of the paper.

Let us close with a remark on W[P], which often is seen as the parameterized analogue of NP. In [3] it is shown that W[P] precisely consists of the problems fpt-reducible to a problem p-WD$_\varphi$, where $\varphi(Z)$ is a formula of fixpoint logic. For such a formula φ the formula 1-max-$\varphi(Z)$ is again in fixpoint logic. Hence: If $\varphi(Z)$ is a formula of fixpoint logic negative in Z, then p-MAXIMAL-WD$_\varphi \in$ W[P].

References

1. V. Arvind and V. Raman. Approximation algorithms for some parameterized counting problems. In P. Bose and P. Morin, editors, *Algorithms and Computation, 13th International Symposium, ISAAC 2002*, volume 2518 of *Lecture Notes in Computer Science*, 453–464, Springer, 2002.
2. C.C. Chang and H. J. Keisler. *Model Theory*, Studies in Logic and Mathematical Foundations, Volume 73. Amsterdam, the Netherlands, North-Holland, 1990.
3. Y. Chen, J. Flum, and M. Grohe. Bounded nondeterminism and alternation in parameterized complexity theory. In *Proceedings of the 18rd IEEE Conference on Computational Complexity CCC '03*, 13–29, 2003.
4. R.G. Downey and M.R. Fellows. *Parameterized Complexity*. Springer-Verlag, 1999.
5. R.G. Downey, M.R. Fellows, and K. Regan. Descriptive complexity and the W-hierarchy. In *Proof Complexity and Feasible Arithmetic, AMS-DIMACS Volume 39 Series*, 119–134. 1998.
6. T. Eiter and G. Gottlob. Identifying the minimal transversals of a hypergraph and related problems. *SIAM Journal on Computing*, 24(6): 1278-1304, 1995.
7. H. Fernau. Parameterized algorithms: A graph-theoretic approach. Habilitationsschrift, Universität Tübingen, Tübingen, Germany, 2005.
8. J. Flum and M. Grohe. Fixed-parameter tractability, definability, and model checking. *SIAM Journal on Computing*, 31(1):113–145, 2001.
9. J. Flum and M. Grohe. The parameterized complexity of counting problems. *SIAM Journal on Computing*, 33(4):892–922 2005.
10. J. Flum and M. Grohe. Model-checking problems as a basis for parameterized intractability. *Logical Methods in Computer Science*, 1(1), 2004.
11. J. Flum and M. Grohe. *Parameterized Complexity Theory*. Springer-Verlag, 2006.
12. J. Flum, M. Grohe, and M. Weyer Bounded fixed-parameter tractability and $\log^2 n$ nondeterministic bits. *Journal of Computer and System Sciences*, 72: 34–71, 2006.
13. S. Khot and V. Raman. Parameterized complexity of finding subgraphs with hereditary properties. *Theoretical Computer Science*, 289(2): 997–1008, 2002.
14. C. Ritter. Fagin-Definierbarkeit. Diplomarbeit, Universität Freiburg, 2005.

Parameterizing MAX SNP Problems Above Guaranteed Values

Meena Mahajan, Venkatesh Raman, and Somnath Sikdar

The Institute of Mathematical Sciences, C.I.T Campus, Taramani, Chennai 600113
{meena, vraman, somnath}@imsc.res.in

Abstract. We show that every problem in MAX SNP has a lower bound on the optimum solution size that is unbounded and that the above guarantee question with respect to this lower bound is fixed parameter tractable. We next introduce the notion of "tight" upper and lower bounds for the optimum solution and show that the parameterized version of a variant of the above guarantee question with respect to the tight lower bound cannot be fixed parameter tractable unless P = NP, for a class of NP-optimization problems.

1 Introduction

In this paper, we consider the parameterized complexity of NP-optimization problems Q with the following property: for non-trivial instance I of Q, the optimum $opt(I)$, is lower-bounded by an increasing function of the input size. That is, there exists a function $f : \mathbb{N} \to \mathbb{N}$ which is increasing such that for non-trivial instances I, $opt(I) \geq f(|I|)$. For such an optimization problem Q, the standard parameterized version $\tilde{Q}$ defined below is easily seen to be fixed parameter tractable. For if $k \leq f(|I|)$, we answer 'yes'; else, $f(|I|) < k$ and so $|I| < f^{-1}(k)$[1] and we have a kernel.

$$\tilde{Q} = \{(I, k) \ : \ I \text{ is an instance of } Q \text{ and } opt(I) \geq k\}$$

Thus for such an optimization problem it makes sense to define an "above guarantee" parameterized version $\bar{Q}$ as

$$\bar{Q} = \{(I, k) \ : \ I \text{ is an instance of } Q \text{ and } opt(I) \geq f(|I|) + k\}.$$

Such above guarantee parameterized problems were first considered by Mahajan and Raman in [5]. The problems dealt with by them are MAX SAT and MAX CUT. An instance of the MAX SAT problem is a boolean formula ϕ in conjunctive normal form and the standard parameterized version asks whether ϕ has at least k satisfiable clauses, k being the parameter. Since any boolean formula ϕ with m clauses has at least $\lceil m/2 \rceil$ satisfiable clauses (see Motwani and Raghavan [6]), by the above argument, this problem is fixed parameter tractable. The above guarantee MAX SAT question considered in [5] asks whether a given formula ϕ has at least $\lceil m/2 \rceil + k$ satisfiable clauses, with k as parameter. This was shown to be fixed parameter tractable.

The standard parameterized version of the MAX CUT problem asks whether an input graph G has a cut of size at least k, where k is the parameter. This problem is also fixed

---

[1] Assuming f to be invertible; the functions considered in this paper are.

H.L. Bodlaender and M.A. Langston (Eds.): IWPEC 2006, LNCS 4169, pp. 38–49, 2006.

parameter tractable since any graph G with m edges has a cut of size $\lceil m/2 \rceil$. The above guarantee MAX CUT question considered in [5] asks whether an input graph G on m edges has a cut of size at least $\lceil m/2 \rceil + k$, where k is the parameter. This problem was shown to be fixed parameter tractable too.

In this paper, we consider above guarantee questions for problems in the class MAX SNP. This paper is structured as follows. In Section 2, we introduce the necessary ideas about parameterized complexity and state some basic definitions needed in the rest of the paper. In Section 3, we show that every problem in the class MAX SNP has a guaranteed lower bound that is an unbounded function of the input size and that the above guarantee problem with respect to this lower bound is fixed parameter tractable. In Section 4, we define a notion of *tight lower bound* and show that a variant of the above guarantee question with respect to tight lower bounds is hard (unless P = NP) for a number of NP-maximization problems. Finally in Section 5, we end with a few concluding remarks.

2 Preliminaries

We briefly introduce the necessary concepts concerning optimization problems and parameterized complexity.

To begin with, a parameterized problem is a subset of $\Sigma^* \times \mathbb{N}$, where Σ is a finite alphabet and $\mathbb{N}$ is the set of natural numbers. An instance of a parameterized problem is therefore a pair (I, k), where k is the parameter. In the framework of parameterized complexity, the run time of an algorithm is viewed as a function of two quantities: the size of the problem instance *and* the parameter. A parameterized problem is said to be *fixed parameter tractable* (fpt) if there exists an algorithm for the problem with time complexity $O(f(k) \cdot |I|^{O(1)})$, where f is a recursive function of k alone. The class FPT consists of all fixed parameter tractable problems.

A parameterized problem π_1 is *fixed-parameter-reducible* to a parameterized problem π_2 if there exist functions $f, g : \mathbb{N} \to \mathbb{N}$, $\Phi : \Sigma^* \times \mathbb{N} \to \Sigma^*$ and a polynomial $p(\cdot)$ such that for any instance (I, k) of π_1, $(\Phi(I, k), g(k))$ is an instance of π_2 computable in time $f(k) \cdot p(|I|)$ and $(I, k) \in \pi_1$ if and only if $(\Phi(I, k), g(k)) \in \pi_2$.

An NP-optimization problem Q is a 4-tuple $Q = \{\mathscr{I}, S, V, \text{opt}\}$, where

1. $\mathscr{I}$ is the set of input instances. (w.l.o.g., $\mathscr{I}$ can be recognized in polynomial time.)
2. $S(x)$ is the set of feasible solutions for the input $x \in \mathscr{I}$.
3. V is a polynomial-time computable function called the *cost function* and for each $x \in \mathscr{I}$ and $y \in S(x)$, $V(x, y) \in \mathbb{N}$.
4. opt $\in \{\max, \min\}$.
5. The following decision problem (called the *underlying decision problem*) is in NP: Given $x \in \mathscr{I}$ and an integer k, does there exist a feasible solution $y \in S(x)$ such that $V(x, y) \geq k$, when Q is a maximization problem (or, $V(x, y) \leq k$, when Q is a minimization problem).

The class MAX SNP was defined by Papadimitriou and Yannakakis [7] using logical expressiveness. They showed that a number of interesting optimization problems such as MAX 3-SAT, INDEPENDENT SET-b, MAX CUT, MAX k-COLORABLE SUBGRAPH etc. lie in this class. They also introduced the notion of completeness for **MAX SNP** by a reduction known as the *L-reduction*. We define this next.

Let Q_1 and Q_2 be two optimization (maximization or minimization) problems. We say that Q_1 *L-reduces* to Q_2 if there exist polynomial-time computable functions f, g, and constants $\alpha, \beta > 0$ such that for each instance I_1 of Q_1:

1. $f(I_1) = I_2$ is an instance of Q_2, such that $\text{opt}(I_2) \leq \alpha \cdot \text{opt}(I_1)$.
2. Given any solution y_2 of I_2, g maps (I_2, y_2) to a solution y_1 of I_1 such that
$$|V(I_1, y_1) - \text{opt}(I_1)| \leq \beta \cdot |V(I_2, y_2) - \text{opt}(I_2)|$$

We call such an *L*-reduction from Q_1 to Q_2 an $\langle f, g, \alpha, \beta \rangle$ reduction.

A problem Q is **MAX SNP**-*hard* if every problem in the class **MAX SNP** *L*-reduces to Q. A problem Q is **MAX SNP**-*complete*, if Q is in **MAX SNP** and is **MAX SNP**-hard. Cai and Chen [1] established that all maximization problems in the class **MAX SNP** are fixed parameter tractable. In the next section, we show that for all problems in **MAX SNP**, a certain above-guarantee question is also fixed parameter tractable.

3   Parameterizing Above Guaranteed Values

Consider the problem MAX 3-SAT which is complete for the class **MAX SNP**. An instance of MAX 3-SAT is a boolean formula f in conjunctive normal form with at most three literals per clause. As already stated, any boolean formula with m clauses has at least $\lceil m/2 \rceil$ satisfiable clauses, and the following above guarantee parameterized problem is fixed parameter tractable.

$$L = \{(f, k) : f \text{ is a MAX 3-SAT instance and } \exists \text{ an assignment satisfying}$$
$$\text{at least } k + \lceil m/2 \rceil \text{ clauses of the formula } f \}.$$

Since MAX 3-SAT is **MAX SNP**-complete and has a guaranteed lower bound, we have

Proposition 1. *If Q is in* **MAX SNP***, then for each instance x of Q there exists a positive number γ_x such that $\gamma_x \leq \text{opt}(x)$. Further, if Q is* NP*-hard, then the function $\gamma : x \to \gamma_x$ is unbounded, assuming* P $\neq$ NP.

Proof. Let Q be a problem in **MAX SNP** and let $\langle f, g, \alpha, \beta \rangle$ be an *L*-reduction from Q to MAX 3-SAT. Then for an instance x of Q, $f(x)$ is an instance of MAX 3-SAT such that $\text{opt}(f(x)) \leq \alpha \cdot \text{opt}(x)$. If $f(x)$ is a formula with m clauses, then $\lceil m/2 \rceil \leq \text{opt}(f(x))$ and therefore $\text{opt}(x)$ is bounded below by $\lceil m/2 \rceil / \alpha$. This proves that each instance x of Q has a lower bound. We can express this lower bound in terms of the parameters of the *L*-reduction. Since $f(x)$ is an instance of MAX 3-SAT, we can take the size of $f(x)$ to be m. Then $\gamma_x = |f(x)|/(2 \cdot \alpha)$. Further, note that if m is not unbounded, then we can solve Q in polynomial time via this reduction. ∎

Note that this lower bound γ_x depends on the complete problem to which we reduce Q. By changing the complete problem, we might construct different lower bounds for the problem at hand. It is also conceivable that there exist more than one *L*-reduction between two optimization problems. Different *L*-reductions should give different lower bounds. Thus the polynomial-time computable lower bound that we exhibit in Proposition 1 is a special lower bound obtained from a specific L-reduction to a specific complete problem (MAX 3-SAT) for the class **MAX SNP**. Call the lower bound of Proposition 1 a MAX 3-SAT-lower bound for the problem Q.

Since the above guarantee parameterized version L of MAX 3-SAT is known to be in FPT, we immediately have the following.

Theorem 1. *For a maximization problem Q in* **MAX SNP**, *let $\langle f, g, \alpha, \beta \rangle$ be an L-reduction from Q to* MAX 3-SAT, *and for an instance x of Q, let γ_x represent the corresponding* MAX 3-SAT-*lower bound. Then the following problem is in FPT:*

$$L_Q = \{\langle x, k \rangle : x \text{ is an instance of } Q \text{ and } opt(x) \geq \gamma_x + k\}$$

Proof. We make use of the fact that there exists a fixed parameter tractable algorithm $\mathscr{A}$ for MAX 3-SAT which takes as input, a pair of the form $\langle \psi, k \rangle$, and in time $O(|\psi| + h(k))$, returns YES if there exists an assignment to the variables of ψ that satisfies at least $\lceil m/2 \rceil + k$ clauses, and NO otherwise. See [5,9] for such algorithms.

Consider an instance $\langle x, k \rangle$ of L_Q. Then $f(x)$ is an instance of MAX 3-SAT. Let $f(x)$ have m clauses. Then the guaranteed lower bound for the instance x of Q, $\gamma_x = \frac{m}{2\alpha}$, and $opt(f(x)) \leq \alpha \cdot opt(x)$. Apply algorithm $\mathscr{A}$ on input $\langle f(x), k\alpha \rangle$. If $\mathscr{A}$ outputs YES, then $opt(f(x)) \geq m/2 + k \cdot \alpha$, implying $opt(x) \geq \frac{m}{2 \cdot \alpha} + k = \gamma_x + k$. Thus $\langle x, k \rangle \in L_Q$.

If $\mathscr{A}$ answers NO, then $\lceil \frac{m}{2} \rceil \leq opt(f(x)) < \lceil \frac{m}{2} \rceil + k\alpha$. Apply algorithm $\mathscr{A}$ $k\alpha$ times on inputs $(f(x), 1), (f(x), 2), \ldots, (f(x), k\alpha)$ to obtain $opt(f(x))$. Let $c' = opt(f(x))$. Then use algorithm g of the L-reduction to obtain a solution to x with cost c. By the definition of L-reduction, we have $|c - opt(x)| \leq \beta \cdot |c' - opt(f(x))|$. But since $c' = opt(f(x))$, it must be that $c = opt(x)$. Therefore we simply need to compare c with $\gamma_x + k$ to check whether $\langle x, k \rangle \in L_Q$.

The total time complexity of the above algorithm is $O(k\alpha \cdot (|f(x)| + h(k\alpha)) + p_1(|x|) + p_2(|f(x)|))$, where $p_1(\cdot)$ is the time taken by algorithm f to transform an instance of Q to an instance of MAX 3-SAT, and $p_2(\cdot)$ is the time taken by g to output its answer. Thus the algorithm that we outlined is indeed an FPT algorithm for L_Q. ∎

Note that the proof of Proposition 1 also shows that every minimization problem in **MAX SNP** has a MAX 3-SAT-lower bound. For minimization problems whose optimum is lower bounded by some function of the input, it makes sense to ask how far removed the optimum is with respect to the lower bound. The parameterized question asks whether for a given input x, $opt(x) \leq \gamma_x + k$, with k as parameter. The following result can be proved similarly to Theorem 1.

Theorem 2. *For a minimization problem Q in* **MAX SNP**, *let $\langle f, g, \alpha, \beta \rangle$ be an L-reduction from Q to* MAX 3-SAT, *and for an instance x of Q, let γ_x represent the corresponding* MAX 3-SAT-*lower bound. Then the following problem is in FPT:*

$$L_Q = \{\langle x, k \rangle : x \text{ is an instance of } Q \text{ and } opt(x) \leq \gamma_x + k\}$$

Examples of minimization problems in **MAX SNP** include VERTEX COVER-B and DOMINATING SET-B which are, respectively, the restriction of the VERTEX COVER and the DOMINATING SET problems to graphs whose vertex degree is bounded by B.

4 Hardness Results

For an optimization problem, the question of whether the optimum is at least lower bound + k, for some lower bound and with k as parameter, is not always interesting because if the lower bound is "loose" then the problem is trivially fixed parameter tractable. For instance, for the MAX CUT problem, the question of whether an input graph has a cut of size at least $\frac{m}{2} + k$ is fpt since any graph G with m edges, n vertices and c components has a cut of size at least $\frac{m}{2} + \lceil \frac{n-c}{4} \rceil$ [8]. Thus if $k \leq \lceil \frac{n-c}{4} \rceil$, we answer YES; else, $\lceil \frac{n-c}{4} \rceil < k$ and we have a kernel.

We therefore examine the notion of a *tight lower bound* and the corresponding above guarantee question. A tight lower bound is essentially the best possible lower bound on the optimum solution size. For the MAX SAT problem, this lower bound is $m/2$: if ϕ is an instance of MAX SAT, then $\mathrm{opt}(\phi) \geq m/2$, and there are infinitely many instances for which the optimum is *exactly* $m/2$. This characteristic motivates the next definition.

Definition 1 (Tight Lower Bound). Let $Q = \{\mathscr{I}, S, V, \mathrm{opt}\}$ be an NP-optimization problem and let $f : \mathbb{N} \to \mathbb{N}$. We say that f is a **tight lower bound** for Q if the following conditions hold:

1. $f(|I|) \leq \mathrm{opt}(I)$ for all $I \in \mathscr{I}$.
2. There exists an infinite family of instances $\mathscr{I}' \subseteq \mathscr{I}$ such that $\mathrm{opt}(I) = f(|I|)$ for all $I \in \mathscr{I}'$.

Note that we define the lower bound to be a function of the *input size* rather than the input itself. This is in contrast to the lower bound of Proposition 1 which depends on the input instance. We can define the notion of a tight *upper* bound analogously.

Definition 2 (Tight Upper Bound). Let $Q = \{\mathscr{I}, S, V, \mathrm{opt}\}$ be an NP-optimization problem and let $g : \mathbb{N} \to \mathbb{N}$. We say that g is a **tight upper bound** for Q if the following conditions hold:

1. $\mathrm{opt}(I) \leq g(|I|)$ for all $I \in \mathscr{I}$.
2. There exists an infinite family of instances $\mathscr{I}' \subseteq \mathscr{I}$ such that $\mathrm{opt}(I) = g(|I|)$ for all $I \in \mathscr{I}'$.

Some example optimization problems which have tight lower and upper bounds are given below. The abbreviations TLB and TUB stand for tight lower bound and tight upper bound, respectively.

1. MAX EXACT c-SAT

INSTANCE   A boolean formula F with n variables and m clauses with each clause having *exactly* c distinct literals.

QUESTION   Find the maximum number of simultaneously satisfiable clauses.

BOUNDS   TLB $= (1 - \frac{1}{2^c})m$;   TUB $= m$.

The expected number of clauses satisfied by the random assignment algorithm is $(1 - \frac{1}{2^c})m$; hence the lower bound. To see tightness, note that if $\phi(x_1, \ldots, x_c)$ denotes the EXACT c-SAT formula comprising of all possible combinations of c variables, then ϕ has 2^c clauses of which exactly $2^c - 1$ clauses are satisfiable. By taking disjoint copies of this formula one can construct EXACT c-SAT instances of arbitrary size with exactly $(1 - \frac{1}{2^c})m$ satisfiable clauses.

2. CONSTRAINT SATISFACTION PROBLEM (CSP)

INSTANCE   A system of m linear equations modulo 2 in n variables, together with positive weights w_i, $1 \le i \le m$.

QUESTION   Find an assignment to the variables that maximizes the total weight of the satisfied equations.

BOUNDS   TLB $= \frac{W}{2}$, where $W = \sum_{i=1}^{m} w_i$;   TUB $= W$.

If we use $\{+1, -1\}$-notation for boolean values with -1 corresponding to true then we can write the ith equation of the system as $\prod_{j \in \alpha_i} x_j = b_i$, where each α_i is a subset of $[n]$ and $b_i \in \{+1, -1\}$. To see that we can satisfy at least half the equations in the weighted sense, we assign values to the variables sequentially and simplify the system as we go along. When we are about to give a value to x_j, we consider all equations reduced to the form $x_j = b$, for a constant b. We choose a value for x_j satisfying at least half (in the weighted sense) of these equations. This procedure of assigning values ensures that we satisfy at least half the equations in the weighted sense. A tight lower bound instance, in this case, is a system consisting of pairs $x_j = b_i, x_j = \bar{b}_i$, with each equation of the pair assigned the same weight. See [3] for more details.

3. MAX INDEPENDENT SET-B

INSTANCE   A graph G with n vertices such that the degree of each vertex is bounded by B.

QUESTION   Find a maximum independent set of G.

BOUNDS   TLB $= \frac{n}{B+1}$;   TUB $= n$.

A graph whose vertex degree is bounded by B can be colored using $B + 1$ colors, and in any valid coloring of the graph, the vertices that get the same color form an independent set. By the pigeonhole principle, there exists an independent set of size at least $n/(B + 1)$. The complete graph K_{B+1} on $B + 1$ vertices has an independence number of $\frac{n}{B+1}$. By taking disjoint copies of K_{B+1} one can construct instances of arbitrary size with independence number exactly $\frac{n}{B+1}$.

4. MAX PLANAR INDEPENDENT SET

INSTANCE   A planar graph G with n vertices and m edges.

QUESTION   Find a maximum independent set of G.

BOUNDS   TLB $= \frac{n}{4}$;   TUB $= n$.

A planar graph is 4-colorable, and in any valid 4-coloring of the graph, the vertices that get the same color form an independent set. By the pigeonhole principle, there exists an independent set of size at least $\frac{n}{4}$. A disjoint set of K_4's can be use to construct arbitrary sized instances with independence number exactly $\frac{n}{4}$.

5. MAX ACYCLIC DIGRAPH

INSTANCE   A directed graph G with n vertices and m edges.

QUESTION   Find a maximum acyclic subgraph of G.

BOUNDS   TLB $= \frac{m}{2}$;   TUB $= m$.

To see that any digraph with m arcs has an acyclic subgraph of size $\frac{m}{2}$, place the vertices $v_1, \ldots, v_n$ of G on a line in that order with arcs (v_i, v_j), $i < j$, drawn above the line and arcs (v_i, v_j), $i > j$, drawn below the line. Clearly, by deleting all arcs either above or below the line we obtain an acyclic digraph. By the pigeonhole principle, one of these two sets must have size at least $\frac{m}{2}$. To see that this bound is tight, consider the digraph D on n vertices: $v_1 \leftrightarrows v_2 \leftrightarrows v_3 \leftrightarrows \ldots \leftrightarrows v_n$ which has a maximum acyclic digraph of size exactly $\frac{m}{2}$. Since n is arbitrary, we have an infinite set of instances for which the optimum matches the lower bound exactly.

6. Max Planar Subgraph

INSTANCE   A connected graph G with n vertices and m edges.
QUESTION   Find an edge-subset E' of maximum size such that $G[E']$ is planar.
BOUNDS   TLB $= n - 1$;   TUB $= 3n - 6$.

Any spanning tree of G has $n - 1$ edges; hence any maximum planar subgraph of G has at least $n - 1$ edges. This bound is tight as the family of all trees achieves this lower bound. An upper bound is $3n - 6$ which is tight since for each n, a maximal planar graph on n vertices has exactly $3n - 6$ edges.

7. Max Cut

INSTANCE   A graph G with n vertices, m edges and c components.
QUESTION   Find a maximum cut of G.
BOUNDS   TLB $= \frac{m}{2} + \lceil \frac{n-c}{4} \rceil$;   TUB $= m$.

The lower bound for the cut size was proved by Poljak and Turzík [8]. This bound is tight for complete graphs. The upper bound is tight for bipartite graphs.

A natural question to ask in the above-guarantee framework is whether the language

$$L = \{\langle I, k\rangle : \text{opt}(I) \geq \text{TLB}(I) + k\}$$

is in **FPT**. The parameterized complexity of such a question is not known for most problems. To the best of our knowledge, this question has been resolved only for the Max Sat and Max c-Sat problems [5] and, very recently, for the Linear Arrangement problem [2].

In this section, we study a somewhat different, but related, parameterized question: Given an **NP**-maximization problem Q which has a tight lower bound (TLB) a function of the input size, what is the parameterized complexity of the following question?

$$Q(\epsilon) = \{\langle I, k\rangle : \text{opt}(I) \geq \text{TLB}(I) + \epsilon \cdot |I| + k\}$$

Here $|I|$ denotes the input size, ϵ is some fixed positive rational and k is the parameter. We show that this question is not fixed parameter tractable for a number of problems, unless **P = NP**.

Theorem 3. *For any problem Q in the following, the $Q(\epsilon)$ problem is not fixed parameter tractable unless* **P = NP**:

Problem	TLB$(I) + \epsilon \cdot |I| + k$	Range of ϵ
1. Max Sat	$(\frac{1}{2} + \epsilon)m + k$	$0 < \epsilon < \frac{1}{2}$
2. Max c-Sat	$(\frac{1}{2} + \epsilon)m + k$	$0 < \epsilon < \frac{1}{2}$

3. MAX EXACT c-SAT	$(1 - \frac{1}{2^c} + \epsilon)m + k$	$0 < \epsilon < \frac{1}{2^c}$
4. CSP	$(\frac{1}{2} + \epsilon)m + k$	$0 < \epsilon < \frac{1}{2}$
5. PLANAR INDEPENDENT SET	$(\frac{1}{4} + \epsilon)n + k$	$0 < \epsilon < \frac{3}{4}$
6. INDEPENDENT SET-B	$(\frac{1}{B+1} + \epsilon)n + k$	$0 < \epsilon < \frac{B}{B+1}$
7. MAX ACYCLIC SUBGRAPH	$(\frac{1}{2} + \epsilon)m + k$	$0 < \epsilon < \frac{1}{2}$
8. MAX PLANAR SUBGRAPH	$(1 + \epsilon)n - 1 + k$	$0 < \epsilon < 2$
9. MAX CUT	$\frac{m}{2} + \lceil \frac{n-c}{4} \rceil + \epsilon n + k$	$0 < \epsilon < \frac{1}{4}$
10. MAX DICUT	$\frac{m}{4} + \sqrt{\frac{m}{32} + \frac{1}{256}} - \frac{1}{16} + \epsilon m + k$	$0 < \epsilon < \frac{3}{4}$

The proof, in each case, follows this outline: Assume that for some ϵ in the specified range, $Q(\epsilon)$ is indeed in FPT. Now consider an instance $\langle I, s \rangle$ of the underlying decision version of Q. Here is a P-time procedure for deciding it. If $s \leq$ TLB, then the answer is trivially YES. If s lies between TLB and TLB $+ \epsilon|I|$, then "add" a gadget of suitable size corresponding to the TUB, to obtain an equivalent instance $\langle I', s' \rangle$. This increases the input size, but since we are adding a gadget whose optimum value matches the upper bound, the increase in the optimum value of I' is more than proportional, so that now s' exceeds TLB $+ \epsilon|I'|$. If s already exceeds TLB $+ \epsilon|I|$, then "add" a gadget of suitable size corresponding to the TLB, to obtain an equivalent instance $\langle I', s' \rangle$. This increases the input size faster than it boosts the optimum value of I', so that now s' exceeds TLB $+ \epsilon|I'|$ by only a constant, say c_1. Use the hypothesized fpt algorithm for $Q(\epsilon)$ with input $\langle I', c_1 \rangle$ to correctly decide the original question.

Rather than proving the details for each item separately, we use this proof sketch to establish a more general theorem (Theorem 4 below) which automatically implies items 1 through 10 above. We first need some definitions.

Definition 3 (Dense Set). Let $Q = \{\mathscr{I}, S, V, \text{opt}\}$ be an NPO problem. A set of instances $\mathscr{I}' \subseteq \mathscr{I}$ is said to be *dense with respect to a set of conditions* $\mathcal{C}$ if there exists a constant $c \in \mathbb{N}$ such that for all closed intervals $[a, b] \subseteq \mathbb{R}^+$ of length $|b - a| \geq c$, there exists an instance $I \in \mathscr{I}'$ with $|I| \in [a, b]$ such that I satisfies all the conditions in $\mathcal{C}$. Further, if such an I can be found in polynomial time (polynomial in b), then $\mathscr{I}'$ is said to be *dense poly-time uniform with respect to* $\mathcal{C}$.

For example, for the MAXIMUM ACYCLIC SUBGRAPH problem, the set of all oriented digraphs is dense (poly-time uniform) with respect to the condition: $\text{opt}(G) = |E(G)|$.

We also need the notion of a partially additive NP-optimization problem.

Definition 4 (Partially Additive Problems). An NPO problem $Q = \{\mathscr{I}, S, V, \text{opt}\}$ is said to be *partially additive* if there exists an operator $+$ which maps a pair of instances I_1 and I_2 to an instance $I_1 + I_2$ such that

1. $|I_1 + I_2| = |I_1| + |I_2|$, and
2. $\text{opt}(I_1 + I_2) = \text{opt}(I_1) + \text{opt}(I_2)$.

A partially additive NPO problem that also satisfies the following condition is said to be additive in the framework of Khanna, Motwani et al [4]: there exists a polynomial-time computable function f that maps any solution s of $I_1 + I_2$ to a pair of solutions s_1 and s_2 of I_1 and I_2, respectively, such that $V(I_1 + I_2, s) = V(I_1, s_1) + V(I_1, s_2)$.

For many graph-theoretic optimization problems, the operator $+$ can be interpreted as disjoint union. Then the problems MAX CUT, MAX INDEPENDENT SET-B, MINIMUM VERTEX COVER, MINIMUM DOMINATING SET, MAXIMUM DIRECTED ACYCLIC SUBGRAPH, MAXIMUM DIRECTED CUT are partially additive. For other graph-theoretic problems, one may choose to interpret $+$ as follows: given graphs G and H, $G + H$ refers to a graph obtained by placing an edge between some (possibly arbitrarily chosen) vertex of G and some (possibly arbitrarily chosen) vertex of H. The MAX PLANAR SUBGRAPH problem is partially additive with respect to both these interpretations of $+$. For boolean formulae ϕ and ψ in conjunctive normal form with disjoint sets of variables, define $+$ as the conjunction $\phi \wedge \psi$. Then the MAX SAT problem is easily seen to be partially additive.

Let $Q = \{\mathscr{I}, S, V, \max\}$ be an NP-maximization problem with tight lower bound $f : \mathbb{N} \to \mathbb{N}$ and tight upper bound $g : \mathbb{N} \to \mathbb{N}$. We assume that both f and g are increasing and satisfy the following conditions

P1 For all $a, b \in \mathbb{N}$, $f(a + b) \leq f(a) + f(b) + c^*$, where c^* is a constant (positive or negative),

P2 There exists $n_0 \in \mathbb{N}$ and $r \in \mathbb{Q}^+$ such that $g(n) - f(n) > rn$ for all $n \geq n_0$.

Property $P1$ is satisfied by linear functions ($f(n) = an + b$) and by some sub-linear functions such as $\sqrt{n}, \log n, \frac{1}{n}$. Note that a super-linear function cannot satisfy $P1$. Define $\mathscr{R}$ to be the set

$$\mathscr{R} = \{r \in \mathbb{Q}^+ : g(n) - f(n) > rn \text{ for all } n \geq n_0\},$$

and $p = \sup \mathscr{R}$. For $0 < \epsilon < p$, define $Q(\epsilon)$ as follows

$$Q(\epsilon) = \{(I, k) : I \in \mathscr{I} \text{ and } \max(I) \geq f(|I|) + \epsilon|I| + k\}.$$

Note that for $0 < \epsilon < p$, the function h defined by $h(n) = g(n) - f(n) - \epsilon n$ is strictly increasing, and $h(n) > 0 \; \forall n \geq n_0 \in \mathbb{N}$.

Theorem 4. *Let $Q = \{\mathscr{I}, S, V, \max\}$ be a polynomially bounded NP-maximization problem such that the following conditions hold.*

1. *Q is partially additive.*
2. *Q has a tight lower bound (TLB) f, which is increasing and satisfies condition $P1$. The infinite family of instances $\mathscr{I}'$ witnessing the tight lower bound is dense poly-time uniform with respect to the condition $\max(I) = f(|I|)$.*
3. *Q has a tight upper bound (TUB) g, which with f satisfies condition $P2$. The infinite family of instances $\mathscr{I}'$ witnessing the tight upper bound is dense poly-time uniform with respect to the condition $\max(I) = g(|I|)$.*
4. *The underlying decision problem $\tilde{Q}$ of Q is NP-hard.*

For $0 < \epsilon < p$, define $Q(\epsilon)$ to be the following parameterized problem

$$Q(\epsilon) = \{(I, k) : \max(I) \geq f(|I|) + \epsilon|I| + k\}$$

where $p = \sup \mathscr{R}$. If $Q(\epsilon)$ is FPT for any $0 < \epsilon < p$, then P = NP.

Proof. Suppose that for some $0 < \epsilon < p$, the parameterized problem $Q(\epsilon)$ is fixed parameter tractable and let $\mathscr{A}$ be an fpt algorithm for it with run time $O(t(k)\text{poly}(|I|))$. We will use $\mathscr{A}$ to solve the underlying decision problem of Q in polynomial time.

Let (I, s) be an instance of the decision version of Q. Then (I, s) is a YES-instance if and only if $\max(I) \geq s$. We consider three cases and proceed as described below.

Case 1: $s \leq f(|I|)$.

Since $\max(I) \geq f(|I|)$, we answer YES.

Case 2: $f(|I|) < s < f(|I|) + \epsilon|I|$.

In this case, we claim that we can transform the input instance (I, s) into an 'equivalent' instance (I', s') such that

1. $f(|I'|) + \epsilon|I'| \leq s'$.
2. $|I'| = \text{poly}(|I|)$.
3. $\text{opt}(I) \geq s$ if and only if $\text{opt}(I') \geq s'$.

This will show that we can, without loss of generality, go to Case 3 below directly. Add a TUB instance I_1 to I. Define $I' = I + I_1$ and $s' = s + g(|I_1|)$. Then it is easy to see that $\max(I) \geq s$ if and only if $\max(I') \geq s'$. We want to choose I_1 such that $f(|I'|) + \epsilon|I'| \leq s'$. Since $|I'| = |I| + |I_1|$ and $s' = s + g(I_1)$, and since $f(|I|) < s$, it suffices to choose I_1 satisfying

$$f(|I| + |I_1|) + \epsilon|I| + \epsilon|I_1| \leq f(|I|) + g(|I_1|)$$

By Property $P1$, we have $f(|I| + |I_1|) \leq f(|I|) + f(|I_1|) + c^*$, so it suffices to satisfy

$$f(|I_1|) + c^* + \epsilon|I| + \epsilon|I_1| \leq g(|I_1|)$$

By Property P2 we have $g(|I_1|) > f(|I_1|) + p|I_1|$, so it suffices to satisfy

$$c^* + \epsilon|I| \leq (p - \epsilon)|I_1|$$

Such an instance I_1 (of size polynomial in $|I|$) can be chosen because $0 < \epsilon < p$, and because the tight upper bound is polynomial-time uniform dense.

Case 3: $f(|I|) + \epsilon|I| \leq s$

In this case, we transform the instance (I, s) into an instance (I', s') such that

1. $f(|I'|) + \epsilon|I'| + c_1 = s'$, where $0 \leq c_1 \leq c_0$ and c_0 is a fixed constant.
2. $|I'| = \text{poly}(|I|)$.
3. $\max(I') \geq s'$ if and only if $\max(I) \geq s$.

We then run algorithm $\mathscr{A}$ with input (I', c_1). Algorithm $\mathscr{A}$ answers YES if and only if $\max(I') \geq s'$. By condition 3 above, this happens if and only if $\max(I) \geq s$. This takes time $O(t(c_1) \cdot \text{poly}(|I'|))$.

We want to obtain I' by adding a TLB instance I_1 to I. What if addition of any TLB instance yields an I' with $s' < f(I') + \epsilon|I'|$? In this case, s must already be very close to $f(|I|) + \epsilon|I|$; the difference $k \triangleq s - f(|I|) - \epsilon|I|$ must be at most $\epsilon d + c^*$, where d is the size of the smallest TLB instance I_0. (Why? Add I_0 to I to get $s + f(d) < f(|I| + d) + \epsilon(|I| + d)$; applying property P1, we get $s + f(d) < f(|I|) + f(d) + c^* + \epsilon|I| + \epsilon d$, and so $k < c^* + \epsilon d$.) In such a case, we can use the fpt algorithm $\mathscr{A}$ with input (I, k) directly to answer the question "Is $\max(I) \geq s$?" in time $O(t(\epsilon d + c^*) \cdot \text{poly}(|I|))$.

So now assume that $k \geq c^* + \epsilon d$, and it is possible to add TLB instances to $|I|$. Since f is an increasing function, there is a *largest* TLB instance I_1 we can add to I to get I' while still satisfying $s' \geq f(I') + \epsilon|I'|$. The smallest TLB instance bigger than I_1 has size at most $|I_1| + c$, where c is the constant that appears in the definition of density. We therefore have the following inequalities

$$f(|I'|) + \epsilon|I'| \leq s' < f(|I'| + c) + \epsilon(|I'| + c).$$

Since f is increasing and satisfies property P1, we have $[f(|I'| + c) + \epsilon(|I'| + c)] - [f(|I'|) + \epsilon|I'|] \leq f(c) + c^* + \epsilon c \triangleq c_0$, and hence $s' = f(|I'|) + \epsilon|I'| + c_1$, where $0 \leq c_1 \leq c_0$. Note that c_0 is a constant independent of the input instance (I, s). Also, since Q is a polynomially bounded problem, $|I_1|$ is polynomially bounded in $|I|$. ∎

Remark. Note that there are some problems, notably MAX 3-SAT, for which the constant c_0 in Case 3 of the proof above, is 0. For such problems, the proof of Theorem 4 actually proves that the problem $Q' = \{(I, k) : \max(I) \geq f(|I|) + \epsilon|I|\}$ is **NP**-hard. But in general, the constant $c_0 \geq 1$ and so this observation cannot be generalized.

We can extend Theorem 4 to minimization problems. For a minimization problem $Q = \{\mathscr{I}, S, V, \min\}$, we need the tight lower bound $f : \mathbb{N} \to \mathbb{N}$ and tight upper bound $g : \mathbb{N} \to \mathbb{N}$ to be increasing functions and satisfy the following conditions

P3 For all $a, b \in \mathbb{N}$, $g(a + b) \leq g(a) + g(b) + c^*$, where c^* is a constant,
P4 There exists $r \in \mathbb{Q}^+$ such that $g(n) - f(n) > rn$ for all $n \geq n_0$ for some $n_0 \in \mathbb{N}$.

Define $\mathscr{R}$ to be the set

$$\mathscr{R} = \{r \in \mathbb{Q}^+ : g(n) - f(n) > rn \text{ for all } n \geq n_0\},$$

and $p = \sup \mathscr{R}$. For $0 < \epsilon < p$, define $Q(\epsilon)$ as follows

$$Q(\epsilon) = \{(I, k) : I \in \mathscr{I} \text{ and } \min(I) \leq g(|I|) - \epsilon|I| - k\}.$$

For minimization problems, we have the following

Theorem 5. *Let $Q = \{\mathscr{I}, S, V, \min\}$ be a polynomially bounded* **NP**-*minimization problem such that the following conditions hold.*

1. *Q is partially additive.*
2. *Q has a tight lower bound (TLB) f such that the infinite family of instances $\mathscr{I}'$ witnessing the tight lower bound is dense poly-time uniform with respect to the condition $\min(I) = f(|I|)$.*
3. *Q has a tight upper bound (TUB) g which is increasing, satisfies condition P3, and with f satisfies P4. The infinite family of instances $\mathscr{I}'$ witnessing the tight upper bound is dense poly-time uniform with respect to the condition $\min(I) = g(|I|)$.*

4. *The underlying decision problem $\tilde{Q}$ of Q is NP-hard.*

For $0 < \epsilon < p$, define $Q(\epsilon)$ to be the following parameterized problem

$$Q(\epsilon) = \{(I, k) : \; I \in \mathscr{I} \text{ and } \min(I) \leq g(|I|) - \epsilon|I| - k\}$$

where $p = \sup \mathscr{R}$. If $Q(\epsilon)$ is FPT for any $0 < \epsilon < p$, then P = NP.

The proof of this is similar to that of Theorem 4 and is omitted.

5   Conclusion

We have shown that every problem in MAX SNP has a lower bound on the optimal solution size that is unbounded and that the above guarantee question with respect to that lower bound is in FPT. We have also shown that the $\text{TLB}(I) + \epsilon \cdot |I| + k$ question is hard for a general class that includes a number of NP-maximization problems. However we do not know the parameterized complexity of tight lower bound $+ k$ questions for most NPO problems. In particular, apart from MAX SAT, MAX c-SAT and LINEAR ARRANGEMENT, this question is open for the rest of the problems stated in Theorem 3. It would be interesting to explore the parameterized complexity of these problems and above guarantee problems in general.

References

1. L. CAI AND J. CHEN. *On Fixed-Parameter Tractability and Approximability of NP Optimization Problems.* Jour. Comput. Sys. Sci. 54(3): 465-474 (1997).
2. G. GUTIN, A. RAFIEY, S. SZEIDER AND A. YEO. *The Linear Arrangement Problem Parameterized Above Guaranteed Value.* Available at:
   http://arxiv.org/abs/cs.DS/0511030
3. J. HÅSTAD AND S. VENKATESH. *On the Advantage Over a Random Assignment.* Proceedings of the 34th Annual ACM Symposium on Theory of Computing, 2002, pages 43-52.
4. S. KHANNA, R. MOTWANI, M. SUDAN AND U. VAZIRANI. *On Syntactic Versus Computational Views of Approximability.* SIAM Jour. Computing. Vol. 28, No. 1, pp 164-191.
5. M. MAHAJAN AND V. RAMAN. *Parameterizing above Guaranteed Values: MaxSat and Max-Cut,* Journal of Algorithms **31**, 335-354 (1999).
6. R. MOTWANI AND P. RAGHAVAN. *Randomized Algorithms.* Cambridge University Press, 1995.
7. C. H. PAPADIMITRIOU AND M. YANNAKAKIS. *Optimization, Approximation, and Complexity Classes,* JCSS **43**, 425-440 (1991).
8. S. POLJAK AND D. TURZÍK. *A Polynomial Algorithm for Constructing a Large Bipartite Subgraph with an Application to a Satisfiability Problem.* Canadian Jour. Math., Vol. 34, No. 3, 1982, pp. 519-524.
9. P. ROSSMANITH AND R. NIEDERMEIER. *New Upper Bounds for Maximum Satisfiability.* Journal of Algorithms, 36: 63-88, 2000.

Randomized Approximations of Parameterized Counting Problems

Moritz Müller

Abteilung für mathematische Logik,
Albert-Ludwigs-Universität Freiburg, Eckerstr.1, 79104 Freiburg, Germany
moritz.mueller@math.uni-freiburg.de

Abstract. We prove that each parameterized counting problem in the class #W[P] has a randomized fpt approximation algorithm using a W[P] oracle. Analoguous statements hold for #W[t] and #A[t] for each $t \geq 1$. These results are parameterized analogues of a theorem due to O.Goldreich and L.Stockmeyer.

1 Introduction

It is common to consider besides decision problems also other types of problems such as search, listing or counting problems. They ask respectively to examplify, list all or count solutions. For a wide class of problems the decision, search and listing versions all have the same complexity[1]. In particular this holds for all so-called self-reducible problems.

In contrast counting seems in general to be harder than decision. As a prominent example the perfect matching problem (given a bipartite graph, decide if it contains a perfect matching) is tractable [11] and by self-reducibility so are the associated search and listing problems. L.G.Valiant introduced the intractable class of counting problems #P and proved (see [14,8]) the famous.

Theorem 1 (Valiant 1979). *To compute the number of perfect matchings in a bipartite graph is #P complete under polynomial time Turing reductions.*

This is not a sole standing phenomenon: later other tractable decision problems were shown to have intractable counting versions (see e.g. [14]). Twelve years later S.Toda reveiled [13] the surprising power of counting:

Theorem 2 (Toda 1991). *Each problem in the polynomial hierarchy can be decided in polynomial time using a #P oracle.*[2]

This apparent intractability of (exact) counting suggests the quest for feasible approximations. A precise notion is that of a fully polynomial time randomized approximation scheme (fpras). Randomized approximation turns out to be related to almost uniform sampling: here again self-reducibility implies equitractability [7,8], a fact opening the door to the rich theory of Markov chains. This finally enabled M.Jerrum et al. to prove (see [8]).

---

[1] Concepts of tractability for listing problems have been introduced in [9].
[2] As a matter of fact one needs only one call to the oracle.

H.L. Bodlaender and M.A. Langston (Eds.): IWPEC 2006, LNCS 4169, pp. 50–59, 2006.

Theorem 3 (Jerrum, Sinclair, Vigoda 2001). *The counting problem for perfect matchings has a fpras.*

This examplifies that we have a chance to find fast randomized approximation algorithms for hard counting problems. From a theoretical point of view the complexity of randomized almost correct counting is much lower than that of exact counting. While by Toda's theorem the latter is at least as hard as PH, the former is at most as hard as NP [7,12][3]:

Theorem 4 (Goldreich 2001, Stockmeyer 1985). *Any counting problem in #P has a fpras using a NP oracle.*

In the parameterized world a theory of counting complexity has been started (see [6] for a survey). Consider the parameterized decision problem p-CYCLE and its counting version p-#CYCLE (given a graph and a parameter $k \in \mathbb{N}$ compute the number of length k cycles). While p-CYCLE is fixed-parameter tractable[4] J.Flum and M.Grohe proved [5]

Theorem 5 (Flum, Grohe 2004). *p-#CYCLE is complete for #W[1] under fpt Turing reductions.*

So as in the classical setting we are faced with natural tractable parameterized decision problems having an intractable counting version. The quest is again to find fast (randomized) approximations. V.Arvind and V.Raman [1] introduced the notion of a fixed-parameter tractable randomized approximation scheme (fptras) and proved a theorem implying

Theorem 6 (Arvind, Raman 2002). *p-#CYCLE has a fptras.*

This paper is concerned with parameterized analogues of theorem 4. Both W[P] and W[1] can be viewed as parameterized analogues of NP. It turns out that for both of them we find analogues of theorem 4. Even more we find such analogues for any class of the W- and the A-hierarchy. Here we focus on an analogue for W[P]. The main result of this paper is

Theorem 7. *Each parameterized counting problem (F, κ) in #W[P] has a W[P]-fptras using a κ-balanced oracle for W[P].*

In section 2 some standard terminology is recalled. Section 3 discusses parameterized randomization. There we introduce so-called W[P]-fptrases. Section 4 contains a proof of theorem 7. The last section 5 states further results and problems.

2   #W[P]

This section recalls some definitions. In general we follow [6]. Fix a finite alphabet Σ containing at least two elements. As in [6] we view *parameterized decision*

---

[3] This was proved by O.Goldreich [7] using „a variant of a procedure" in [12].

[4] In general the embedding problem for relational structures, where the structure to be embedded has bounded treewidth, is fixed-parameter tractable (see [6]).

problems as pairs (Q, κ) for $Q \subseteq \Sigma^*$ and polynomial time computable *parameterizations* $\kappa : \Sigma^* \to \mathbb{N}$.

Let κ be a parameterization. An algorithm is *fpt* with relation to κ if and only if for a computable function $f : \mathbb{N} \to \mathbb{N}$ and a polynomial p its running time on any input $x \in \Sigma^*$ is bounded by $f(\kappa(x))p(|x|)$.[5] We sloppily allow $Q \subseteq \Sigma^* \times \mathbb{N}$ or the like and consider natural numbers as encoded in unary.

An oracle algorithm uses a κ-*balanced* oracle to a parameterized decision problem (Q, κ') if and only if there is a computable function g such that for all $x \in \Sigma^*$ and all oracle queries „$y \in Q$?" posed by the algorithm on any run on x we have $\kappa'(y) \le g(\kappa(x))$.

A *parameterized counting problem* is a pair (F, κ) for a function $F : \Sigma^* \to \mathbb{N}$ and a parameterization κ.

An *fpt parsimonious reduction* from one parameterized counting problem (F, κ) to another (F', κ') is a function $r : \Sigma^* \to \Sigma^*$ fpt computable with relation to κ such that $F = F' \circ r$ and $\kappa' \circ r \le g \circ \kappa$ for some computable function g.

We define W[P] by its machine characterization [2]. The machine model is that of a *nondeterministic RAM*, NRAM for short. These are usual RAMs [11] using registers $0, 1, \ldots$, whose contents are natural numbers $r_0, r_1, \ldots$; additionally to the usual instructions NRAMs have the instruction GUESS:

"guess a natural number $< r_0$ and store it in register 0."

A *program* is a finite sequence of instructions. Runs and acceptance are defined as usual. An execution of GUESS is a *nondeterministic step*. We use the uniform cost measure.

Let κ be a parameterization. A program $\mathbb{P}$ is κ-*restricted* if and only if there are computable functions f, g and a polynomial p such that for all $x \in \Sigma^*$ and each run of $\mathbb{P}$ on x the program $\mathbb{P}$ performs at most $g(\kappa(x))$ many nondeterministic steps and the number $f(\kappa(x))p(|x|)$ upper bounds the number of steps, the registers used and the numbers stored in any register at any time.

Definition 8. W[P] *is the class of all parameterized decision problems* (Q, κ) *decidable by some κ-restricted program.*

With a κ-restricted program $\mathbb{P}$ we associate the parameterized counting problem $(F_{\mathbb{P}}, \kappa)$, where $F_{\mathbb{P}}$ is given by

$$F_{\mathbb{P}}(x) := \text{the number of accepting runs of } \mathbb{P} \text{ on } x .$$

Definition 9. #W[P] *is the class of all parameterized counting problems of the form* $(F_{\mathbb{P}}, \kappa)$ *for some κ-restricted program $\mathbb{P}$.*

This parallels Valiant's definition [14] of #P if we use W[P] as an analogue for NP. The proof [2] of the machine characterization of W[P] shows.

---

[5] In [3] such a running time corresponds to *strongly uniform* FPT.

Theorem 10 (Chen, Flum, Grohe 2005). *$\#W[P]$ is the class of all parameterized counting problems fpt parsimoniously reducible to $p\text{-}\#\mathrm{WSAT}(\mathrm{CIRC})$.*

Here $p\text{-}\#\mathrm{WSAT}(\mathrm{CIRC})$ is the counting version of parameterized weighted circuit satisfiability $p\text{-}\mathrm{WSAT}(\mathrm{CIRC})$: given a circuit $\mathcal{C}$ and a parameter $k \in \mathbb{N}$, compute the number of assignments of weight k satisfying $\mathcal{C}$.[6]

3   W[P]-Randomization

Classically a randomized polynomial time algorithm can be viewed as a binary ,,NP-machine'', where a run on some input is determined by a sequence in $\{0,1\}$. This sequence can be interpreted as the outcome of independent ,,coin tosses''. The algorithm is then analyzed by means of probability statements concerning events of runs. The probability measure concerned is the uniform measure provided the machine is *exact*, that is for each input $x \in \Sigma^*$ it performs the same number of nondeterministic steps on every run on x.

We get different concepts of parameterized randomized computation by replacing ,,NP-machine'' in the classical definition by ,,paraNP-machine''[7] or ,,W[P]-machine''. Instead of flipping coins programs for NRAMs ,,roll dices'' (execute GUESS). In order to get these rolls induce the uniform measure on runs, the program should besides being exact ,,always use the same dice'': in analogy to the above restriction to binary machines we define a program $\mathbb{P}$ to have *uniform guess bounds* if and only if for all $x \in \Sigma^*$ the content r_0 of register 0 is the same for any two nondeterministic steps in runs of $\mathbb{P}$ on x.

Definition 11. *Let κ be a parameterization. An exact binary nondeterministic algorithm which is fpt with relation to κ is* paraNP-randomized *with relation to κ. An exact κ-restricted program having uniform guess bounds is* W[P]-randomized *with relation to κ.*

Notation: let κ be a parameterization, $\mathbb{P}$ a κ-restricted program and $x \in \Sigma^*$; for the probability space given by the uniform measure P_x on the set of runs of $\mathbb{P}$ on x we let $\mathbb{P}(x)$ denote the random variable mapping such a run to the output of that run (say r_0 of its last configuration).

For reals r and $\epsilon > 0$ we write $(1 \pm \epsilon) \cdot r$ for the open interval $(r - \epsilon \cdot r, r + \epsilon \cdot r)$.

Definition 12. *A* W[P]-fptras *(fixed parameter tractable W[P]-randomized approximation scheme) for a parameterized counting problem (F, κ) is a program $\mathbb{P}$ expecting inputs (x, l, l') for $x \in \Sigma^*$ and positive $l, l' \in \mathbb{N}$ which is* W[P]-randomized *with relation to $\Sigma^* \times \mathbb{N} \times \mathbb{N} \to \mathbb{N} : (x, l, l') \mapsto \kappa(x)$ such that for all $(x, l, l') \in \Sigma^* \times \mathbb{N} \times \mathbb{N}$*

$$P_{(x,l,l')}\left(\mathbb{P}((x,l,l')) \in (1 \pm 1/l) \cdot F(x)\right) > 1 - 1/l' \ .$$

---

[6]   We always formulate counting problems sloppily like above. It will always be clear how to make it precise by defining a function and a parameterization.

[7]   See [6] for a definition of paraNP.

The following simple characterization is useful.

Definition 13. *A parameterized counting problem (F, κ) is fpt paddable if and only if there is a function $r : \Sigma^* \times \mathbb{N} \to \Sigma^*$ fpt computable with relation to $\Sigma^* \times \mathbb{N} \to \mathbb{N} : (x, l) \mapsto \kappa(x)$ such that for some computable g we have for all $(x, l) \in \Sigma^* \times \mathbb{N}$ that $F(r(x, l)) = F(x)$, $\kappa(r(x, l)) \leq g(\kappa(x))$ and $|r(x, l)| \geq |x| + l$.*

Proposition 14 *Let (F, κ) be a fpt paddable parameterized counting problem and let $c, c' \in \mathbb{N}$ positive. Then (F, κ) has a W[P]-fptras $\mathbb{P}$ if and only if there is a W[P]-randomized program $\mathbb{P}'$ such that for all $x \in \Sigma^*$*

$$P_x \left(\mathbb{P}'(x) \in (1 \pm |x|^{-c}) \cdot F(x) \right) > 1 - |x|^{-c'} \ .$$

Sketch of proof: For necessity define $\mathbb{P}'$ on x to simulate the given $\mathbb{P}$ on input $(x, |x|^c, |x|^{c'})$. For sufficiency define $\mathbb{P}$ on (x, l, l') to simulate the given $\mathbb{P}'$ on x in case $|x| \geq \max\{l, l'\}$; else run $\mathbb{P}'$ on $r(x, \max\{l, l'\} - |x|)$ for r witnessing fpt paddability of (F, κ). $\qquad\square$

4   Proof of the Main Theorem

In this section we prove theorem 7. We need some facts on hashing [10]. A technical trick which has proved useful also in other contexts is to use a sufficiently large finite field for encoding solutions and to use hash families on these encodings.

Let $p, k, i \in \mathbb{N}$ be positive, $k \geq i$ and p prime. The field with p elements is denoted by $\mathbb{F}_p$. We don't distinguish $\mathbb{F}_p$ from $\{0, \ldots, p - 1\}$ notationally. $\mathbb{F}_p^k$ is the k-dimensional vector space over $\mathbb{F}_p$. Vectors are represented with relation to the standard base as k-tuples over $\{0, \ldots, p - 1\}$.

We let $H_{k,i}^p$ denote the set of all affine transformations h from $\mathbb{F}_p^k$ to $\mathbb{F}_p^i$, i.e. mappings of the form $\bar{x} \mapsto A\bar{x} + \bar{b}$ for $\bar{x} \in \mathbb{F}_p^k$, where A is an $i \times k$-matrix with entries in $\mathbb{F}_p$, and $\bar{b} \in \mathbb{F}_p^i$. Set $h^{-1}(\bar{a}) := \{\bar{x} \in \mathbb{F}_p^k \mid h(\bar{x}) = \bar{a}\}$ for $\bar{a} \in \mathbb{F}_p^i$.

It can be shown that $H_{k,i}^p$ is a so-called 2-universal family of hash functions [7]. It then easily follows:

Lemma 15 (Hashing Lemma). *Let $p, k \in \mathbb{N}$ and p prime. For $S \subseteq \mathbb{F}_p^k$ and $i \in [k] = \{1, \ldots, k\}$ define*

$$Y_i^S : H_{k,i}^p \to \mathbb{N} : h \mapsto |S \cap h^{-1}(0^i)| \ ,$$

where 0^i is the zero vector in $\mathbb{F}_p^i$. Then for all $\epsilon > 0$ we have for $\rho_i := |S| \cdot p^{-i}$

$$P \left(|Y_i^S - \rho_i| \geq \epsilon \rho_i \right) \leq \frac{1}{\epsilon^2 \rho_i} \ ,$$

where P is the uniform measure on $H_{k,i}^p$.

Proof of theorem 7: The proof is similar to the one given in [7] for theorem 4. Write $p\text{-}\#\text{WSAT}(\text{CIRC})$ as (F, κ). By the machine characterization theorem 10 it suffices to prove that (F, κ) has a W[P]-fptras using a κ-balanced oracle for the decision problem $p\text{-}\text{WSAT}(\text{CIRC})$.

We may restrict attention to circuits $\mathcal{C}$ of size polynomial in n, say $\leq n^2$, where n is the number of input nodes of $\mathcal{C}$: if necessary we can pass in polynomial time to a new circuit by adding sufficiently many new input nodes and conjunct the negations of all of them with the old output node. This does not change the number of satisfying weight k assignments. The same argument shows that (F, κ) is fpt paddable.

Let $\mathcal{C}$ have n input nodes. We identify the set of weight k assignments for $\mathcal{C}$ with the set of tuples $(i_1, \ldots, i_k) \in [n]^k$ with $i_1 < \ldots < i_k$ by stipulating that $(i_1, \ldots, i_k)$ corresponds to the assignment setting exactly the i_1th and the i_2th ... and the i_kth input node to TRUE. Let S be the set of weight k assignments satisfying $\mathcal{C}$. Then $S \subseteq [n]^k \subseteq \mathbb{F}_p^k$ for any prime $p \geq n$.

l is a natural number we are going to fix during the proof. In fact l will be ≥ 6. We use the notation from the Hashing Lemma 15. Additionally we set

$$\min \emptyset := 0 \text{ and } Y_0^S \text{ constantly } 0 \ .$$

The program $\mathbb{P}$ on instance $(\mathcal{C}, k)$ of $p\text{-}\#\text{WSAT}(\text{CIRC})$ does the following:

1: compute $p :=$ the smallest prime $> n$
2: **if** $|S| < p^l$ **then** out $|S|$
3: **else for all** $i \in [k]$: guess $h_i \in H_{k,i}^p$
4: compute $i := \min \left\{ j \in [k] \mid Y_j^S(h_j) < p^l \right\}$
5: out $p^i \cdot Y_i^S(h_i)$

Probability analysis. Before explaining how $\mathbb{P}$ manages to do that, we first prove that it works. Let p be the prime computed in line 1. $\mathbb{P}$ gives with probability one the correct answer if $|S| < p^l$, so let's assume that $|S| \geq p^l$. For $m := \lceil \log_p |S| \rceil$ clearly

$$p^{m-1} < |S| \leq p^m \text{ and } l \leq m \leq k \ ,$$

since $|S| \geq p^l$ and $S \subseteq \mathbb{F}_p^k$ (wlog we assume $k \geq l$). Instead of the uniform measure P_x (for $x \in \Sigma^*$ encoding $(\mathcal{C}, k)$) on the set of runs of $\mathbb{P}$ on x consider the uniform measure P on the set of all tuples $(h_1, \ldots, h_k)$ as guessed by $\mathbb{P}$ in line 3. These tuples correspond bijectively to the runs of $\mathbb{P}$ on $(\mathcal{C}, k)$. Let $i_{\mathbb{P}}((h_1, \ldots, h_k))$ be the i computed by $\mathbb{P}$ in line 4 in the run determined by $(h_1, \ldots, h_k)$. $i_{\mathbb{P}}$ is a random variable with values in $\mathbb{N}$ defined on the set of the $(h_1, \ldots, h_k)$ endowed with the measure P.

For $i \in [k]$ consider the projection π_i mapping $(h_1, \ldots, h_k)$ to h_i. Because the random variable $Y_i^S \circ \pi_i$ is distributed as Y_i^S, we have the Hashing Lemma 15 for $Y_i^S \circ \pi_i$ and P as defined here. For notational simplicity we denote $Y_i^S \circ \pi_i$ again by Y_i^S.

For $\epsilon > 0$ consider the following events:

$$A_\epsilon := \left\{ \left| p^{i_{\mathbb{P}}} \cdot Y_{i_{\mathbb{P}}}^S - |S| \right| < \epsilon |S| \right\} \text{ and } B := \{ i_{\mathbb{P}} \in [m - l + 1] \} \ .$$

The reason of line 2 is to get $m - l + 1 \geq 1$ (because $m \geq l$) enabling us (note $m - l + 1 < k$) to apply the Hashing Lemma on Y^S_{m-l+1}: because

$$p^{l-1} = p^m / p^{m-l+1} \geq \rho_{m-l+1} > p^{m-1}/p^{m-l+1} = p^{l-2}$$

we have for the complement $\overline{B}$ of B (note $\overline{B} \subseteq \{Y^S_{m-l+1} \geq p^l\}$)

$$P\left(\overline{B}\right) \leq P\left(Y^S_{m-l+1} \geq p^l\right) \leq P\left(\left|Y^S_{m-l+1} - \rho_{m-l+1}\right| \geq (p-1)\rho_{m-l+1}\right)$$
$$\underset{\text{Hashing Lemma}}{\leq} \frac{1}{(p-1)^2 \rho_{m-l+1}} < \frac{1}{(p-1)^2 p^{l-2}} < n^{-l} .$$

For all $j \in [m - l + 1]$ we have $\rho_j > p^{m-1}/p^{m-l+1} = p^{l-2}$. Using the Hashing Lemma again we get for all $\epsilon > 0$ and all $j \in [m - l + 1]$

$$P\left(\left|Y^S_j - \rho_j\right| \geq \epsilon \rho_j\right) \leq 1/(\epsilon^2 \rho_j) < 1/(\epsilon^2 p^{l-2}) < 1/(\epsilon^2 n^{l-2}) .$$

Note $\overline{A}_\epsilon \cap \{i_\mathbb{P} = j\} = \{|p^j \cdot Y^S_j - |S|| \geq \epsilon|S|\} = \{|Y^S_j - \rho_j| \geq \epsilon \rho_j\}$. Thus

$$P(\overline{A}_\epsilon) \leq P(\overline{B}) + P(\overline{A}_\epsilon \cap B)$$
$$= P(\overline{B}) + \sum_{j=1}^{m-l+1} P(\overline{A}_\epsilon \cap \{i_\mathbb{P} = j\})$$
$$< n^{-l} + \frac{m - l + 1}{\epsilon^2 n^{l-2}} .$$

Now let $c, c' \in \mathbb{N}$ be arbitrary positive constants. For $\epsilon := n^{-c}$ the above is $\leq n^{-l} + (k-1)n^{2c}n^{-l+2} < kn^{2c+2-l}$ since $m - l + 1 < k$ and hence $< n^{2c+3-l}$ since $k \leq n$ (else we would have $S = \emptyset$). Choose $l := 2c + 3 + c'$ and get

$$P\left(A_{n^{-c}}\right) = P\left(\mathbb{P}((\mathcal{C}, k)) \in (1 \pm n^{-c}) \cdot |S|\right) > 1 - n^{-c'} .$$

This suffices by our assumption that the size of $\mathcal{C}$ is $\leq n^2$ and by proposition 14 (we already noted that (F, κ) is fpt paddable).

We now describe the subroutines of $\mathbb{P}$ and analyze the complexity.

Random complexity. For $i \in [k]$ to guess $h_i \in H^p_{k,i}$ the program guesses the entries of a $i \times k$- matrix and a length i vector over $\mathbb{F}_p$. In total $\mathbb{P}$ performs on every run on $(\mathcal{C}, k)$ exactly $\sum_{i=1}^k (ik + i)$ many guesses - especially $\mathbb{P}$ is exact and performs a number of nondeterministic steps recursiveley bounded in terms of the parameter. $\mathbb{P}$ has uniform guess bounds because it always guesses numbers $< p$. Provided time, registers used and the numbers stored are fpt bounded, this shows that $\mathbb{P}$ is κ-restricted and hence W[P]-randomized.

Time complexity. That $\mathbb{P}$ uses not too large numbers and not too many registers is clear. We focus on the running time. In line 1 brute force needs only polynomial time since there is a prime between n and $2n$ for any positive natural number n.[8] It follows that p^l is polynomially bounded in n.

---

[8] This is Bertrand's Postulate (1845) proved by Chebyshev in 1850.

We'll later look at line 2. The remaining time needed is dominated by the computations of i in line 4 and $Y_i^S(h_i)$ in line 5. We have to solve the following problem:

Input: *a circuit* $\mathcal{C}, k \in \mathbb{N}, h \in H_{k,i}^p$ *with* $i \in [k]$
(where p is the smallest prime larger than the number of input nodes of $\mathcal{C}$ and i is determined by h)
Parameter: k
Problem: *decide if* $Y_i^S(h) < p^l$ *and in this case compute* $Y_i^S(h)$
(where S is the set of weight k assignments satisfying $\mathcal{C}$)

Consider the program $\mathbb{P}'$ which on input $(\mathcal{C}, h, k)$ as above first guesses $\bar{a} \in [n]^k$ and then checks if both $\bar{a}$ satisfies $\mathcal{C}$ and $h(\bar{a}) = 0^i$.
Clearly $\mathbb{P}'$ is κ'-restricted for κ' the parameterization mapping $(\mathcal{C}, h, k)$ to k. The run of $\mathbb{P}'$ determined by $\bar{a} \in [n]^k$ is accepting if and only if $\bar{a} \in S \cap h^{-1}(0^i)$, especially

$$F_{\mathbb{P}'}((\mathcal{C}, h, k)) = Y_i^S(h) \ .$$

By the machine characterization theorem 10 there is a fpt parsimonious reduction from $(F_{\mathbb{P}'}, \kappa')$ to (F, κ). Using this reduction we can compute given $(\mathcal{C}, h, k)$ a circuit $\mathcal{C}'$ and a parameter $k' \in \mathbb{N}$ such that

$$F_{\mathbb{P}'}((\mathcal{C}, h, k)) = \text{the number of weight } k' \text{ assignments satisfying } \mathcal{C}'.$$

We have thus reduced our problem to the parameterized counting problem:

$$\text{Given } x \in \Sigma^* \text{ and a prime } p, \text{ compute } \min\{F(x), p^l\},$$

with the parameterization mapping (x, p) to $\kappa(x)$. It suffices to show that this is fixed-parameter tractable using a balanced oracle for the decison problem p-WSAT(CIRC) thereby explaining also how to serve line 2.

But p-WSAT(CIRC) is *self-reducible*! It follows by general means that there is an algorithm $\mathbb{A}$ using a κ-balanced oracle for p-WSAT(CIRC) solving the associated listing problem with *fixed parameter tractable delay* with relation to κ[9]: $\mathbb{A}$ on an instance $(\mathcal{C}', k')$ of p-WSAT(CIRC) puts out without repetitions all weight k' assignments satisfying $\mathcal{C}'$ such that the delay of $\mathbb{A}$ is fpt bounded with relation to κ; *delay* means the maximum number of steps until the first output, between any two outputs and from the last output to the end (see [9]).

We proceed straightforwardly: simulate $\mathbb{A}$ on $(\mathcal{C}', k')$; increase a counter (initialized by 0) for each output of $\mathbb{A}$; stop the simulation in case the counter becomes p^l; return the counter. The time needed here is at most p^l times the delay of $\mathbb{A}$ and that obeys a fpt bound.

For completeness we describe $\mathbb{A}$. We now view assignments as strings over $\{0, 1\}$. $\mathbb{A}$ on $(\mathcal{C}', k')$ is $\mathbb{A}'$ on $(\mathcal{C}', k', \lambda)$ for the empty string λ. $\mathbb{A}'$ takes inputs $(\mathcal{C}, k, \bar{a})$ for circuits $\mathcal{C}$, $k \in \mathbb{N}$ the parameter, and $\bar{a} \in \{0, 1\}^*$. For $i \in \{0, 1\}$ let $\mathcal{C}^i$ be the circuit obtained from $\mathcal{C}$ by replacing its first input node by the constant i.

---

[9] Actually we get *polynomial delay* [9].

$\mathbb{A}'$ on $(\mathcal{C}, k, \bar{a})$ behaves as follows:

1: **if** $\mathcal{C}$ has no satisfying assignment of weight k **then** stop
2: **else if** $\mathcal{C}$ has no input nodes **then** out $\bar{a}$
3:   **else for all** $i \in \{0,1\}$:
4:   **if** $\mathcal{C}^i$ has a satisfying assignment of weight $k - i$
5:   **then** $\mathbb{A}'$ on $(\mathcal{C}^i, k - i, \bar{a}i)$

$\mathbb{A}'$ uses an oracle for $p\text{-}\mathrm{WSAT}(\mathrm{CIRC})$ to check the „if"-conditions in lines 1 and 4. It is easy to see that this is a listing algorithm as desired.   □

5   Further Results and Questions

Let κ be a parameterization. A κ-restricted program is *tail-nondeterministic* if and only if for some computable g in any run on any input $x \in \Sigma^*$ each nondeterministic step is any of the last $g(\kappa(x))$ steps of the computation.

W[1] is the class of all parameterized decision problems (Q, κ) decidable by some κ-restricted tail-nondeterministic program [2].

Definition 16. *#W[1] is the class of all parameterized counting problems of the form $(F_{\mathbb{P}}, \kappa)$ for some κ-restricted tail-nondeterministic program $\mathbb{P}$.*

By a proof similar to the one given for theorem 7 you can get

Theorem 17. *Each parameterized counting problem (F, κ) in #W[1] has a W[P]-fptras using a κ-balanced oracle for W[1].*

A shortcoming of this result is that it does not settle the question for *W[1]-fptrases*. These are defined using the concept of parameterized randomization you get by use of W[1] as an analogue for NP, that is

Definition 18. *An exact κ-restricted tail-nondeterministic program having uniform guess bounds is* W[1]-*randomized with relation to κ.*

This is our third notion of parameterized randomized computation. Theorem 6 refers to paraNP-fptrases. [1] and [6] use different bounds on the „probability to fail" in the definition of paraNP-fptrases ($1/l'$ in our definition 12). These definitions can be seen to be equivalent [7] since for paraNP-randomized algorithms classical methods of probability amplification apply. For W[P]- and W[1]-fptrases such bounds are to be handled with care [4].

Theorem 17 can be generalized to all classes of the W- and the A-hierarchy:

Theorem 19. *For all $t \geq 1$ each parameterized counting problem (F, κ) in #A[t] has a W[P]-fptras using a κ-balanced oracle for A[t].*

Theorem 20. *For all $t \geq 1$ each parameterized counting problem (F, κ) in #W[t] has a W[P]-fptras using a κ-balanced oracle for W[t].*

A second shortcoming of these results is that the query complexity is bad. Our program in the proof of theorem 7 uses polynomially many oracle queries.

Under what conditions do we have these theorems „problemwise"? That is: when approximating a counting problem coming from a decision problem (Q, κ), say $(Q, \kappa) \in W[t]$, but not $W[t]$-hard, are we really in need of an oracle for the full class $W[t]$ or could we do with (Q, κ)? This is particularly interesting for problems in FPT with a $\#W[1]$-hard counting version. An answer would be a step towards a theoretical understanding of what makes certain problems have fast randomized approximations.

One of the main gaps in parameterized counting complexity theory is that we don't have [4,6] an analogue of Toda's theorem 2 such as $\bigcup_{t \geq 1} A[t] \subseteq FPT^{\#A[1]}$ or at least $\bigcup_{t \geq 1} W[t] \subseteq FPT^{\#W[1]}$ or $\bigcup_{t \geq 1} A[t] \subseteq FPT^{\#W[P]}$. The main problem here seems to be probability amplification.

Acknowledgements. The idea of looking for parameterized analogues of theorem 4 came from J.A. Montoya. I thank J.Flum for his advices.

References

1. V.Arvind and V.Raman, *Approximation Algorithms for Some Parameterized Counting Problems.* In I.Bose and P.Morin, ed., Proceedings of the 13th Annual International Symposium on Algorithms and Computation, LNCS 2518, pp. 453-464. Springer, 2002.
2. Y.Chen, J.Flum and M.Grohe, *Machine-Based Methods in Parameterized Complexity Theory*, Theoretical Computer Science 339, pp. 167-199, 2005.
3. R.G.Downey and M.R.Fellows, *Parameterized Complexity.* Springer, 1999.
4. R.G.Downey, M.R.Fellows ans K.W.Regan, *Parameterized Circuit Complexity and the W Hierachy*, Theoretical Computer Science 191(1-2), pp. 97-115, 1998.
5. J.Flum and M.Grohe, *The Parameterized Complexity of Counting Problems.* SIAM Journal on Computing 33(4), pp. 892-922, 2004.
6. J.Flum and M.Grohe, *Parameterized Complexity Theory.* Springer, 2006.
7. O.Goldreich, *Randomized Methods in Computation - Lecture Notes*, 2001. http://www.wisdom.weizmann.ac.il/~oded/homepage.html
8. M.Jerrum, *Counting, Sampling and Intergrating: Algorithms and Complexity.* Birkhäuser, 2003.
9. D.S.Johnson, C.H.Papadimitriou and M.Yannakakis, *On Generating All Maximal Independent Sets*, Information Processing Letters 27, pp. 119-123, 1988.
10. M.Luby and A.Wigderson, *Pairwise Independence and Derandomization*, International Computer Science Institute, TR-95-035, 1995.
11. C.H.Papadimitriou, *Computational Complexity.* Addison Wesley, 1994.
12. L.Stockmeyer, *On Approximation Algorithms for #P*, SIAM Journal on Computing 14(4), pp. 849-861,1985.
13. S.Toda, *PP is as Hard as the Polynomial Hierarchy*, SIAM Journal on Computing 20(5) , pp. 865-877, 1991.
14. L.G.Valiant, *The Complexity of Enumeration and Reliability Problems*, SIAM Journal on Computing 8(3), pp. 410-421, 1979.

Fixed-Parameter Complexity of Minimum Profile Problems

Gregory Gutin[1,2,*], Stefan Szeider[3,**], and Anders Yeo[1]

[1] Department of Computer Science,
Royal Holloway University of London,
Egham, Surrey TW20 OEX, England, United Kingdom
{gutin, anders}@cs.rhul.ac.uk
[2] Department of Computer Science, University of Haifa, Israel
[3] Department of Computer Science, Durham University,
Durham DH1 3LE, England, United Kingdom
stefan.szeider@durham.ac.uk

Abstract. An ordering of a graph $G = (V, E)$ is a one-to-one mapping $\alpha : V \to \{1, 2, \ldots, |V|\}$. The profile of an ordering α of G is $\mathrm{prf}_\alpha(G) = \sum_{v \in V} (\alpha(v) - \min\{\alpha(u) : u \in N[v]\})$; here $N[v]$ denotes the closed neighborhood of v. The profile $\mathrm{prf}(G)$ of G is the minimum of $\mathrm{prf}_\alpha(G)$ over all orderings α of G. It is well-known that $\mathrm{prf}(G)$ equals the minimum number of edges in an interval graph H that contains G as a subgraph. We show by reduction to a problem kernel of linear size that deciding whether the profile of a connected graph $G = (V, E)$ is at most $|V| - 1 + k$ is fixed-parameter tractable with respect to the parameter k. Since $|V| - 1$ is a tight lower bound for the profile of a connected graph $G = (V, E)$, the parameterization above the guaranteed value $|V| - 1$ is of particular interest.

1 Introduction

The *profile* $\mathrm{prf}(G)$ of a graph G is an integer-valued graph parameter defined via vertex orderings (see Section 2). Fomin and Golovach [4] established the equivalence of the profile and other parameters including one that is important in graph searching. Further areas of application of the profile and equivalent parameters include computational biology [2,6], archaeology [9] and clone fingerprinting [8]. The following is a well-known NP-complete problem [3,10].

Minimum Profile Problem (MPP)
Instance: A graph $G = (V, E)$ and a positive integer k.
Question: Is the profile of G at most k?

In fact this problem is equivalent to the following problem that was proved to be NP-complete even earlier (see [5]).

* Research supported in part by the IST Programme of the European Community, under the PASCAL Network of Excellence, IST-2002-506778.
** Research supported in part by the Nuffield Foundation, NAL/01012/G.

H.L. Bodlaender and M.A. Langston (Eds.): IWPEC 2006, LNCS 4169, pp. 60–71, 2006.

Interval Graph Completion (IGC)
Instance: A graph $G = (V, E)$ and a positive integer $k \geq |E|$.
Question: Is there a supergraph H of G such that H is an interval graph and contains at most k edges?

Recall that a graph $G = (V, E)$ is an *interval graph* if we can associate each vertex $v \in V$ with a closed interval I_v in the real line such that two distinct vertices x and y are adjacent in G if and only if $I_x \cap I_y \neq \emptyset$. The equivalence of MPP and IGC follows from the next result:

Theorem 1 ([1]). *For any graph G, $\mathrm{prf}(G)$ equals the smallest number of edges in an interval supergraph of G.*

Consequently, $\mathrm{prf}(G) \geq |E|$ holds for every graph $G = (V, E)$, and so the following parameterized problem is fixed-parameter tractable (FPT); that is, it can be solved in time $O(f(k)(|V| + |E|)^{O(1)})$ for some computable function $f(k)$.

Profile Problem (PP)
Instance: A graph $G = (V, E)$. *Parameter:* A positive integer k.
Question: Is the profile of G at most k?

Several authors consider the following much more interesting problem.

Profile Above Guaranteed Value (PAGV)
Instance: A graph $G = (V, E)$. *Parameter:* A positive integer k.
Question: Is the profile of G at most $|E| + k$?

It is unknown whether this problem is FPT (private communications with L. Cai, F. Fomin and H. Kaplan). Clearly, PAGV is equivalent to the problem of checking whether a graph can be made interval by adding at most k edges. In this paper, we restrict ourselves to connected graphs (the case of general graphs can be reduced to connected graphs) and consider the following somewhat weaker version of PAGV (note that $|E| \geq |V| - 1$ holds for connected graphs $G = (V, E)$).

Profile Above Vertex Guaranteed Value (PAVGV)
Instance: A connected graph $G = (V, E)$. *Parameter:* A positive integer k.
Question: Is the profile of G at most $|V| - 1 + k$?

This problem is of interest also because of the problem VAP considered by Serna and Thilikos [11] (see Section 5). We prove by means of a kernelization scheme that the problem PAVGV is fixed-parameter tractable.

2 Definitions and Preliminary Results

Let $G = (V, E)$ be a graph. An *ordering* of G is a one-to-one mapping $\alpha : V \to \{1, 2, \ldots, |V|\}$. We denote the set of orderings of G by $\mathrm{OR}(G)$. For a vertex v in G, its *neighborhood* is $N(v) = \{u \in V : uv \in E\}$ and its *closed neighborhood* is $N[v] = N(v) \cup \{v\}$. The *profile of a vertex* z of G in an ordering α of G is $\mathrm{prf}_\alpha(G, z) = \alpha(z) - \min\{\alpha(w) : w \in N[z]\}$. The *profile of a set* $Z \subseteq V$ in an ordering α of G is $\mathrm{prf}_\alpha(G, Z) = \sum_{z \in Z} \mathrm{prf}_\alpha(G, z)$. The *profile of an ordering* α of G is $\mathrm{prf}_\alpha(G) = \mathrm{prf}_\alpha(G, V)$. An ordering α of G is *optimal* if $\mathrm{prf}_\alpha(G) =$

$\min\{\text{prf}_\beta(G) : \beta \in \text{OR}(G)\}$. If α is optimal, then $\text{prf}(G) = \text{prf}_\alpha(G)$ is called the *profile* of G. If $X \subseteq V$ and α is an ordering of G, then let α_X denote the ordering of $G - X$ in which $\alpha_X(u) < \alpha_X(v)$ if and only if $\alpha(u) < \alpha(v)$ for all $u, v \in V(G) - X$. If $X = \{x\}$, then we simply write α_x instead of $\alpha_{\{x\}}$.

Lemma 1. *Let $G = (V, E)$ be a graph of order n and let X be a set of vertices such that $G - X$ is connected. If an optimal ordering α has $\{\alpha^{-1}(1), \alpha^{-1}(n)\} \subseteq V(G - X)$ then $\text{prf}_\alpha(G, V - X) \geq \text{prf}_{\alpha_X}(G - X) + |X|$.*

Proof. Let $X = \{x_1, x_2, \ldots, x_r\}$ and define $X_i = \{x_1, x_2, \ldots, x_i\}$ for all $0 \leq i \leq r$. We will by induction show the following: (*) $\text{prf}_{\alpha_{X_i}}(G - X_i, V - X) \geq \text{prf}_{\alpha_X}(G - X) + |X| - i$. The above is clearly true when $i = r$ as $X_r = X$ and $|X| = r$. If we can show that (*) is true for $i = 0$, then we are done. We will assume that (*) is true for some $i > 0$. Since $G - X$ is connected and $\{\alpha^{-1}(1), \alpha^{-1}(n)\} \subseteq V(G - X)$, there is an edge $uv \in E(G - X)$ such that $\alpha_{X_{i-1}}(u) > \alpha_{X_{i-1}}(x_i) > \alpha_{X_{i-1}}(v)$. This implies that the profile of u is one larger in $\alpha_{X_{i-1}}$ than it is in α_{X_i}. This implies $\text{prf}_{\alpha_{X_{i-1}}}(G - X_{i-1}, V - X) \geq \text{prf}_{\alpha_{X_i}}(G - X_i, V - X) + 1 \geq \text{prf}_{\alpha_X}(G - X) + |X| - i + 1$. We are now done by induction. $\square$

Lemma 2 ([10]). *(i) If G is a connected graph with n vertices, then $\text{prf}(G) \geq n - 1$. (ii) For a cycle C_n with n vertices we have $\text{prf}(C_n) = 2n - 3$.*

For a vertex x, $d(x)$ denotes its degree, i.e., $d(x) = |N(x)|$. A slightly weaker version of the following lemma is stated in [10] without a proof.

Lemma 3. *If G is an arbitrary graph of order n, $x \in V(G)$ and α is an optimal ordering of G, then $\text{prf}_\alpha(G) \geq \text{prf}_{\alpha_x}(G - x) + d(x)$.*

Proof. Let α be an optimal ordering of G and let

$$X = \{\alpha^{-1}(1), \alpha^{-1}(2), \ldots, \alpha^{-1}(\alpha(x) - 1)\}.$$

Note that for all $a \in N(x) - X$ we have $\text{prf}_\alpha(G, a) \geq \text{prf}_{\alpha_x}(G - x, a) + 1$. Furthermore, $\text{prf}_\alpha(G, x) \geq |N(x) \cap X|$. Thus, $\text{prf}_\alpha(G) - \text{prf}_{\alpha_x}(G - x) \geq \text{prf}_\alpha(G, x) + \sum_{a \in N(x) - X}(\text{prf}_\alpha(G, a) - \text{prf}_{\alpha_x}(G - x, a)) \geq |N(x) \cap X| + |N(x) - X| = d(x)$. Hence, $\text{prf}_\alpha(G) \geq \text{prf}_{\alpha_x}(G - x) + d(x)$. $\square$

Theorem 3 gives a lower bound for the profile of a 2-edge-connected graph, which is important for our FPT algorithm. Lin and Yuan [10] used a concise and elegant argument to show that $\text{prf}(G) \geq k(2n - k - 1)/2$ for every k-connected graph G of order n. Their argument uses Menger's Theorem in a clever way, yet the argument cannot be used to prove our bound. Instead of Menger's Theorem we will apply the following well-known decomposition of 2-edge-connected graphs (see, e.g., Theorem 4.2.10 in [12]) called a *closed-ear decomposition*.

Theorem 2. *Any 2-edge-connected graph G has a partition of its edges $E_1, E_2, \ldots, E_r$, such that $G_i = G[E_1 \cup E_2 \cup \ldots \cup E_i]$ is 2-edge-connected for* all $i = 1, 2, 3, \ldots, r$. Furthermore, E_j induces either a path with its endpoints in $V(G_{j-1})$ but all other vertices in $V(G_j) - V(G_{j-1})$ or a cycle with one vertex in $V(G_{j-1})$ but all other vertices in $V(G_j) - V(G_{j-1})$ for every $j = 2, 3, \ldots, r$. Moreover, G_1 is a cycle and every cycle of G can be G_1.

Theorem 3. If G is a 2-edge-connected graph of order n, then $\mathrm{prf}(G) \geq \frac{3n-3}{2}$.

Proof. Let α be an optimal ordering of $V(G)$ and let y be the vertex with $\alpha(y) = n$. Since G is 2-edge-connected, y is contained in a cycle C. By Theorem 2, G has an ear-decomposition $E_1, E_2, \ldots, E_r$ such that $G[E_1] = C$. Let $G_i = G[E_1 \cup E_2 \cup \ldots \cup E_i]$, which by Theorem 2 are 2-edge-connected for all $i = 1, 2, \ldots, r$. We will prove this theorem by induction. If $r = 1$ then the Theorem holds by Lemma 2 (ii), as $n \geq 3$. So assume that $r \geq 2$. Let $n_i = |V(G_i)|$ for all $i = 1, 2, \ldots, r$ and note that by induction we know that $\mathrm{prf}(G_{r-1}) \geq \frac{3n_{r-1}-3}{2}$. If $n_r = n_{r-1}$ then E_r is just one edge and we are done as $\mathrm{prf}(G_r) \geq \mathrm{prf}(G_{r-1})$. So assume that $a = n_r - n_{r-1} > 0$. If $a = 1$ and $V(G_r) - V(G_{r-1}) = \{x\}$, then by Lemma 3 we obtain

$$\mathrm{prf}(G) \geq \mathrm{prf}(G_{r-1}) + d(x) \geq \frac{3n_{r-1}-3}{2} + 2 > \frac{3n-3}{2}.$$

So we may assume that $a \geq 2$. Let P be the path $G_r - V(G_{r-1})$, let x and z be the endpoints of P such that $\alpha(x) < \alpha(z)$, and let u be a neighbor of x in G_{r-1}. Let $j = \min\{\alpha(q) : q \in V(G_{r-1})\}$, and let $Q = \{p \in V(P) : \alpha(p) > j\}$ and $M = \{p \in V(P) : \alpha(p) < j\}$, which is a partition of $V(P)$. (Note that $\alpha^{-1}(j) \in V(G_{r-1})$ and recall that $\alpha^{-1}(n) = y \in V(G_{r-1})$.) Furthermore let β denote the ordering α restricted to P (i.e., $\beta = \alpha_{V(G_{r-1})}$) and let $H = G - M$. By Lemma 1 (with $X = Q$) we obtain

$$\mathrm{prf}_{\alpha_M}(H, V(H) - Q) \geq \mathrm{prf}_{\alpha_{(M \cup Q)}}(H - Q) + |Q| = \mathrm{prf}_{\alpha_{V(P)}}(G_{r-1}) + |Q|.$$

Now assume that $\alpha(x) < j$ and note that $\mathrm{prf}_\alpha(G, u) \geq \mathrm{prf}_{\alpha_M}(H, u) + j - \alpha(x)$, as $\mathrm{prf}_{\alpha_M}(H, u) \leq \alpha(u) - j$ and $\mathrm{prf}_\alpha(G, u) = \alpha(u) - \alpha(x)$. As $|Q| = |V(P)| - j + 1$ and $\mathrm{prf}_\alpha(G, V(P)) \geq \mathrm{prf}_\beta(P)$ we obtain

$$\begin{aligned}
\mathrm{prf}_\alpha(G) &= \mathrm{prf}_\alpha(G, V(H) - Q) + \mathrm{prf}_\alpha(G, V(P)) \\
&\geq \mathrm{prf}_{\alpha_M}(H, V(H) - Q) + j - \alpha(x) + \mathrm{prf}_\beta(P) \\
&\geq \mathrm{prf}_{\alpha_{V(P)}}(G_{r-1}) + |Q| + j - \alpha(x) + \mathrm{prf}_\beta(P) \\
&= \mathrm{prf}_{\alpha_{V(P)}}(G_{r-1}) + |V(P)| - \alpha(x) + 1 + \mathrm{prf}_\beta(P)
\end{aligned}$$

Now assume that $\alpha(x) > j$. Analogously to the above we get the following:

$$\begin{aligned}
\mathrm{prf}_\alpha(G) &= \mathrm{prf}_\alpha(G, V(H) - Q) + \mathrm{prf}_\alpha(G, V(P)) \\
&\geq \mathrm{prf}_{\alpha_M}(H, V(H) - Q) + \mathrm{prf}_\beta(P) \\
&\geq \mathrm{prf}_{\alpha_{V(P)}}(G_{r-1}) + |Q| + \mathrm{prf}_\beta(P) \\
&\geq \mathrm{prf}_{\alpha_{V(P)}}(G_{r-1}) + |V(P)| - \alpha(x) + 1 + \mathrm{prf}_\beta(P)
\end{aligned}$$

So, we always have $\mathrm{prf}_\alpha(G) \geq \mathrm{prf}_{\alpha_{V(P)}}(G_{r-1}) + |V(P)| - \alpha(x) + 1 + \mathrm{prf}_\beta(P)$.

We add an artificial vertex u' to the end of the ordering β and add the edges $u'x$ and $u'z$. This results in an ordering β' of $V(P) \cup \{u'\}$ where $\beta'(u') = |V(P)| + 1$. Since we have created a cycle we note that $\mathrm{prf}_{\beta'}(P \cup u') \geq 2(|V(P)| + 1) - 3$, by Lemma 2 (ii). Since the profile of u' in β' is $|V(P)| + 1 - \alpha(x)$ we note that $\mathrm{prf}_\beta(P) \geq 2(|V(P)| + 1) - 3 - (|V(P)| + 1 - \alpha(x))$. We get $\mathrm{prf}_\alpha(G) \geq \mathrm{prf}_{\alpha_{V(P)}}(G_{r-1}) + |V(P)| - \alpha(x) + 1 + 2(|V(P)| + 1) - 3 - (|V(P)| + 1 - \alpha(x))$.

By reducing this formula and using the fact that $\mathrm{prf}_{\alpha_{V(P)}}(G_{r-1}) \geq \frac{3n_{r-1}-3}{2}$, we get $\mathrm{prf}(G) = \mathrm{prf}_\alpha(G) \geq \frac{3n_{r-1}-3}{2} + 2|V(P)| - 1$. Since $|V(P)| = a \geq 2$ we note that $2|V(P)| - 1 \geq \frac{3a}{2}$, which implies the desired result. □

Theorem 4. *Let $G = (V, E)$ be a connected graph of order n, let $\mathrm{prf}(G) \leq n - 1 + k$ and let α be an optimal ordering of G. Let $V_1, V_2, \ldots V_t$ be a partition of V such that $|V_1|, |V_t| \geq k + 2$ and there is only one edge $x_i y_i$ between $G[V_1 \cup V_2 \cup \cdots \cup V_i]$ and $G[V_{i+1} \cup V_{i+2} \cup \cdots \cup V_t]$ for each $i = 1, 2, \ldots t - 1$. Let $x_i \in V_i$ and $y_i \in V_{i+1}$ for each $i = 1, 2, \ldots t - 1$ and let $\alpha^{-1}(1) \in V_1$ or $\alpha^{-1}(n) \in V_t$. Let an ordering α' of G be defined as follows: $\alpha'_{V-V_i} = \alpha_{V-V_i}$ for each $i = 1, 2, \ldots t$, and $\alpha'(v_i) < \alpha'(v_{i+1})$ for each $i = 1, 2, \ldots, t - 1$. Then α' is optimal.*

Proof. Consider first the case of $t = 2$. Let $xy = x_1 y_1$, $X = V_1$, $Y = V_2$. Let α be an optimal ordering of G and let $\alpha^{-1}(n) = y' \in Y$ (the case $\alpha^{-1}(1) \in X$ is treated similarly). Let x' be the vertex with $\alpha(x') = 1$. If $x' \in Y$, then Lemma 1 implies that $\mathrm{prf}_\alpha(G, Y) \geq \mathrm{prf}_{\alpha_X}(G - X, Y) + |X|$. Since $\mathrm{prf}_\alpha(G, X) \geq \mathrm{prf}_{\alpha_Y}(G[X]) \geq |X| - 1$ and $\mathrm{prf}_{\alpha_X}(G[Y]) \geq |Y| - 1$ (both by Lemma 2 (i)) and $|X| \geq k + 2$, we conclude that $\mathrm{prf}_\alpha(G) \geq |X| + |Y| + k$, a contradiction. Therefore, $x' \in X$.

Let $i = \min\{\alpha(y'') : y'' \in Y\}$ and let $j = \max\{\alpha(x'') : x'' \in X\}$. Assume for the sake of contradiction that $i < j$. Let $I = \alpha^{-1}(\{i, i+1, \ldots, j\})$. Recall that α' is defined as follows: $\alpha'_X = \alpha_X$ and $\alpha'_Y = \alpha_Y$ but $\alpha'(x'') < \alpha'(y'')$ for all $x'' \in X$ and $y'' \in Y$. We will prove that α' is optimal.

Let $H = G[X \cup (Y \cap I)]$ and let $G' = H$ if $xy \notin E(H)$ and $G' = H - xy$, otherwise. Let $\beta = \alpha_{V(G)-V(G')}$ (so β is equal to α, except we have deleted the last $n - j$ vertices in the ordering). Note that by Lemma 1 (used with the set $Y \cap I$) we get that $\mathrm{prf}_\beta(G', V(G') - (Y \cap I)) \geq \mathrm{prf}_{\beta_{Y \cap I}}(V(G') - (Y \cap I)) + |Y \cap I|$. This implies the following:

$$\mathrm{prf}_\alpha(G - xy, X) \geq \mathrm{prf}_{\alpha_Y}(G[X]) + |Y \cap I|.$$

Analogously we obtain that $\mathrm{prf}_\alpha(G - xy, Y) \geq \mathrm{prf}_{\alpha_X}(Y) + |X \cap I|$, which implies

$$\mathrm{prf}_\alpha(G - xy) \geq \mathrm{prf}_{\alpha_Y}(X) + \mathrm{prf}_{\alpha_X}(Y) + |I| = \mathrm{prf}_{\alpha'}(G - xy) + (j - i + 1). \quad (1)$$

If $\alpha(x) > \alpha(y)$, then the above implies the following contradiction, as $\alpha'(y) - \alpha'(x) < j - i + 1$.

$$\mathrm{prf}_\alpha(G) \geq \mathrm{prf}_\alpha(G - xy) \geq \mathrm{prf}_{\alpha'}(G - xy) + (j - i + 1) > \mathrm{prf}_{\alpha'}(G).$$

Therefore we may assume that $\alpha(x) < \alpha(y)$. Let $l = \min\{\alpha(z) : z \in N[y] - \{x\}\}$ and let $L = \alpha^{-1}(\{\alpha(x), \alpha(x)+1, \alpha(x)+2, \ldots, l-1\})$. Note that $L = \emptyset$ if $l < \alpha(x)$. By the definition of L and the inequality in (1), we get the following:

$$\mathrm{prf}_\alpha(G) = \mathrm{prf}_\alpha(G - xy) + |L| \geq \mathrm{prf}_{\alpha'}(G - xy) + |I| + |L|.$$

When we add the edge xy to $G - xy$, we observe that, in the ordering α', the profile of y will increase by one for every vertex from Y with an α-value less then l and every vertex in X with an α-value larger than $\alpha(x)$. This is exactly the set $R_1 \cup R_2 \cup R_3 \cup R_4$, where

$$R_1 = \{y'' \in Y : \alpha(y'') < l \text{ and } \alpha(x) < \alpha(y'')\},$$
$$R_2 = \{x'' \in X : \alpha(x) < \alpha(x'') \text{ and } \alpha(x'') < l\},$$
$$R_3 = \{y'' \in Y : \alpha(y'') < l \text{ and } \alpha(y'') < \alpha(x)\},$$
$$R_4 = \{x'' \in X : \alpha(x) < \alpha(x'') \text{ and } l < \alpha(x'')\}.$$

Since $R_1 \cup R_2 \subseteq L$ and $R_3 \cup R_4 \subseteq I$ (as $\alpha^{-1}(l) \in Y$), we conclude that $\mathrm{prf}_\alpha(G) \geq \mathrm{prf}_{\alpha'}(G) + |I| + |L| - |R_1| - |R_2| - |R_3| - |R_4| \geq \mathrm{prf}_{\alpha'}(G)$.

Now let $t \geq 3$. Let $X = \bigcup_{i=1}^{t-1} V_i$ and $Y = V_t$. By the case $t = 2$, the following ordering β is optimal: $\beta_X = \alpha_X$, $\beta_Y = \alpha_Y$, and $\beta(x) < \beta(y)$ for each $x \in X, y \in Y$. Now let $X' = \bigcup_{i=1}^{t-2} V_i$, $Y' = V_{t-1} \cup V_t$. By the case $t = 2$, the following ordering β' is optimal: $\beta'_{X'} = \beta_{X'}$, $\beta'_{Y'} = \beta_{Y'}$, and $\beta'(x') < \beta'(y')$ for each $x' \in X', y \in Y'$. Combining the properties of β and β', we obtain that $\beta'_{Y'} = \alpha_{Y'}$, $\beta'_{V-V_{t-1}} = \alpha_{V-V_{t-1}}$, $\beta'_{V-V_t} = \alpha_{V-V_t}$, and $\beta'(x') < \beta'(v_{t-1}) < \beta'(v_t)$ for each $x' \in X', v_{t-1} \in V_{t-1}, v_t \in V_t$. Continuation of this argument allows us to show that α' is an optimal ordering. $\square$

A *bridgeless component* of a graph G is a maximal induced subgraph of G with no bridges. We call a connected graph G a *chain of length t* if the following holds: (a) G has bridgeless components C_i, $1 \leq i \leq t$ such that $V(G) = \bigcup_{i=1}^t V(C_i)$, and (b) C_i is linked to C_{i+1} by a bridge, $1 \leq i \leq t - 1$. A component C_i is *nontrivial* if $|V(C_i)| > 1$, and *trivial*, otherwise. An ordering α of G is *special* if for any two vertices $x, y \in V(G)$ and $x \in V(C_i), y \in V(C_j)$, $i < j$ implies $\alpha(x) < \alpha(y)$.

Lemma 4. *Let G be a chain of order n and let η be the total number of vertices in the nontrivial bridgeless components of G. Let α be a special ordering of G with $\mathrm{prf}_\alpha(G) \leq n - 1 + k$. Then $\eta \leq 3k$.*

Proof. We show $\eta \leq 3k$ by induction on n. Suppose that G has a trivial component. If C_1 is trivial, then $G - C_1$ is a chain with $\mathrm{prf}_{\alpha_{V(C_1)}}(G - C_1) \leq n' - 1 + k$, where $n' = n - 1$. Thus, by the induction hypothesis, $\eta \leq 3k$. Similarly, we prove $\eta \leq 3k$ when C_t is trivial. Assume that C_i, $1 < i < t$, is trivial. Let C_i be adjacent to $x \in V(C_{i-1})$ and $y \in V(C_{i+1})$. Consider G' obtained from G by deleting C_i and appending edge xy. Observe that G' is a chain and $\mathrm{prf}_{\alpha_{V(C_i)}}(G') \leq n' - 1 + k$, where $n' = n - 1$. Thus, by induction hypothesis, $\eta \leq 3k$. So, now we may assume that $\eta = n$.

Let $C_1, \ldots, C_t$ denote the bridgeless components of G as in the definition above. Let $n_i = |V(C_i)|$. If $t = 1$, then by Theorem 3 we have $n \leq 2k + 1$ and we are done as $k \geq 1$. Now assume $t \geq 2$. Let $G' = G - V(C_t)$ and $n' = n - n_t$. Observe that G' is a chain and $\alpha_{V(C_t)}$ is a special ordering of G'. Let $k_t = \mathrm{prf}_\alpha(G, V(C_t)) - n_t + 1$ and let $k' = \mathrm{prf}_\alpha(G, V(G')) - n' + 1$. We have $k_t + k' - 1 \leq k$. Theorem 3 implies that

$$n_t - 1 + k_t = \mathrm{prf}_\alpha(G, V(C_t)) \geq \mathrm{prf}(C_t) + 1 \geq \frac{3n_t - 3}{2} + 1 = \frac{3n_t - 1}{2},$$

and thus $k_t \geq \frac{n_t+1}{2}$ and $n_t \leq 2k_t - 1$. Since $n_t \geq 3$, we have $k_t \geq 2$. By induction hypothesis, $n' \leq 3k'$. Thus $n = n' + n_t \leq 3(k - k_t + 1) + 2k_t - 1 \leq 3k$. □

A connected component of a graph G is called *nontrivial* if it has more than one vertex.

Lemma 5. *Let $G = (V, E)$ be a connected graph of order n, let $X \subseteq V$ such that $G[X]$ is connected. Let $G_1, \ldots, G_r$ denote the nontrivial connected components of $G - X$. Assume that $|V(G_i)| \leq |V(G_{i+1})|$ for $1 \leq i \leq r-1$. If $k+n-1 \geq \mathrm{prf}(G)$, then $k + 2 \geq r$ and $2k \geq \sum_{i=1}^{r-2} |V(G_i)|$.*

Proof. The result holds vacuously true if $r < 3$, hence assume $r \geq 3$. Let α be an optimal ordering of G. Let $I = \{1 \leq i \leq r : V(G_i) \cap \{\alpha^{-1}(1), \alpha^{-1}(n)\} = \emptyset\}$. Clearly $|I| \geq r - 2$. Let $Y = X \cup \bigcup_{i \notin I} V(G_i)$ and $Z = V \setminus Y$. Observe that $G[Y] = G - Z$ is connected and G_i, $i \in I$, are exactly the nontrivial components of $G - Y$. Since also $\{\alpha^{-1}(1), \alpha^{-1}(n)\} \subseteq Y$, Lemma 1 applies. Thus we get $\mathrm{prf}(G) = \mathrm{prf}_\alpha(G) \geq \mathrm{prf}_\alpha(G, V - Z) + \sum_{i \in I} \mathrm{prf}_\alpha(G, V(G_i)) \geq \mathrm{prf}(G - Z) + |Z| + \sum_{i \in I} \mathrm{prf}(G_i)$. Furthermore, by Lemma 2 (i), $k \geq \mathrm{prf}(G) - n + 1 \geq \mathrm{prf}(G - Z) + |Z| - |Y| + (\sum_{i \in I} \mathrm{prf}(G_i) - |V(G_i)|) + 1 \geq (\mathrm{prf}(G[Y]) - |Y|) + |Z| - |I| + 1 \geq -1 + |Z| - |I| + 1$. Hence $k \geq |Z| - |I|$. However, since the components G_i are nontrivial, $|Z| \geq 2|I|$. Thus, $|I| \leq k$ and $|Z| \leq k + |I| \leq 2k$. □

3   Dealing with Vertices of Degree 1

In this section, G denotes a connected graph of order n. For an ordering α of G let $E_\alpha(G)$ denote the set of edges uv of G such that $\alpha(u) = \min_{w \in N[v]} \alpha(w)$ and $u \neq v$. The *length* $\ell_\alpha(uv)$ *of an edge* $uv \in E(G)$ *relative to* α is $|\alpha(u) - \alpha(v)|$ if $uv \in E_\alpha(G)$, and 0 if $uv \notin E_\alpha(G)$. Observe that $\mathrm{prf}_\alpha(G) = \sum_{e \in E(G)} \ell_\alpha(e)$.

Let X, Y be two disjoint sets of vertices of G and let α be an ordering of G. We say that (X, Y) is an α-*consecutive pair* if there exist integers a, b, c with $1 \leq a < b < c \leq n$ so that $X = \{x \in V(G) : a \leq \alpha(x) \leq b - 1\}$ and $Y = \{y \in V(G) : b \leq \alpha(y) \leq c\}$. By $\mathrm{swap}_{Y,X}(\alpha)$ we denote the ordering obtained from α by swapping the α-consecutive pair (X, Y). For a set $X \subseteq V(G)$ let $E_\alpha^r(X)$ (respectively, $E_\alpha^l(X)$) denote the set of edges $uv \in E_\alpha$ with $u \in X$, $v \in V(G) \setminus X$, and $\alpha(u) < \alpha(v)$ (respectively, $\alpha(u) > \alpha(v)$).

Lemma 6. *Let α be an ordering of G and (X, Y) an α-consecutive pair such that there are no edges between X and Y. If $|E_\alpha^l(X)| \leq |E_\alpha^r(X)|$ and $|E_\alpha^l(Y)| \geq |E_\alpha^r(Y)|$, then for $\beta = \mathrm{swap}_{Y,X}(\alpha)$ we have $\mathrm{prf}_\beta(G) \leq \mathrm{prf}_\alpha(G)$.*

Proof. Observe that $E_\alpha(G) = E_\beta(G)$. Moreover, the only edges of $E_\alpha(G)$ that have different length in α and in β are the edges in $E_\alpha^l(Y) \cup E_\alpha^r(Y) \cup E_\alpha^l(X) \cup E_\alpha^r(X)$. Observe that $\ell_\beta(e) = \ell_\alpha(e) + |Y|$, $\ell_\beta(e') = \ell_\alpha(e') - |Y|$, $\ell_\beta(f) = \ell_\alpha(f) - |X|$, $\ell_\beta(f') = \ell_\alpha(f') + |X|$ for each $e \in E_\alpha^l(X)$, $e' \in E_\alpha^r(X)$, $f \in E_\alpha^l(Y)$ and $f' \in E_\alpha^r(Y)$. Using these relations and the inequalities $|E_\alpha^l(X)| \leq |E_\alpha^r(X)|$ and $|E_\alpha^l(Y)| \geq |E_\alpha^r(Y)|$, we obtain $\mathrm{prf}_\beta(G) \leq \mathrm{prf}_\alpha(G)$. □

Lemma 7. *Let α be an ordering of G and $(\{x\}, Y)$ an α-consecutive pair such that x has a neighbor z of degree 1 with $\alpha(z) > \alpha(y)$ for all $y \in Y$. If $|E_\alpha^l(Y)| \geq |E_\alpha^r(Y)|$, then for $\beta = \mathrm{swap}_{Y,\{x\}}$ we have $\mathrm{prf}_\beta(G) \leq \mathrm{prf}_\alpha(G)$.*

Proof. If there are no edges between x and vertices in Y then the result follows from Lemma 6 since $|E_\alpha^l(\{x\})| \leq 1 \leq |E_\alpha^r(\{x\})|$. Now consider the case where $E_\alpha^l(\{x\}) = \{wx\}$ for a vertex w. It follows that $E_\beta(G) \subseteq E_\alpha(G)$. Moreover, we have $\sum_{e \in E_\beta^l(Y) \cup E_\beta^r(Y)} \ell_\beta(e) \leq \sum_{e \in E_\alpha^l(Y) \cup E_\alpha^r(Y)} \ell_\alpha(e)$ and $\ell_\beta(wx) + \ell_\beta(xz) \leq \ell_\alpha(wx) + \ell_\alpha(xz)$. Hence the result also holds true in that case. It remains to consider the case where x has neighbors in Y and $E_\alpha^l(\{x\}) = \emptyset$. Let y, y' be the neighbors of x in Y with largest $\alpha(y)$ and smallest $\alpha(y')$. Now $E_\beta(G) \setminus E_\alpha(G) = \{xy'\}$ and $\ell_\beta(xy') + \ell_\beta(xz) \leq \ell_\alpha(xz)$. Thus, $\mathrm{prf}_\beta(G) \leq \mathrm{prf}_\alpha(G)$. $\qquad\square$

Lemma 8. *Let α be an ordering of G and let $(\{x\}, Y)$ be an α-consecutive pair. Let all vertices in Y be of degree 1 and adjacent with x. Then for $\beta = \mathrm{swap}_{Y,\{x\}}(\alpha)$ we have $\mathrm{prf}_\beta(G) \leq \mathrm{prf}_\alpha(G)$.*

Proof. Let y, y' denote the vertex in Y with largest $\alpha(y)$ and smallest $\alpha(y')$. Observe $\ell_\alpha(yx) = |Y|$. First assume that $E_\alpha^l(\{x\})$ contains an edge zx. We have $E_\beta(G) \subseteq E_\alpha(G) \setminus \{xy\}$, and $\ell_\beta(e) \leq \ell_\alpha(e)$ holds for all $e \in E_\beta(G) \setminus \{xz\}$. Since $\ell_\beta(zx) = \ell_\alpha(zx) + \ell_\alpha(xy)$, the result follows. Next assume that $E_\alpha^l(\{x\}) = \emptyset$. We have $E_\beta(G) \subseteq (E_\alpha(G) \setminus \{xy\}) \cup \{xy'\}$, and $\ell_\beta(e) \leq \ell_\alpha(e)$ holds for all $e \in E_\beta(G) \setminus \{xy'\}$. Since $\ell_\beta(xy') = \ell_\alpha(xy)$, the result follows. $\qquad\square$

For $x \in V(G)$ let $N_1(X)$ denote the set of neighbors of x that have degree 1. We say that an ordering α of G is *conformal for a vertex* x of G if $\{\alpha(w) : w \in N_1(x)\}$ forms a (possibly empty) interval and $\alpha(w) < \alpha(x)$ holds for all $w \in N_1(x)$. We say that α is *conformal for a graph* G if it is conformal for all vertices of G.

Theorem 5. *For every connected graph G there exists an optimal ordering which is conformal.*

Proof. Let α be an optimal ordering of G. Let x be a vertex of G for which α is not conformal. We apply the following steps to α, until we end up with an optimal ordering which is conformal for x. In each step we transform α into an optimal ordering β in such a way that whenever α is conformal for a vertex x', so is β. Hence, we can repeat the procedure for all the vertices one after the other, and we are finally left with an optimal ordering which is conformal.

Let $w_1, w_2 \in N_1(x) \cup \{x\}$ with minimal $\alpha(w_1)$ and maximal $\alpha(w_2)$. We call a set $B \subseteq N_1(x)$ a *block* if $\{\alpha(b) : b \in B\}$ is a nonempty interval of integers. A block is *maximal* if it is not properly contained in another block.

Step 1. Assume that there exist α-consecutive pairs $(\{x\}, Y)$, (Y, Z) with the following properties: (a) Y and Z are nonempty; (b) $Y \cap N_1(x) = \emptyset$; (c) Z is a maximal block. By assumption, there is a $z \in Z$ such that $xz \in E(G)$ and $\alpha(z) > \alpha(y)$ holds for all $y \in Y$. Moreover, there are no edges between Y and Z and $E_\alpha^r(Z) = \emptyset$. If $|E_\alpha^l(Y)| \geq |E_\alpha^r(Y)|$, then we put $\beta = \mathrm{swap}_{Y,\{x\}}(\alpha)$, otherwise we put $\beta = \mathrm{swap}_{Z,Y}(\alpha)$. It follows from Lemmas 7 and 6, respectively, that β is optimal.

Step 2. Assume that there exists an α-consecutive pair $(\{x\}, Y)$ such that Y is a maximal block. We put $\beta = \mathrm{swap}_{Y, \{x\}}(\alpha)$. It follows by Lemma 8 that β is optimal. (*Remark:* If neither Step 1 nor Step 2 can be applied, then $\alpha(w_2) < \alpha(x)$.)

Step 3. Assume that there exist α-consecutive pairs (X, Y), (Y, Z) with the following properties: (a) X and Z are maximal blocks; (b) $Y \subseteq V(G) \setminus N_1(X)$; (c) $w_1 \in X$. Note that there are no edges between X and Y and no edges between Y and Z. Furthermore, we have $E_\alpha^l(X) = \emptyset$ and $E_\alpha^r(Z) = \emptyset$ (the former follows from Property (c)). If $|E_\alpha^l(Y)| \geq |E_\alpha^r(Y)|$, then we put $\beta = \mathrm{swap}_{Y,X}(\alpha)$, otherwise we put $\beta = \mathrm{swap}_{Z,Y}(\alpha)$. In both cases it follows from Lemma 6 that β is optimal. (*Remark:* If none of the above Steps 1, 2, or 3, applies, then α is conformal for x.) $\square$

4   Kernelization

For technical reasons, in this section we will deal with a special kind of weighted graphs, but they will be nothing else but compact representations of (unweighted) graphs. We consider a *weighted graph* $G = (V, E, \rho)$ whose vertices v of degree 1 have an arbitrary positive integral weight $\rho(v)$, vertices u of degree greater than one have weight $\rho(u) = 1$. The *weight* $\rho(G)$ of $G = (V, E, \rho)$ is the sum of weights of all vertices of G. An *ordering* of a weighted graph $G = (V, E, \rho)$ is an injective mapping $\alpha : V \to \{1, \ldots, \rho(G)\}$ such that for every vertex $v \in V$ of degree 1 we have $\alpha(v) \neq \rho(G)$ and for all $u \in V$ we have $\alpha(u) \notin \{\alpha(v) + 1, \ldots, \alpha(v) + \rho(v) - 1\}$. The *profile* $\mathrm{prf}(G)$ of a weighted graph is defined exactly as the profile of an unweighted graph. A weighted graph $G = (V, E, \rho)$ corresponds to an unweighted graph G^u, which is obtained from G by replacing each vertex v of degree 1 (v is adjacent to a vertex w) with $\rho(v)$ vertices adjacent to w. By Theorem 5 and the definitions above, $\mathrm{prf}(G) = \mathrm{prf}(G^u)$ and an optimal ordering of G can be effectively transformed into an optimal ordering of G^u. Also, $\rho(G) = |V(G^u)|$. The correspondence between G and G^u allows us to use the results given in the previous sections.

Kernelization Rule 1. *Let G be a weighted graph and x a vertex of G with $N_1(x) = \{v_1, \ldots, v_r\}$, $r \geq 2$. We obtain the weighted graph $G_0 = (V_0, E_0, \rho_0)$, where $G_0 = G - \{v_2, \ldots, v_r\}$ and $\rho_0(u) = \rho(u)$ for $u \in V_0 \setminus \{v_1\}$ and $\rho_0(v_1) = \sum_{i=1}^r \rho(v_i)$.*

The next lemma follows from Theorem 5.

Lemma 9. *Let G be a weighted connected graph and G_0 the weighted graph obtained from G by Kernelization Rule 1. Then $\mathrm{prf}(G) = \mathrm{prf}(G_0)$, and an optimal ordering α_0 of G_0 can be effectively transformed into an optimal ordering α of G.*

Let e be a bridge of a weighted connected graph G and let G_1, G_2 denote the connected components of $G - e$. We define the *order* of e as $\min\{\rho(G_1), \rho(G_2)\}$. Let v be a vertex of a (weighted) graph G. We say that v is *k-suppressible* if the following conditions hold: (a) v forms a trivial bridgeless component of G; (b) v is of degree 2 or 3; (c) there are exactly two bridges e_1, e_2 of order at least $k + 2$ incident with v; (d) if there is a third edge $e_3 = vw$ incident with v, then w is a vertex of degree 1.

Kernelization Rule 2 (w.r.t. parameter k). *Let v be a k-suppressible vertex of a weighted graph $G = (V, E, \rho)$ and let xv, yv be the bridges of order at least $k + 2$. From G we obtain a weighted graph by removing $\{v\} \cup N_1(v)$ and adding the edge xy.*

Lemma 10. *Let $G = (V, E, \rho)$ be a weighted connected graph with $\mathrm{prf}(G) \leq \rho(G) - 1 + k$ and G' the weighted graph obtained from G by means of Kernelization Rule 2 with respect to parameter k. Then $\mathrm{prf}(G) - \rho(G) = \mathrm{prf}(G') - \rho(G')$, and an optimal ordering α' of G' can be effectively transformed into an optimal ordering α of G.*

Proof. Let v be a k-suppressible vertex of G^u and let xv, yv be the bridges of order at least $k + 2$. We consider the case when $N_1(v) = \{w_1, \ldots, w_r\} \neq \emptyset$; the proof for the case when $N_1(v) = \emptyset$ is similar. Let $G^u[X]$ and $G^u[Y]$ denote the components of $G^u - v$ that contain x and y, respectively. Consider an optimal ordering α of G^u and assume that $\alpha^{-1}(n) \in Y$. By Theorem 5, we may assume that $\alpha(w_i) < \alpha(v)$ for every $1 \leq i \leq r$. Now by Theorem 4, we can find an optimal ordering α' of G^u such that $\alpha'(x') < \alpha'(w_i) < \alpha'(v) < \alpha'(y')$ for each $x' \in X$, $y' \in Y$ and $i = 1, 2, \ldots, r$. Now it will be more convenient to argue using the weighted graphs G and G'. Using Kernelization Rule 1, we transform α' into the corresponding optimal ordering of G. For simplicity we denote the new ordering α' as well. Observe that $\mathrm{prf}_{\alpha'_{\{v,w\}}}(G', y) = \mathrm{prf}_{\alpha'}(G, y) + \mathrm{prf}_{\alpha'}(G, v) - 1 - \rho(w)$. Hence, $\mathrm{prf}(G') - \rho(G') \leq \mathrm{prf}_{\alpha'_{\{v,w\}}}(G') - \rho(G') = \mathrm{prf}(G) - \rho(G)$.

Conversely, let α' be an optimal ordering of G'. Since the bridge xy of G' is of order at least $k + 2$, we may assume by Theorem 4 that either for all $x' \in X$ and $y' \in Y$ we have $\alpha'(x') < \alpha'(y')$. It is straightforward to extend α' into an ordering α of G such that $\alpha_{\{v,w\}} = \alpha'$ and $\mathrm{prf}_\alpha(G) = \mathrm{prf}_{\alpha'}(G') + 1 + \rho(w)$. Hence $\mathrm{prf}(G) - \rho(G) \leq \mathrm{prf}_\alpha(G) - \rho(G) = \mathrm{prf}_{\alpha'}(G') - \rho(G')$. Thus, $\mathrm{prf}(G') - \rho(G') = \mathrm{prf}(G) - \rho(G)$. □

Theorem 6. *Let $G = (V, E, \rho)$ be a weighted connected graph with $n = |V|$ and $m = |E|$. Let k be a positive integer such that $\mathrm{prf}(G) \leq \rho(G) - 1 + k$. One of the Kernelization Rules 1 and 2 can be applied with respect to parameter k, or $n \leq 12k + 6$ and $m \leq 13k + 5$.*

Proof. For a weighted graph $G = (V, E, \rho)$ let G^* be an unweighted graph with $V(G^*) = V$ and $E(G^*) = E$. Observe that $\mathrm{prf}(G^*) \leq \mathrm{prf}(G)$. Thus, in the rest of the proof we consider G^* rather than G, but for the simplicity of notation we use G instead of G^*. Assume that none of the Kernelization Rules 1 and 2 can be applied with respect to parameter k. We will show that the claimed bounds on n and m hold. By Theorem 1 we have $m \leq \mathrm{prf}(G) \leq n - 1 + k$. Thus, $n \leq 12k + 6$ implies $m \leq 13k + 5$. Therefore, it suffices to prove that $n \leq 12k + 6$. If G is bridgeless, then by Theorem 3, we have $n - 1 + k \geq \mathrm{prf}(G) \geq \frac{3n-3}{2}$ and, thus, $n \leq 2k + 1$. Hence, we may assume that G has bridges. Let C_i, $i = 1, \ldots, t$, denote the bridgeless components of G such that at least one vertex in C_i is incident with a bridge of order at least $k + 2$. We put $X = \bigcup_{i=1}^{t} V(C_i)$.

Suppose that there is a component C_i incident with three or more bridges of order at least $k+2$. Then, we may assume that there are three bridges e_2, e_3, e_4 of order at least $k+2$ that connect a subgraph F_1 of G with subgraphs F_2, F_3, F_4, respectively, and $V = \bigcup_{i=1}^4 V(F_i)$. Let α be an optimal ordering of G. Assume without loss of generality that $\alpha^{-1}(1) \notin V(F_2)$ and $\alpha^{-1}(n) \notin V(F_2)$. Let $X = V(F_2)$ and note that $G - X$ is connected. Therefore Lemmas 1 and 2 (i) imply $\mathrm{prf}(G) = \mathrm{prf}_\alpha(G, X) + \mathrm{prf}_\alpha(G, V - X) \geq |X| - 1 + (|V| - |X| - 1) + |X| \geq n + k$, a contradiction. Since G is connected, it follows that $G[X]$ is connected. Thus, $G[X]$ is a chain and we may assume that C_i and C_{i+1} are linked by a bridge b_i for each $i = 1, 2, \ldots, t - 1$. Notice that each b_i is of order at least $k + 2$ in G.

Let $G_1, \ldots, G_r$ be the connected components of $G - X$. Observe that each G_i ($1 \leq i \leq r$) is linked with exactly one C_j ($1 \leq j \leq t$) with a bridge e_{ij}. The bridge e_{ij} must be of order less than $k + 2$, since otherwise $V(G_i) \cap X \neq \emptyset$. Hence (**) $|V(G_i)| \leq k + 1$ follows for all $i \in \{1, \ldots, r\}$. For each j, let $IG(j)$ be the set of indices i such that G_i is linked to C_j. Let $N = \{1 \leq i \leq t : |V(C_i)| > 1\}$ and $T = \{1 \leq i \leq t : |V(C_i)| = 1\}$. For $i \in T_i$ let x_i denote the single vertex in C_i. Similarly, let $N' = \{1 \leq i \leq r : |V(G_i)| > 1\}$ and $T' = \{1 \leq i \leq r : |V(G_i)| = 1\}$. Let $H_j = G[\bigcup_{i \in IG(j)} V(G_i) \cup V(C_j)]$ for each $j = 1, 2, \ldots, t$. By Theorem 4, we may assume that there exists an optimal ordering β such that $\beta(h_i) < \beta(h_j)$ for all $i < j$, $h_i \in V(H_i)$, $h_j \in V(H_j)$. Let $\gamma = \beta_{V(G)-X}$. Clearly, γ is a special ordering of the chain $G[X]$, i.e., $\gamma(c_i) < \gamma(c_j)$ for all $i < j$, $c_i \in V(C_i)$, $c_j \in V(C_j)$.

If G_i is nontrivial, then it has a vertex z such that $G_i - z$ is connected and z is not incident to the bridge between G_i and $G[X]$. If G_i is trivial, let $z = V(G_i)$. In both cases, by Lemma 3, $\mathrm{prf}_{\beta_z}(G-z) \leq (n-1) - 1 + k$. Repeating this argument, we conclude that $\mathrm{prf}_\gamma(G[X]) \leq |X| - 1 + k$. Now by Lemma 4, $\sum_{i \in N} |V(C_i)| \leq 3k$. Lemma 5 yields that $|N'| \leq k + 2$. Observe that for each $i \in T$, x_i is linked by a bridge $x_i y_{\pi(i)}$ to at least one nontrivial $G_{\pi(i)}$, where $\pi(i) \neq \pi(i')$ whenever $i \neq i'$. Hence, $|T| \leq k + 2$. Thus, $|X| = \sum_{i \in N} |V(C_i)| + |T| \leq 3k + (k + 2) = 4k + 2$. Using (**) and Lemma 5, we have that $\sum_{i \in N'} |V(G_i)| \leq 2(k+1) + 2k = 4k + 2$.

Let $Y = \bigcup_{i=1}^r V(G_i)$. Since Kernelization Rule 1 cannot be applied, every vertex in X is adjacent with at most one G_i with $i \in T'$. Hence $|T'| \leq |X| \leq 4k + 2$. Consequently $|Y| \leq 2(4k + 2) = 8k + 4$. Hence $n = |X| + |Y| \leq 4k + 2 + 8k + 4 = 12k + 6$ follows. $\square$

Corollary 1. *The problem PAVGV is fixed-parameter tractable.*

Remark 1. *We see that PAVGV can be solved in time $O(|V|^2 + f(k))$, where $f(k) = (12k + 6)!$. It would be interesting to significantly decrease $f(k)$, but even as it is now our algorithm is of practical interest because the kernel produced by the two kernelization rules can be solved using fast heuristics.*

5   Vertex Average Profile Problem

Serna and Thilikos [11] asked whether the following problem is FPT.

Vertex Average Profile (VAP)
Instance: A graph $G = (V, E)$. *Parameter:* A positive integer k.
Question: Is the profile of G at most $k|V|$?

The following result, announced in [7] without a proof, implies that VAP is not fixed-parameter tractable unless $P = NP$.

Theorem 7. *Let $k \geq 2$ be a fixed integer. Then it is NP-complete to decide whether $\mathrm{prf}(H) \leq k|V(H)|$ for a graph H.*

Proof. Let G be a graph and let r be an integer. We know that it is NP-complete to decide whether $\mathrm{prf}(G) \leq r$. Let $n = |V(G)|$. Let k be a fixed integer, $k \geq 2$. Define G' as follows: G' contains k copies of G, j isolated vertices and a clique with i vertices (all of these subgraphs of G' are vertex disjoint). We have $n' = |V(G')| = kn + i + j$. Observe that $\mathrm{prf}(K_i) = \binom{i}{2}$. By the definition of G', $k \cdot \mathrm{prf}(G) = \mathrm{prf}(G') - \mathrm{prf}(K_i) = \mathrm{prf}(G') - \binom{i}{2}$. Therefore, $\mathrm{prf}(G) \leq r$ if and only if $\mathrm{prf}(G') \leq kr + \binom{i}{2}$. If there is a positive integer i such that $kr + \binom{i}{2} = kn'$ and the number of vertices in G' is bounded from above by a polynomial in n, then G' provides a reduction from to VAP with the fixed k. Observe that $kr + \binom{i}{2} \geq k(kn+i)$ for $i = 2kn$. Thus, by setting $i = 2kn$ and $j = r + \frac{1}{k}\binom{i}{2} - kn - i$, we ensure that G' exists and the number of vertices in G' is bounded from above by a polynomial in n. $\qquad\square$

References

1. A. Billionnet, On interval graphs and matrix profiles. RAIRO Tech. Oper. 20 (1986), 245–256.
2. H.L. Bodlaender, R.G. Downey, M.R. Fellows, M.T. Hallett and H.T. Wareham, Parameterized complexity analysis in computational biology. Comput. Appl. Biosci. 11 (1995), 49–57.
3. J. Diaz, A. Gibbons, M. Paterson and J. Toran, The minsumcut problem. Lect. Notes Comput. Sci. 519 (1991), 65-79.
4. F.V. Fomin and P.A. Golovach, Graph searching and interval completion. SIAM J. Discrete Math. 13 (2000), 454–464.
5. M. R. Garey and D. R. Johnson, *Computers and Intractability*, Freeman, 1979.
6. P.W. Goldberg, M.C. Golumbic, H. Kaplan and R. Shamir, Four strikes against physical mapping of DNA. J. Comput. Biol. 2 (1995), 139–152.
7. G. Gutin, A. Rafiey, S. Szeider and A. Yeo, The Linear Arrangement Problem Parameterized Above Guaranteed Value. To appear in Theory of Comput. Systems.
8. R.M. Karp, Mapping the genome: some combinatorial problems arising in molecular biology. In Proc. 25th Annual Symp. Theory Comput. (1993), 278–285.
9. D.G. Kendall, Incidence matrices, interval graphs, and seriation in archeology. Pacific J. Math. 28 (1969), 565–570.
10. Y. Lin and J. Yuan, Profile minimization problem for matrices and graphs. Acta Math. Appl. Sinica, English-Series, Yingyong Shuxue-Xuebas 10 (1994), 107-112.
11. M. Serna and D.M. Thilikos, Parameterized complexity for graph layout problems. EATCS Bulletin 86 (2005), 41–65.
12. D.B. West, Introduction to Graph Theory, Prentice Hall, 2001.

On the OBDD Size for Graphs of Bounded Tree- and Clique-Width

Klaus Meer[1] and Dieter Rautenbach[2]

[1] Department of Mathematics and Computer Science
Syddansk Universitet, Campusvej 55, 5230 Odense M, Denmark
Fax: 0045 6593 2691
meer@imada.sdu.dk
[2] Forschungsinstitut für Diskrete Mathematik
Universität Bonn, Lennéstrasse 2, 53113 Bonn, Germany
Fax: 0049 228 738771
rauten@or.uni-bonn.de

Abstract. We study the size of OBDDs (ordered binary decision diagrams) for representing the adjacency function f_G of a graph G on n vertices. Our results are as follows:

-) For graphs of bounded tree-width there is an OBDD of size $O(\log n)$ for f_G that uses encodings of size $O(\log n)$ for the vertices;

-) For graphs of bounded clique-width there is an OBDD of size $O(n)$ for f_G that uses encodings of size $O(n)$ for the vertices;

-) For graphs of bounded clique-width such that there is a *reduced term* for G (to be defined below) that is balanced with depth $O(\log n)$ there is an OBDD of size $O(n)$ for f_G that uses encodings of size $O(\log n)$ for the vertices;

-) For cographs, i.e. graphs of clique-width at most 2, there is an OBDD of size $O(n)$ for f_G that uses encodings of size $O(\log n)$ for the vertices. This last result improves a recent result by Nunkesser and Woelfel [14].

1 Introduction

The most usual way to represent graphs certainly is by adjacency lists or adjacency matrices. However, for applications like traffic scheduling, Travelling Salesman, and many more such representations can easily become very large. A way of trying to circumvent the storing problems related to such large graphs is to consider data structures for Boolean functions, hoping for a more succinct representation of the graph's adjacency function by such structures. In data structure literature almost tight bounds for designing a shortest string encoding a graph (with fast en- and decoding) are known for general graphs [13] and for many subclasses such as planar graphs [10].

One of the most important data structures for Boolean functions that has been used quite successfully are OBDDs (*Ordered Binary Decision Diagrams*). For a comprehensive introduction into the theory of OBDDs we refer to [17].

The content of this paper is to study succinct graph representations using OBDDs. By cardinality arguments we cannot represent any arbitrary graph by

H.L. Bodlaender and M.A. Langston (Eds.): IWPEC 2006, LNCS 4169, pp. 72–83, 2006.

OBDDs of small size. Thus it is a reasonable question whether succinct OBDD representations can be found at least for significant graph classes. We shall look for such classes among graphs whose local structure is somehow controlled. For some graph classes including cographs and unit interval graphs this kind of problem was recently studied by Nunkesser and Woelfel [14], where several upper and lower bounds for the sizes of OBDDs representing such graphs were given. Other works relating algorithmic graph problems with OBDD representations are [9,15,16,18].

Our paper takes the work of [14] as starting point and analyzes further important graph classes with respect to their representability by small sized OBDDs. Two of the most important parameters for graphs are the tree-width and the clique-width. For graphs of bounded tree- or bounded clique-width many otherwise intractable algorithmic problems are known to be efficiently solvable [1,6,7]. Note that bounded tree-width implies bounded clique-width [8].

We study the representation of graphs belonging to one of the above classes by OBDDs. Section 2 recalls the main concepts used in the paper. In Section 3 it is shown that for each graph with n vertices and of bounded tree-width there exists an OBDD of size $O(\log n)$ that uses vertex encodings of length $O(\log n)$. Section 4 is devoted to graphs of bounded clique-width. First, we show that when using vertex encodings of length $O(n)$, then each graph with n vertices having a bounded clique-width allows an OBDD representation of size $O(n)$. Thereafter, we improve the above result for cographs, i.e. graphs of clique-width at most 2. Here, representing OBDDs of size $O(n)$ exist which use vertex encodings of length $O(\log n)$. This last result improves a related one in [14] by a factor of $\log n$.

2   Basic Concepts

In order to make the paper self-contained we briefly recall the main notions we are dealing with in this paper, i.e. ordered binary decision diagrams as well as the tree- and the clique-width of a graph. For more elaborate treatments refer to [17] and [5]. At the end of this section we collect some results from [11] which are important later on.

Definition 1 (BDD and OBDD). *a) A binary decision diagram or BDD is a directed acyclic graph having two kinds of nodes. Output nodes are nodes with no outgoing edge and are labeled with a Boolean constant from $\{0, 1\}$. Inner nodes are labeled with an element from some variable set $\{x_1, \ldots, x_n\}$. They have exactly two outgoing edges one of which is labeled by 0 and the other by 1.*

b) Each node v of a BDD computes a Boolean function $f_v : \{0, 1\}^n \to \{0, 1\}$ in the following way: Given an assignment for the boolean variables $x_1, \ldots, x_n$ one follows, starting at v, the edges according to the value of the corresponding label until an output node is reached whose label gives $f_v(x_1, \ldots, x_n)$. If the underlying graph has a root r, then f_r is also called the function computed *by the BDD.*

c) The size *of a BDD is the number of nodes of the underlying graph.*

d) A BDD is an ordered binary decision diagram *with respect to the variable ordering* $x_1 < \ldots < x_n$, *or OBDD for short, if for every edge of the BDD from an inner node with label* x_i *to an inner node with label* x_j *the indices satisfy* $j > i$.

In an OBDD the variables are read at most once and always in the same order.

Let $G = (V, E)$ be a graph. We are interested in OBDDs computing the adjacency function $f_G : V^2 \rightarrow \{0, 1\}$ with $f_G(i, j) = 1$ iff $(i, j) \in E$. Here, we first have to specify how a node is encoded in order to use it as (part of) the argument of a Boolean function.

Definition 2. *Let* $G = (V, E)$ *be a graph with* $n := |V|$ *nodes and let* name : $V \rightarrow \{0, 1\}^L$ *be an encoding of the nodes by binary strings of length* L. *An OBDD* O *is said to* represent G *if* O *computes a Boolean function* f_O *from* $\{0, 1\}^{2L} \rightarrow \{0, 1\}$ *such that for all* $i, j \in V$ *we have*

$$f_O(name(i), name(j)) = f_G(i, j).$$

A few words concerning the used encoding are appropriate. The following upper bound is well known.

Theorem 1. *([3]) Let* $f : \{0, 1\}^N \rightarrow \{0, 1\}$. *There is an OBDD of size* $(2 + o(1)) \cdot \frac{2^N}{N}$ *that computes* f.

For the adjacency function of a graph with n nodes this implies that we can always find a representing OBDD of size at most $O(\frac{n^2}{\log n})$ by choosing the binary representations of the elements in $\{1, \ldots, n\}$ as names for the vertices. The adjacency function then takes $2 \cdot \lceil \log n \rceil$ many arguments.

Thus, if below we shall use encodings of a different length we have to keep in mind this upper bound when comparing the sizes of the OBDDs constructed.

Definition 3 (Tree-width of a graph). *A* tree decomposition *of width* $k \in \mathbb{N}$ *of a (simple and undirected) graph* $G = (V, E)$ *is a pair* $(T, (X_t)_{t \in V_T})$ *where* $T = (V_T, E_T)$ *is a rooted tree and* $(X_t)_{t \in V_T}$ *is a collection of subsets* $X_t \subseteq V$ *of cardinality at most* $k + 1$ *such that for every edge* $uv \in E$ *there is a* $t \in V_T$ *with* $u, v \in X_t$ *and for every vertex* $u \in V$ *the set* $\{t \in V_T \mid u \in X_t\}$ *induces a subtree of* T *having at least one vertex.*

The tree-width *of* G *is the minimum* k *such that there is a tree decomposition of width* k *of* G.

The final concept we recall is that of the clique-width of a graph. For $k \in \mathbb{N}$ a k-graph is a graph $G = (V, E)$ whose vertices are labeled with a label from $\{1, \ldots, k\}$. We consider four operations on k-graphs: First, for two k-graphs $G = (V, E)$ and $H = (W, F)$ with disjoint sets of vertices the k-graph $G \oplus H$ is obtained by taking $V \cup W$ as new vertex set and $E \cup F$ as new edge set. The original labels are maintained. Secondly, for $i, j \in \{1, \ldots, k\}, i \neq j$ and a k-graph G the k-graph $\eta_{i,j}(G)$ is obtained from G by connecting all vertices labeled i with those labeled j in G. Thirdly, for $i, j \in \{1, \ldots, k\}, i \neq j$ and a k-graph G

the k-graph $\rho_{i \to j}(G)$ is obtained from G by relabeling all nodes having label i with label j. Finally, for $i \in \{1, \ldots, k\}$ the operation $i(v)$ creates a k-graph with one vertex v that carries label i.

Definition 4 (Clique-width of a graph).

a) Let $k \in \mathbb{N}$ be fixed. A k-expression is a well formed term using the above mentioned operations that represents a graph G in the obvious way. (The term $t = \eta_{1,2}(2(z) \bigoplus (\rho_{2 \to 1}(\eta_{1,2}(1(x) \bigoplus 2(y)))))$ for instance represents a clique with the three vertices x, y and z.)

b) The clique-width $cw(G)$ of a graph G is the minimal $k \in \mathbb{N}$ such that there exists a k-expression representing G.

It is well known that a graph of tree-width k has a clique-width bounded by $2^{k+1} + 1$, see [8]. Cographs, which will be important below in Section 4 can be characterized as being exactly the graphs of clique-width at most 2.

The main concern of this paper is to find small OBDD representations of adjacency functions of graphs for which one of the above parameters is bounded. Note that the converse cannot be achieved, i.e. the existence of a small sized OBDD representing the adjacency function of a graph does not imply any significant bounds on the tree- or clique-width of the graph as shown by the square-grid.

The construction of OBDDs representing graphs of bounded clique-width that we perform in Section 4 relies on results established in [11]. We recall these results before we turn to our constructions.

To each k-expression t representing a graph G there is a naturally related *reduced term* $red(t)$. It is basically obtained from t by deleting all information about the labeling. More precisely, define $\mathcal{R}(V)$ to be the set of well-formed terms r written with the nullary symbols v for $v \in V$, the binary symbol $\bigoplus$ and such that for every $v \in V$ the term r contains exactly once the symbol v. Then for a k-expression t we have that $red(t) \in \mathcal{R}(V)$. (For the example in Definition 4 for instance we have $red(t) = z \bigoplus (x \bigoplus y)$.)

In what follows subterms of an $r \in \mathcal{R}(V)$ and subtrees of T_r are important. For $r \in \mathcal{R}(V)$ and a subterm s of r denote by V_s those vertices occuring in s. Define a graph $H_s := (V_s, E_s)$, where for $u, v \in V_s$ there is an edge $(u, v) \in E_s$ if and only if u and v have at least one different neighbor in G outside V_s, i.e. $(u, v) \in E_s \Leftrightarrow N_G(u) \setminus V_s \neq N_G(v) \setminus V_s$, where $N_G(u)$ is the set of neighbors of u in G.

The following result from [11] collects those properties of the graphs H_s that will be important below.

Theorem 2. *Let G be a graph of clique-width k, t a k-expression representing G, $r = red(t)$ a reduced term and s a subterm of r.*

a) The graph H_s is a complete multipartite graph of at most k different partite sets.

b) If $s = s_1 \bigoplus s_2$, then every partite set of H_s arises as the union of partite sets of H_{s_1} and H_{s_2}.

This result implies in particular that in order to figure out whether two different vertices in V_{s_1} and V_{s_2} are adjacent in G it is sufficient to analyze the relation between the at most k many partite sets in each of the graphs H_{s_1} and H_{s_2}.

3 Succinct Representations for Graphs of Bounded Tree-Width

Let $G = (V, E)$ be a graph with $V = \{1, \ldots, n\}$ of tree-width at most k for some fixed $k \in \mathbb{N}$. In this section we prove that the adjacency function f_G has a succinct representation by an OBDD. More precisely, we shall assign to each vertex $i \in V$ a name $name(i)$ of length $O(\log n)$ and construct an OBDD of size $O(\log n)$ that represents f_G in the sense of Definition 2.

Suppose that G has a tree-decomposition $(T, (X_\ell)_{\ell \in V_T})$ such that T is a binary tree of depth d and each X_ℓ contains at most $\tilde{k}$ vertices for some $\tilde{k} \in \mathbb{N}$. The values d and $\tilde{k}$ will be specified below.

The main idea for constructing an OBDD of size $O(\log n)$ is as follows. Let r denote the root of T. We shall assign to each vertex i an encoding $name(i)$ that consists of three parts. The entire length of $name(i)$ is $O(\log n)$. By the definition of a tree-decomposition there is a unique node $\ell(i)$ of T with $i \in X_{\ell(i)}$ that is closest to the root r of T. The first part of the encoding for i is the binary string encoding the unique path from r to $\ell(i)$. Its length is at most d. For reasons that become clear below we double each bit and add a pair 01 at the end. For example, a path 011 is encoded as 00111101. Denote this first part of $name(i)$ by $p(i)$.

The second part of $name(i)$ is simply taken as the binary representation $bin(i)$ of i of length $\lceil \log n \rceil$.

The third part of $name(i)$, denoted by $adj(i)$, consists of the binary representation of at most $\tilde{k} - 1$ other vertices. More precisely, $adj(i)$ is the concatenation of $bin(s)$ for all those vertices $s \in X_{\ell(i)}$ that are adjacent to i. If $X_{\ell(i)} \setminus \{i\}$ has less than $\tilde{k} - 1$ many vertices we add dummy 0's at the end of $adj(i)$ such that it has precisely length $(\tilde{k} - 1) \cdot \lceil \log n \rceil$. We order the $\tilde{k} - 1$ strings in $adj(i)$ with respect to the numerical value of the integers they represent from the largest to the smallest.

Altogether, $name(i)$ is the concatenation $p(i)bin(i)adj(i)$ and has length at most $L := 2d + 2 + \tilde{k} \cdot \lceil \log n \rceil$.

For arbitrary tree-decompositions of tree-width k we use the following theorem by Bodlaender [2] to obtain a situation as above with values $d = O(\log n)$ and $\tilde{k} = O(k)$.

Theorem 3. *([2]) Let $G = (V, E)$ be a graph of tree-width k with n vertices. Then there exists a tree-decomposition $(T, (X_\ell)_{\ell \in V_T})$ of G of width $3k + 2$ such that T is a binary tree of depth at most $2 \cdot \lceil \log_{\frac{5}{4}}(2n) \rceil$.*

In the present section we only need the statement of the theorem. In Section 4 below we also have to study its proof more closely in the case where G is a tree.

Applying the theorem to our above reasoning gives a balanced tree-decomposition and an encoding of the vertices of G by names of length $L := 4 \cdot \lceil \log_{\frac{5}{4}} (2n) \rceil + 2 + (3k + 2 + 1) \cdot \lceil \log n \rceil$, which is $O(\log n)$ for fixed k.

We are now ready to state the main theorem of this section.

Theorem 4. *Let $k \in \mathbb{N}$ be fixed and let $G = (V, E)$ be a graph of tree-width k. Define L as above. Then there is an OBDD O of size $O(\log n)$ which computes a Boolean function f_O of $2L$ input bits such that for all $i, j \in V$ we have*

$$f_O(name(i), name(j)) = f_G(i, j).$$

Proof. Let $(T, (X_\ell)_{\ell \in V_T})$ be a tree-decomposition of G according to Theorem 3 with tree-width $\tilde{k} - 1 = 3k + 2$ and depth logarithmic in n.

We construct an OBDD using the above encoding as names for the vertices of G. The variable ordering requirements for the OBDD will be obvious from the description below. For vertices $i, j \in V$ the OBDD works in two steps.

Step 1: First, the paths $p(i)$ and $p(j)$ are inspected alternately in blocks of two bits each until we reach the end of one of the two paths. This is indicated by reading a consecutive block 01. There are three different cases to treat: Either none of the two paths is a prefix of the other. Then i and j cannot be adjacent in G due to the properties of a tree-decomposition. Or $p(i)$ is a prefix of $p(j)$ (of course when considering $p(i)$ without the final 01 block), or vice versa. Both cases are handled using the same idea in Step 2.

Step 2: Without loss of generality suppose $p(i)$ is a prefix of $p(j)$ (the reverse situation is treated similarly in a parallel part of the OBDD). If $p(i) = p(j)$, then i and j occur in the same set $X_{\ell(i)} = X_{\ell(j)}$ for the first time due to our convention about the tree-decomposition. They are adjacent in G iff this is already visible in the adjacency list related to i and j. If $p(i) \neq p(j)$, then by the same argument adjacency has to be visible in the adjacency list $adj(j)$. By ignoring intermediate inputs the OBDD continues by reading the input part for $bin(i)$ and $adj(j)$. Thus, in both cases the OBDD has to pattern match $bin(i)$ in $adj(j)$. This can be done by parallel bitwise comparison of the string $bin(i)$ with each of the $\tilde{k} - 1$ many strings in $adj(j)$. Since the strings in $adj(j)$ are ordered numerically after reading each new bit the OBDD can maintain two numbers in $\{1, \ldots, \tilde{k}\}$ that indicate where a string in $adj(j)$ has to be looked for in order to pattern match $bin(i)$ - if at all such a string exists in $\{1, \ldots, \tilde{k}\}$. This way, we remember a subset of $\{1, \ldots, k\}$ of still possible matches. There occur at most $O(\tilde{k}^2 \cdot \log(n))$ different situations with respect to how the two numbers look like and which components in $bin(i)$ still have to be read. Similarly if $p(j)$ is a prefix of $p(i)$. Thus, the variable ordering can be chosen as taking alternately bitwise the strings $bin(i)$ and the $\tilde{k} - 1$ strings in $adj(j)$ followed by the corresponding ordering for $bin(j)$ and $adj(i)$. Depending on which path is a prefix of the other the OBDD ignores the corresponding other half of the variables.

The size of this OBDD can be estimated as follows: Step 1 can be done by an OBDD of size $O(\min\{|p(i)|, |p(j)|\}) = O(\log n)$. The pattern matching as described above in Step 2 can be implemented by an OBDD of size $O(\log n)$

for fixed k. Note that the strings $bin(i)$ and $bin(j)$ are compared to $O(k)$ many strings each. Thus, the number of potential subcases that might occur with respect to the question whether $bin(i)$ is one of the strings in $adj(j)$ is bounded as a function in k. The total size of the OBDD therefore is of order $O(\log n)$ as k is fixed.   □

4   OBDDs for Graphs of Bounded Clique-Width

Whereas graphs of bounded tree-width allow succinct representations by small OBDDs, this is in general not the case for graphs of bounded clique-width. Actually, the following lower and upper bounds are known for cographs, which are precisely the graphs of clique-width at most 2.

Proposition 1. *([14]) a)   There exist cographs G with n nodes such that each OBDD that computes the adjacency function of G and uses names of size $\lceil \log n \rceil$ has size at least $1.832 \cdot \frac{n}{\log n} - O(1)$.*

b) For each cograph G with n nodes there exists an OBDD of size at most $3n \cdot \log n + 2n - \log n + \frac{1}{2}$ that computes the adjacency function of G using names of length $\lceil \log n \rceil$ for the nodes.

Thus, for OBDDs representing graphs of bounded clique-width we can guarantee a size at most of order $O(\frac{n^2}{\log n})$ and in the worst case we have at least a size of order $\Omega(\frac{n}{\log n})$.

The purpose of this section is to come closer to this lower bound. We first show in Section 4.1 that each graph of clique-width k can be represented by an OBDD of size $O(n \cdot f(k))$ for some function f only depending on k when using vertex encodings of length $O(n \cdot \log k)$. The result can be improved into the direction of using shorter encodings in case there exists a reduced term $red(t)$ for a k-expression t representing G such that $red(t)$ is balanced. Unfortunately, we do not know a general result guaranteeing such a reduced term to exist. For cographs, however, we can prove more using once again Theorem 3. This will be done in Section 4.2.

4.1   Representations for Graphs of Bounded Clique-Width Using Long Vertex Encodings

Our constructions in this section are based on Theorem 2.

Theorem 5. *Let $k \in \mathbb{N}$ be fixed. Let $G = (V, E)$ be a graph with $n := |V|$ vertices and let t be a k-expression representing G. There is an OBDD O of size $O(n)$ representing G. The encodings of G's nodes used by O have length at most $O(d \cdot \log k)$, where d is the depth of the binary tree T_r encoded by $r = red(t)$.*

Proof. Let $G = (V, E)$ be a graph of clique-width k, t a k-expression representing G, and $r = red(t)$ its reduced term. Denote the depth of the binary tree T_r related to r by d; clearly $d \le n - 1$. We first explain the encoding we use for each $i \in V$. Since all the nodes of G occur as leaves in T_r there is a unique path $p(i) \in \{0,1\}^*$ from T_r's root to leaf i. The length of $p(i)$ is at most d. Given two different graph nodes $i, j \in V$ the first task is to find in T_r the final common tree node s for $p(i)$ and $p(j)$.

The OBDD reads alternately the bits of $p(i)$ and $p(j)$ until s is found. Thereafter, it proceeds with a different subprogram for each s. The first part causes an OBDD size of $O(n)$ since there are $n - 1$ many internal nodes in T_r.

Consider the subtrees of T_r rooted at s_1 and s_2, where s is the final common node on $p(i)$ and $p(j)$ and s_1 is the left and s_2 the right son of s.

Now the following observation is crucial with respect to the question whether i and j are adjacent in G: According to Theorem 2 both H_{s_1} and H_{s_2} are multipartite with $\leq k$ partite sets; and the question whether a node i in one of the partite sets of H_{s_1} is adjacent in G to a node j in one of the partite sets of H_{s_2} only depends on those sets. Thus, we can code each of the $\leq k$ sets by a string of length $\lceil \log k \rceil$. We then include the information to which set the vertex i belongs in H_{s_1} as part of the encoding of i; more precisely, each bit of $p(i)$ is followed by a string of length $\lceil \log k \rceil$ coding the partite sets of the subsequent subgraph to which i belongs. The encoding of each node therefore has length $O(d \cdot \log k)$. Finally, for deciding adjacency of nodes i and j the OBDD has to find the splitting node as explained above and then to compute a Boolean function depending on $2\lceil \log k \rceil$ variables that represents the adjacency relations between the partite sets of H_{s_1} and H_{s_2}. Using the upper bound from Theorem 1 this can be achieved with a sub-OBDD of size $(2 + o(1)) \cdot 4 \cdot \frac{k^2}{\log k}$. Thus, we obtain an OBDD of size $O(n \cdot \frac{k^2}{\log k})$ using encodings of each node of length at most $O(d \cdot \log k)$.   □

If we do not know more about an optimal (with respect to the depth d of T_r) reduced term r for G we can only conclude $d < n$ and obtain vertex names of length $O(n \cdot \log k)$.

Recall how in the above proof the size of the OBDD depends on the depth of T_r and the number of its internal nodes. The depth of T_r enters into the length of the node encoding whereas the number $n-1$ of internal nodes always enters into the size estimate. If the tree T_r related to $r := red(t)$ has depth $O(\log n)$ we thus obtain OBDDs of size $O(n)$ using encodings of length $O(\log n)$. Currently, we do not know whether such a balanced term always exists for graphs of bounded clique-width. Balancing even at the cost of increasing the number of used labels might give better results. We thus pose the

Problem 1. Let G be a graph of order n and clique-width k. Is there always a $\tilde{k}$-expression t with $\tilde{k} = O(f(k))$ for some function f representing G such that $red(t)$ encodes a binary tree of depth $O(\log n)$?

4.2   OBDDs Representing Cographs

The goal of this section is to show that for cographs we can construct small sized OBDDs using vertex encodings of length $O(\log n)$. More precisely, the sizes of the OBDDs we design are of order $O(n)$, thus improving the result of Proposition 1,

b) by a factor $\log n$. The proof relies on an application of the balancing algorithm behind Theorem 3 to the so called *cotree* that is related to each cograph.

Definition 5. *A cograph $G = (V, E)$ is a graph of clique-width at most 2. (This is equivalent to saying that G contains no induced P_4, i.e. no chordless path with four vertices and three edges.)*

Proposition 2. *([4]) To each cograph G there exists an associated tree $T(G)$ called* cotree *representing G as follows. The leaves of $T(G)$ are precisely the vertices of G. The internal nodes of $T(G)$ are labeled with either 0 or 1. Two vertices i, j of V are adjacent in G iff their least common ancestor $lca(i, j)$ in $T(G)$ is labeled with 1. Without loss of generality $T(G)$ can be chosen to be a binary tree.*

Theorem 6. *For every cograph $G = (V, E)$ with n vertices there exists an OBDD O representing G that has size $O(n)$ and uses encodings of the vertices of G of length $O(\log n)$.*

Proof. Let $G = (V, E)$ be a cograph and $T(G)$ a binary cotree of G as explained in Proposition 2. According to a version of Theorem 3 dealing with trees and also proved in [2] $T(G)$ has a tree decomposition of tree-width at most 3 whose underlying rooted tree T' is of depth $d := 2\lceil \log_{\frac{5}{4}}(n) \rceil$. In order to use T' for the design of an OBDD its construction from $T(G)$ has to be studied more carefully. We thus recall the latter from [2], giving special emphasis to some additional information we encode in T'.

Balancing $T(G)$ is based on the so called *parallel tree contraction* of Miller and Reif [12]. This contraction uses two basic operations RAKE and COMPRESS in order to contract a tree to a single vertex. The intermediate steps during the contraction process are used by Bodlaender to arrive at the desired tree decomposition.

Starting from $T(G) =: T_0 = (V_0, E_0)$ the contraction process recursively constructs trees $T_i = (V_i, E_i)$ for $1 \le i \le r$ such that $|V_r| = 1$ and $r \le 2\lceil \log_{\frac{5}{4}}(n) \rceil$ as follows. Each $v \in V_i$ represents a subset $\rho(v, i) \subseteq V$ of nodes of $T(G)$ that after i steps are contracted to the one node v. Starting with $\rho(v, 0) := \{v\}$ the tree $T_{i+1} = (V_{i+1}, E_{i+1})$ and the new set $\rho(v, i + 1)$ are obtained from T_i by applying in parallel the following operations:

1. RAKE removes from each $v \in V_i$ its children that are leaves in T_i. Then $\rho(v, i + 1) = \bigcup \{\rho(w, i) | w = v,$ or w is a child of v in T_i that is a leaf$\}$.

2. COMPRESS contracts pairs of two subsequent nodes in a *chain* of T_i. A sequence of nodes $v_1, \ldots, v_k$ is a *chain* if v_{j+1} is the only child of v_j for $1 \le j \le k - 1$ and v_k has exactly one child in T_i not being a leaf. Then for each odd j the nodes v_j and v_{j+1} in a maximal chain are compressed to a single node w_j in T_{i+1}. It represents all the nodes previously contracted to either v_j or v_{j+1}, i.e. $\rho(w_j, i + 1) = \rho(v_j, i) \cup \rho(v_{j+1}, i)$.

A tree decomposition of $T(G)$ with underlying rooted tree T' is obtained by Bodlaender from the above procedure as follows. Here, we modify a bit Bodlaender's construction for our purposes. T' has $r + 1$ levels numbered bottom up from 0 to r. On the bottom level (corresponding to T_0 with the nodes V_0 of $T(G)$) tree T' has for each node of $T(G)$ a box containing this node. Thus, the nodes of $T(G)$ occur in T' as leaves. By convention we order the leaves of $T(G)$ itself (which are the vertices of the given graph G) in such a way that they occur as the first n leaves of T' from left to right. This will make it easy for the OBDD to detect inputs not encoding a vertex of G.

Since later on we need to mark one node in each box occuring on the upper levels of T' we mark each node of $T(G)$ on the starting level 0 of T' in its corresponding box. The further levels of T' are now recursively determined according to the contracting operations used in Miller's and Reif's algorithm. Each level i (counted bottom up) in T' contains as many boxes as the tree T_i has nodes. Those boxes are connected in T' as described below:

- Suppose X_1, X_2, X_3 are boxes on level i of T' that correspond to nodes in T_i such that X_2, X_3 are children of X_1 in T_i and at the same time X_2, X_3 are leaves in T_i. Suppose furthermore that the nodes marked in these boxes are x_1, x_2, and x_3, respectively. Then on level $i+1$ of T' we take a box X containing the nodes x_1, x_2, x_3 and mark x_1 in this box. In T' we let X be a father (on level $i+1$) of X_1, X_2 and X_3;
- similarly, if X_1, X_2, X_3 are boxes on level i of T' as above, X_2, X_3 children of X_1 in T_i, but only X_2 is a leaf in T_i, then on level $i+1$ of T' we still have box X_3 with x_3 marked as well as a new box $X := \{x_1, x_2\}$ with x_1 marked. X (on level $i+1$) will be father of X_1 and X_2 (on level i) and X_3 (on level $i+1$) father of X_3 (on level i);
- thirdly, if X_1, X_2 are boxes on level i of T' corresponding to two nodes in T_i that are compressed in T_{i+1}, x_1 marked in X_1 and x_2 marked in X_2, and if X_1 is father of X_2, then on level $i+1$ of T' we include a box $X = \{x_1, x_2\}$ with x_1 marked such that X in T' is father of X_1 and X_2;
- finally, if a node is not changed when going from T_i to T_{i+1}, then the same box is used on both levels of T' and connected by an edge.

For designing the desired OBDD we encode each node i in the vertex set V of G by a bitstring of length $O(\log n)$. One part of i's encoding name will be the path $path(i)$ from T''s root to the leaf i. Since T' has depth $d \leq 2\lceil \log_{\frac{5}{4}}(n) \rceil$ this quantity bounds as well the length of this first part of the encoding.

In order to make T' providing all the information about G's vertices that we need, we have to attach some additional information to each box occuring along $path(i)$ for each $i \in V$. It is here where the labels of the original cotree representing the cograph G are important. The allover idea to decide whether two vertices i, j of G are adjacent in G is to find the final common box along $path(i)$ and $path(j)$ in T' and then include some additional information obtained from $T(G)$'s labeling.

Towards this aim first note that each node (box) X in T' corresponds uniquely to a subtree of $T(G)$ that is contracted to this node. Therefore, the label of the root of that subtree in $T(G)$ is naturally related to the node of T' (this is not yet the information we are looking for!). We denote this original label by $ol(X)$

and use those labels in order to compute bottom up a further label $lab(i, \ell)$ for each box X_ℓ in T' that occurs along $path(i)$. These new labels will be included as part of the encoding of vertex i. The first operation that involves leaf i in T' is a RAKE operation that contracts leaf i with an internal node v of $T(G)$. If X_1 is the corresponding node on level 1 in T' (including the nodes i and v with v marked) we take the original label $ol(X_1)$ as the value for $lab(i, 1)$. Now recursively we define each component of the string $label(i) \in \{0, 1\}^{r+1}$ in relation with the unique box along $path(i)$ on the corresponding level of T'. If the label $lab(i, \ell)$ on level ℓ is determined and the corresponding box $X_{\ell+1}$ on level $\ell + 1$ was obtained from a RAKE operation, then we put $lab(i, \ell + 1) = ol(X_{\ell+1})$. This is justified because if $path(i)$ and $path(j)$ have $X_{\ell+1}$ as the final common component, then $ol(X_{l+1})$ is the label of their least common ancestor in $T(G)$. If label $lab(i, \ell)$ on level ℓ was determined and the corresponding box $X_{\ell+1}$ on level $\ell + 1$ was obtained from a COMPRESS operation, then we put $lab(i, \ell+1) = lab(i, \ell)$. In this situation, we add one additional bit of information denoted by $upper(i, \ell + 1)$. It gets value 1 if box X_ℓ was the upper box among the two boxes compressed, otherwise we put $upper(i, \ell + 1) = 0$. Again, if for two vertices i, j of G $path(i)$ and $path(j)$ split in node $X_{\ell+1}$, then the labeling of the upper node is the decisive one for deciding adjacency. If for example $upper(i, \ell+1) = 1, upper(j, \ell+1) = 0$, then $lab(i, \ell + 1)$ is the label of the least common ancestor of i, j in $T(G)$.

This finishes the description of the encoding of the vertices. The entire encoding $name(i)$ is the concatenation of $path(i), lab(i, \ell)$ for all $1 \leq \ell \leq 2\lceil \log_{\frac{5}{4}}(n) \rceil$ and of $upper(i, \ell)$ for the corresponding levels ℓ. For two inputted vertices i, j the OBDD reads the encodings of both alternately top-down until the splitting box X_k of $path(i)$ and $path(j)$ is found. Then it inspects $lab(i, k)$ and $lab(j, k)$. If both are equal the label corresponds to the correct answer. If both are different X_k must have arisen from a COMPRESS operation. Then the OBDD tests whether $upper(i, k) = 1$. If yes it returns $lab(i, k)$, else it returns $lab(j, k)$.

The length of the used encodings is at most $6 \cdot \lceil \log_{\frac{5}{4}}(n) \rceil = O(\log n)$. The size of the OBDD is basically determined by the number of splitting points between two paths $path(i)$ and $path(j)$, which in turn corresponds to the number n of vertices of G. Then, some additional gates are necessary to decide the label and to stop the computation for inputs that do not properly encode a vertex of the original graph. Clearly, the size of the OBDD thus is of order $O(\log n)$.   □

Acknowledgement. This work was done while K. Meer spent a sabbatical at the Forschungsinstitut für Diskrete Mathematik at the University of Bonn, Germany. The hospitality during the stay is gratefully acknowledged. K. Meer is partially supported by the IST Programme of the European Community, under the PASCAL Network of Excellence, IST-2002-506778 and by the Danish Natural Science Research Council SNF. This publication only reflects the authors' views. The authors thank an anonymous referee for pointing them to references [10,13].

References

1. S. Arnborg, J. Lagergren, D. Seese: *Easy problems for tree decomposable graphs*, Journal of Algorithms 12, pp. 308–340 (1991).
2. H.L. Bodlaender: *NC-algorithms for graphs with small tree-width*, In: Proceedings Graph-theoretic concepts in computer science, Lecture Notes in Comput. Sci., vol. 344, Springer, pp. 1–10 (1989).
3. Y. Breitbart, H. Hunt, D. Rosenkrantz: *On the size of binary decision diagrams representing Boolean functions*, Theoretical Computer Science, Vol. 145, Nr. 1, pp. 45–69 (1995).
4. D.G. Corneil, H. Lerchs, L. Stewart Burlingham: *Complement reducible graphs*, Discrete Applied Mathematics 3, pp. 163–174 (1981).
5. B. Courcelle: *Graph grammars, monadic second-order logic and the theory of graph minors*, Contemporary Mathematics 147, pp. 565–590 (1993).
6. B. Courcelle, J.A. Makowsky, U. Rotics: *Linear Time Solvable Optimization Problems on Graphs of Bounded Clique Width*, Theory of Computing Systems, vol. 33(2),pp. 125–150 (2000).
7. B. Courcelle, M. Mosbah: *Monadic second-order evaluations on tree decomposable graphs*, Theoretical Computer Science 109, pp. 49–82 (1993).
8. B. Courcelle, S. Olariu: *Upper bounds to the clique-width of graphs*, Discrete Applied Mathematics 101, pp. 77–114 (2000).
9. J. Feigenbaum, S. Kannan, M.Y. Vardi, M. Viswanathan: *Complexity of Problems on Graphs Represented as OBDDs*, Chicago Journal of Theoretical Computer Science, vol. 1999, issue 5/6, pp. 1–25 (1999)
10. K. Keeler, J. Westbrook: *Short encodings of planar graphs and maps*, Discrete Applied Mathematics 58, pp. 239–252 (1995).
11. V. Lozin, D. Rautenbach: *The Relative Clique-Width of a graph*, Rutcor Research Report RRR 16-2004 (2004).
12. G. Miller, J. Reif: *Parallel tree contraction and its application*, Proc. 26th Foundations of Computer Science FOCS 1985, IEEE, pp. 478–489 (1985).
13. M. Naor: *Succinct representation of general unlabeled graphs*, Discrete Applied Mathematics 28, pp. 303–307 (1990).
14. R. Nunkesser, P. Woelfel: *Representation of Graphs by OBDDs*, Proceedings of ISAAC 2005, X. Deng and D. Du (eds.), LNCS 3827, Springer, pp. 1132–1142 (2005).
15. D. Sawitzki: *Implicit flow maximization by iterative squaring*, Proceedings of 30th SOFSEM, P. van Emde Boas, J. Pokorny, M. Bielikova, and J. Stuller (eds.), LNCS 2932, Springer, pp. 301–313 (2004).
16. D. Sawitzki: *A symbolic approach to the all-pairs shortest paths problem*, Proc. 30th Graph-Theoretic Concepts in Computer Science WG 2004, J. Hromkovic, M. Nagl, B. Westfechtel (eds.), LNCS 3353, Springer, pp. 154 – 167 (2004).
17. I. Wegener: *Branching Programs and Binary Decision Diagrams: Theory and Applications*, SIAM Monographs on Discrete Mathematics and Applications, SIAM (2000).
18. P. Woelfel: *Symbolic topological sorting with OBDDs*, Journal of Discrete Algorithms, to appear.

Greedy Localization and Color-Coding: Improved Matching and Packing Algorithms*

Yang Liu[1], Songjian Lu[1], Jianer Chen[1], and Sing-Hoi Sze[1,2]

[1] Department of Computer Science
[2] Department of Biochemistry & Biophysics
Texas A&M University, College Station, TX 77843, USA
{yangliu, sjlu, chen, shsze}@cs.tamu.edu

Abstract. Matching and packing problems have formed an important class of NP-hard problems. There have been a number of recently developed techniques for parameterized algorithms for these problems, including greedy localization, color-coding plus dynamic programming, and randomized divide-and-conquer. In this paper, we provide further theoretical study on the structures of these problems, and develop improved algorithmic methods that combine existing and new techniques to obtain improved algorithms for matching and packing problems. For the 3-SET PACKING problem, we present a deterministic algorithm of time $O^*(4.61^{3k})$, which significantly improves the previous best deterministic algorithm of time $O^*(12.8^{3k})$. For the 3-D MATCHING problem, we develop a new randomized algorithm of running time $O^*(2.32^{3k})$ and a new deterministic algorithm of running time $O^*(2.77^{3k})$. Our randomized algorithm improves the previous best randomized algorithm of running time $O^*(2.52^{3k})$, and our deterministic algorithm significantly improves the previous best deterministic algorithm of running time $O^*(12.8^{3k})$. Our results also imply improved algorithms for various triangle packing problems in graphs.

1 Introduction

Matching and packing problems have formed an important class of NP-hard problems. In particular, the 3-D MATCHING problem is one of the six "basic" NP-complete problems in terms of Garey and Johnson [10], and the 3-SET PACKING problem is a natural extension of the 3-D MATCHING problem. There has been a remarkable line of research in the study of parameterized algorithms for 3-D MATCHING and 3-SET PACKING problems.

Downey and Fellows [7] proved that the 3-D MATCHING problem is fixed-parameter tractable and gave an algorithm of time $O^*((3k)!(3k)^{9k+1})^1$. Chen et al. [3] improved the time complexity for 3-D MATCHING to $O^*((5.7k)^k)$, and Jia, Zhang, and Chen [11] improved the time complexity for 3-SET PACKING to $O^*((5.7k)^k)$.

---

* This work was supported in part by the National Science Foundation under the Grants CCR-0311590 and CCF-0430683.

[1] Following the recent convention, for a function $f(k)$ of the parameter k, we will use the notation $O^*(f(k))$ to denote the bound $O(f(k)n^{O(1)})$.

H.L. Bodlaender and M.A. Langston (Eds.): IWPEC 2006, LNCS 4169, pp. 84–95, 2006.
© Springer-Verlag Berlin Heidelberg 2006

More progress has been made recently. For the 3-SET PACKING problem, Koutis [13] developed a randomized algorithm of time $O^*(10.88^{3k})$ and a deterministic algorithm of time $O^*(2^{O(k)})$. Koutis [13] did not give the exact constant factor in the exponent of the time complexity $O^*(2^{O(k)})$ for his deterministic algorithm. He used the perfect hashing families proposed by Schmidt and Siegel [16], in which the number of hashing functions to hash n elements into $3k$ colors is larger than $2^{\log\log n + 12k}$. It can be derived that his deterministic algorithm has time complexity of at least $O^*(32000^{3k})$. These algorithms can be applied to the 3-D MATCHING problem without any changes. Fellows et al. [9] studied the complexity of matching and packing problems. They first showed that the 3-D MATCHING problem has a kernel of size $O(k^3)$, and then presented an algorithm of time $O^*(2^{O(k)})$ for the problem, where the term $O(k)$ was also not specified in detail. It can be deduced that the running time of the algorithm given in [9] for 3-D MATCHING is at least $O^*(12.67^{3k}T(k))$, where $T(k)$ is the running time of a dynamic programming algorithm that, on a set of triples whose symbols are colored with $13k$ colors, searches for a matching of k triples in which all symbols are colored with distinct colors ($T(k)$ is at least $O^*(10.4^{3k})$ using currently known techniques). The paper also discussed how these techniques are applied to solve 3-SET PACKING and various graph packing problems.

There are at least two very recent works that give further improved algorithms for 3-D MATCHING and 3-SET PACKING problems. Chen et al. [4] proposed a new technique based on divide-and-conquer that leads to randomized algorithms of time $O^*(2.52^{3k})$ for 3-D MATCHING and 3-SET PACKING problems. Moreover, they proposed a color-coding scheme of $O^*(6.1^k)$ k-colorings which, when combined with standard dynamic programming techniques, gives deterministic algorithms of running time $O^*(12.8^{3k})$ for 3-D MATCHING and 3-SET PACKING problems. We point out that using this new color-coding scheme, the time complexity of the algorithms by Koutis [13] for 3-D MATCHING and 3-SET PACKING can be improved to $O^*(25.6^{3k})$, and the time complexity of the algorithms by Fellows et al. [9] can be improved to $O^*(13.78^{3k})$. In a work performed independently of that in [4], Kneis et al. [12] developed a divide-and-conquer method that leads to randomized algorithms for 3-D MATCHING and 3-SET PACKING problems with time complexity similar to that in [4]. Moreover, a different derandomization method was proposed in [12] based on the work of [1], which leads to deterministic algorithms of running time $O^*(16^{3k})$ for 3-D MATCHING and 3-SET PACKING problems.

The known parameterized algorithms for 3-D MATCHING and 3-SET PACKING have used either the technique of *greedy localization* [3,6,11], the technique of *color-coding* [2] plus dynamic programming [4,9,13], or the *divide-and-conquer* method [4,12]. In this paper, we show how a combination of these techniques and new techniques will yield further improved algorithms for these problems. We start with the 3-SET PACKING problem. In difference from the approach used in [3,11] that constructs a packing of k 3-sets directly from a maximal packing, we concentrate on the construction of a packing of $k + 1$ 3-sets based on a packing of k 3-sets. This slight modification enables us to derive a property for packing that is much stronger than the one given in [11]. Moreover, instead of coloring all elements in an instance of 3-SET PACKING, we color only part of the elements and use either ordering or pre-selected elements to reduce the complexity of the coloring stage in the algorithms. Using these new techniques, we are able to develop a parameterized algorithm of running time $O^*(4.61^{3k})$ for the 3-SET PACKING problem, significantly improving the previous best algorithm of running time $O^*(12.8^{3k})$ for the problem [4]. For the 3-D MATCHING problem, we further show that the complexity of the dynamic programming stage in the algorithms, which seems to have been largely neglected in the previous research, can also be improved using a pre-ordering technique. Combining this new technique and those developed for 3-SET PACKING, we achieve further improved algorithms for the 3-D MATCHING problem. More specifically, our new randomized algorithm for 3-D MATCHING runs in time $O^*(2.32^{3k})$, and our new deterministic algorithm for 3-D MATCHING runs in time $O^*(2.77^{3k})$, both significantly improving the previous best algorithms for the problem.

We would like to point out that all previous parameterized algorithms for 3-D MATCHING and 3-SET PACKING have the same time complexity for both problems, although it is obvious that 3-SET PACKING is a nontrivial generalization of 3-D MATCHING. The results in the current paper seem to give faster algorithms for 3-D MATCHING than for 3-SET PACKING. We also mention that the difference in complexity between our deterministic algorithm (i.e., $O^*(2.77^{3k})$) and our randomized algorithm (i.e., $O^*(2.32^{3k})$, which is also currently the best upper bound) for 3-D MATCHING has been significantly narrowed down, which is remarkable considering the fact that in the previous research on the problem, the difference between these two kinds of algorithms is in general very significant. Table 1 gives a specific comparison of our new algorithms and the previous algorithms for the 3-D MATCHING problem.

Table 1. Comparison of algorithms for 3-D MATCHING

References	Randomized algorithm	Deterministic algorithm
Downey and Fellows [7]		$O^*((3k)!(3k)^{9k+1})$
Chen et al. [3]		$O^*((5.7k)^k)$
Koutis [13]	$O^*(10.88^{3k})$	$> O^*(32000^{3k})$
Fellows et al. [9]*		$> O^*(12.67^{3k}T(k))$
Kneis et al. [12]	$O^*(2.52^{3k})$	$O^*(16^{3k})$
Chen et al. [4]	$O^*(2.52^{3k})$	$O^*(12.8^{3k})$
Our new result	$O^*(2.32^{3k})$	$O^*(2.77^{3k})$

*$T(k)$ is the running time of a dynamic programming process that, on a set of triples whose symbols are colored with $13k$ colors, searches for a matching of k triples in which all symbols are colored with distinct colors. Based on currently known techniques, $T(k)$ is at least $O^*(10.4^{3k})$.

2   Preliminaries and Reformulations

Let X, Y, and Z be three pairwise disjoint symbol sets, and let $U = X \times Y \times Z$ be the product set of X, Y, and Z. Each element $t = (x, y, z)$ in U, where $x \in X$, $y \in Y$, and $z \in Z$, is called a *triple*. For a triple $t = (x, y, z)$ in U, denote by $\text{Val}(t)$ the set $\{x, y, z\}$, and let $\text{Val}^1(t) = \{x\}$, $\text{Val}^2(t) = \{y\}$, $\text{Val}^3(t) = \{z\}$. We say that a triple t_1 *conflicts* with another triple t_2 if $t_1 \neq t_2$ and $\text{Val}(t_1) \cap \text{Val}(t_2) \neq \emptyset$. Let S be a set of triples in U. Denote $\text{Val}(S) = \bigcup_{t \in S} \text{Val}(t)$, and $\text{Val}^i(S) = \bigcup_{t \in S} \text{Val}^i(t)$ for $i = 1, 2, 3$. A *matching* in S is a subset M of triples in S such that no two triples in M conflict with each other. A matching M in S is a *k-matching* if M contains exactly k triples.

Packing problems are a generalization of matching problems. We say that a set ρ_1 *conflicts* with another set ρ_2 if $\rho_1 \neq \rho_2$ and $\rho_1 \cap \rho_2 \neq \emptyset$. Let S be a collection of sets. Denote $\text{Val}(S) = \cup_{\rho \in S} \rho$. A *packing* in S is a sub-collection P of S such that no two sets in P conflict with each other. A packing P in S is a *k-packing* if P contains exactly k sets.

The main problems we study in this paper are formally defined as follows.

(PARAMETERIZED) 3-D MATCHING:
Given a pair (S, k), where S is a set of n triples, and k is an integer, either construct a k-matching in S, or report that no such matching exists.

(PARAMETERIZED) 3-SET PACKING:
Given a pair (S, k), where S is a collection of n sets, each containing at most three elements, and k is an integer, either construct a k-packing in S, or report that no such packing exists.

A set is a *3-set* if it contains exactly three elements. For an instance (S, k) of 3-SET PACKING, we can assume, without loss of generality, that all sets in S are 3-sets (otherwise, we can add new elements, i.e., elements not in S, to convert each set with fewer than three elements to a 3-set). Instead of working on the above problems, we will concentrate on the following related problems.

3-D MATCHING AUGMENTATION:
Given a pair (S, M_k), where S is a set of n triples, and M_k is a k-matching in S, either construct a $(k + 1)$-matching M_{k+1} in S, or report that no such matching exists.

3-SET PACKING AUGMENTATION:
Given a pair (S, P_k), where S is a collection of n 3-sets, and P_k is a k-packing in S, either construct a $(k + 1)$-packing P_{k+1} in S, or report that no such packing exists.

Lemma 1. *For any constant $c > 1$, the 3-D MATCHING AUGMENTATION problem can be solved in time $O^*(c^k)$ if and only if the 3-D MATCHING problem can be solved in time $O^*(c^k)$. Similarly, the 3-SET PACKING AUGMENTATION problem can be solved in time $O^*(c^k)$ if and only if the 3-SET PACKING problem can be solved in time $O^*(c^k)$.*

According to Lemma 1, we only need to concentrate on the 3-D MATCHING AUGMENTATION and 3-SET PACKING AUGMENTATION problems.

3   Improved Packing Algorithms

The method of *greedy localization* has been heavily used in early algorithms for matching and packing problems [3,11]. The method takes advantage of the fact that information for a larger matching/packing can be obtained from a given smaller matching/packing, which narrows down the size of the search space during the construction of the larger matching/packing. We show that this property can be significantly enhanced and more effectively used to develop algorithms for the 3-SET PACKING AUGMENTATION problem.

Lemma 2. *Let* (S, P_k) *be an instance of* 3-SET PACKING AUGMENTATION, *where* P_k *is a* k-*packing in* S. *If* S *also has* $(k+1)$-*packings, then there exists a* $(k+1)$-*packing* P_{k+1} *in* S *such that every set in* P_k *contains at least two elements in* $\mathrm{Val}(P_{k+1})$.

Proof. We prove the lemma by contradiction. Suppose that the lemma does not hold. Then there is a k-packing P_k such that for every $(k+1)$-packing P in S, there is a set in P_k that contains at most one element in $\mathrm{Val}(P)$. Let P_{k+1} be a $(k+1)$-packing in S such that the number of common sets in P_k and P_{k+1} is maximized over all $(k+1)$-packings in S. By our assumption, there is a set ρ in P_k that contains at most one element in $\mathrm{Val}(P_{k+1})$.

   Case 1. Exactly one element a in the set ρ is in $\mathrm{Val}(P_{k+1})$. Then let ρ' be the set in P_{k+1} that contains the element a. Since no other element in ρ is in $\mathrm{Val}(P_{k+1})$, if we replace ρ' in P_{k+1} by ρ, we get a new $(k+1)$-packing that has one more common set (i.e., ρ) with the k-packing P_k (note that ρ' cannot be in P_k because ρ' and ρ share a common element a while ρ contains another two elements not in $\mathrm{Val}(P_{k+1})$). This contradicts our assumption that the $(k+1)$-packing P_{k+1} maximizes the number of common sets with P_k.
   Case 2. No element in ρ is in $\mathrm{Val}(P_{k+1})$. Since P_k contains k sets while P_{k+1} contains $k+1$ sets, there must be a set ρ'' in P_{k+1} that is not in P_k. Since ρ contains no element in $\mathrm{Val}(P_{k+1})$, replacing ρ'' in P_{k+1} by ρ gives a new $(k+1)$-packing that has one more common set (i.e., ρ) with P_k, again contradicting the assumption that the $(k+1)$-packing P_{k+1} maximizes the number of common sets with P_k.
   This contradiction shows that the set ρ in P_k that contains at most one element in $\mathrm{Val}(P_{k+1})$ cannot exist.   □

According to Lemma 2, to construct a $(k+1)$-packing from a given instance (S, P_k) of 3-SET PACKING AUGMENTATION, we can aim at the $(k+1)$-packing P_{k+1} with the property described in the lemma. The advantage of this $(k+1)$-packing P_{k+1} is that at least $2k$ elements in P_{k+1} are already present in the k-packing P_k, and we only need to identify at most $k+3$ other elements in P_{k+1}.

We use the technique of *color-coding*, first introduced by Alon, Yuster, and Zwick [2], to search for these elements that are in P_{k+1} but not in P_k.

Let B be a set of elements. A *coloring* of B is a function mapping B to the natural numbers $\{1, 2, \ldots\}$, and an *h-coloring* of B is a function mapping B to $\{1, 2, \ldots, h\}$. A subset B' of B is *colored properly* by a coloring f if no two elements in B' are colored with the same color under f. A collection C of *h*-colorings of a set B is an *h-color coding scheme* if for every subset B' of h elements in B, there is an *h*-coloring in C that colors B' properly. The following proposition has been proved in [4].

Proposition 1. [4] *For any finite set B and any integer h, there is an h-color coding scheme C of $O^*(6.1^h)$ h-colorings of the set B. Moreover, the h-colorings in C can be constructed and enumerated in time $O^*(6.1^h)$.*

Let S be a collection of 3-sets and let f be a coloring of the set $\mathrm{Val}(S)$. We say that a packing P in S is *colored properly* if the set $\mathrm{Val}(P)$ is colored properly under the coloring f. Let (S, P_k) be an instance of 3-SET PACKING AUGMENTA-TION. Since the set of elements that are in $\mathrm{Val}(P_{k+1})$ but not in $\mathrm{Val}(P_k)$ contains at most $k + 3$ elements, by introducing $3k$ new colors to properly color the $3k$ elements in P_k, P_{k+1} can be colored properly with at most $4k + 3$ colors.

Lemma 3. *Let (S, P_k) be an instance of 3-SET PACKING AUGMENTATION, and let P_{k+1} be a $(k + 1)$-packing in S such that each 3-set in P_k contains at least two elements in $\mathrm{Val}(P_{k+1})$. Then there is a collection C_0 of $O^*(6.1^k)$ $(4k + 3)$-colorings of the set $\mathrm{Val}(S)$ in which at least one properly colors P_{k+1}. Moreover, the collection C_0 can be constructed in time $O^*(6.1^k)$.*

Now we turn to the problem of constructing a properly colored $(k + 1)$-packing P_{k+1}. Alon, Yuster, and Zwick [2] in their seminal work on color-coding suggested a general principle in which a $(3k + 3)$-coloring that properly colors the $3k + 3$ elements is first constructed in $\mathrm{Val}(P_{k+1})$, then a dynamic programming process is applied to find the properly colored $(k + 1)$-packing P_{k+1}. Koutis [13] proposed an algebraic formulation to find the properly colored $(k + 1)$-packing P_{k+1}. Fellows et al. [9] considered a more general approach that first uses g colors to properly color the $(k + 1)$-packing P_{k+1}, where $g \geq 3k + 3$, then perform a dynamic programming algorithm. For completeness, we present such a generalized dynamic programming algorithm in detail, as given in Fig. 1, verify its correctness, and analyze its precise complexity.

Lemma 4. *The algorithm 3SetPack(S, k, f, g) runs in time $O^*(\sum_{j=0}^{k} \binom{g}{3j})$, and constructs a properly colored k-packing in S if such k-packings exist.*

Proof. From steps 4.1-4.2 of the algorithm, it can be seen that every collection P of 3-sets added to the super-collection Q in step 4.4 is a properly colored packing. Therefore, if the algorithm returns a packing in step 5, the packing must be a properly colored k-packing.

For each i, let $S_i = \{\rho_1, \ldots, \rho_i\}$. We prove by induction on i that for all $j \leq k$, if S_i has a properly colored j-packing P_j, then after the i-th execution of the

Algorithm 3SetPack(S, k, f, g)
input: A collection S of 3-sets, an integer k, a g-coloring f of $\mathrm{Val}(S)$
output: A properly colored k-packing if such a packing exists

1.   remove all 3-sets in S in which any two elements have the same color;
2.   let the remaining 3-sets in S be $\rho_1, \rho_2, \ldots, \rho_n$;
3. $\mathcal{Q} = \{\emptyset\}$;
4.   **for** $i = 1$ to n **do**
4.1.   **for** each packing P in $\mathcal{Q}$ such that no element in P is colored with the same
            color as an element in ρ_i **do**
4.2. $P' = P \cup \{\rho_i\}$;
4.3.   **if** P' is a j-packing with $j \leq k$ and $\mathcal{Q}$ contains no packing that uses exactly
                the same colors as that used by P'
4.4.   **then** add P' to $\mathcal{Q}$;
5.   return a k-packing in $\mathcal{Q}$ if such a packing exists.

Fig. 1. Dynamic programming for 3-SET PACKING

for-loop in step 4 of the algorithm, the super-collection $\mathcal{Q}$ contains a properly colored j-packing P'_j such that P_j and P'_j use exactly the same $3j$ colors.

The initial case $i = 0$ is trivial since $\mathcal{Q} = \{\emptyset\}$. Consider $i \geq 1$. Suppose that the collection S_i has a properly colored j-packing $P_j = \{\rho_{i_1}, \rho_{i_2}, \ldots, \rho_{i_j}\}$, where $1 \leq i_1 < i_2 < \cdots < i_j \leq i$. Then the collection S_{i_j-1} contains the properly colored $(j-1)$-packing $P_{j-1} = \{\rho_{i_1}, \rho_{i_2}, \ldots, \rho_{i_{j-1}}\}$. By the inductive hypothesis, after the $(i_j - 1)$-st execution of the **for**-loop in step 4, the super-collection $\mathcal{Q}$ contains a properly colored $(j-1)$-packing P'_{j-1} such that the $(j-1)$-packings P_{j-1} and P'_{j-1} use exactly the same $3(j-1)$ colors. Since $P_j = \{\rho_{i_1}, \rho_{i_2}, \ldots, \rho_{i_j}\}$ is a properly colored j-packing, and $P_{j-1} = \{\rho_{i_1}, \rho_{i_2}, \ldots, \rho_{i_{j-1}}\}$ and P'_{j-1} use exactly the same $3(j-1)$ colors, no element in $\mathrm{Val}(P'_{j-1})$ is colored with the same color as an element in the set ρ_{i_j}. Therefore, in the i_j-th execution of the **for**-loop in step 4, a properly colored j-packing $P'_{j-1} \cup \{\rho_{i_j}\}$ will be added to the super-collection $\mathcal{Q}$ if no properly colored j-packing that uses exactly the same $3j$ colors exists in $\mathcal{Q}$ yet. Note that the j-packing $P'_{j-1} \cup \{\rho_{i_j}\}$ and the j-packing P_j use exactly the same $3j$ colors. Therefore, after the i_j-th execution of the **for**-loop in step 4, a j-packing that uses exactly the same $3j$ colors as P_j will exist in the super-collection $\mathcal{Q}$. Finally, since packings in $\mathcal{Q}$ are never removed from $\mathcal{Q}$ and $i_j \leq i$, we conclude that after the i-th execution of the **for**-loop in step 4, a j-packing that uses exactly the same $3j$ colors as P_j will exist in the super-collection $\mathcal{Q}$. This completes the inductive proof.

Now if we let $i = n$, for any $j \leq k$, if the original collection S contains a properly colored j-packing P_j, then the super-collection $\mathcal{Q}$ contains a j-packing that uses exactly the same $3j$ colors as P_j. In particular, if the collection S contains properly colored k-packings, then the algorithm **3SetPack**(S, k, f, g) must return a properly colored k-packing.

Finally, we analyze the complexity of the algorithm. For each $0 \leq j \leq k$ and for each set of $3j$ colors, the super-collection $\mathcal{Q}$ keeps at most one properly colored j-packing that uses exactly these $3j$ colors. Since there are $\binom{g}{3j}$

different subsets of $3j$ colors over a total of g colors, the total number of packings recorded in $\mathcal{Q}$ is bounded by $\sum_{j=0}^{k} \binom{g}{3j}$. For each i, $1 \leq i \leq k$, we examine each packing P in $\mathcal{Q}$ in step 4.1 and check if we can construct a larger packing by adding the set ρ_i to the packing P. This can be done for each packing P in time $O(k)$. In consequence, the algorithm **3SetPack**(S, k, f, g) runs in time $O(nk \sum_{j=0}^{k} \binom{g}{3j}) = O^*(\sum_{j=0}^{k} \binom{g}{3j})$. $\qquad\square$

Combining Lemmas 2, 3, and 4, the 3-SET PACKING AUGMENTATION problem can be solved in time $O^*(6.1^k 2^{4k}) = O^*(4.61^{3k})$.

Theorem 1. *The* 3-SET PACKING AUGMENTATION *problem can be solved in time* $O^*(4.61^{3k})$.

Corollary 1. *The* 3-SET PACKING *problem can be solved in time* $O^*(4.61^{3k})$.

4   Matching Algorithms Further Improved

All the previous results are applicable to the 3-D MATCHING problem. In fact, if we regard each triple as a 3-set, then each instance S_M of 3-D MATCHING is also an instance S_P of 3-SET PACKING, and a triple set is a matching in S_M if and only if it is a packing in S_P.

As shown in the previous section, a dynamic programming algorithm is used as a second stage in parameterized algorithms for 3-SET PACKING/3-D MATCHING. In this section, we develop a new technique for the dynamic programming stage for the 3-D MATCHING problem so that fewer colors will be needed. This technique has two advantages. First, the use of fewer colors will significantly reduce the time complexity of the coloring stage. Second, since fewer colors are used, the number of different color sets is reduced, which will reduce the time complexity of the dynamic programming stage remarkably.

Let the universal triple set be $U = X \times Y \times Z$, where X, Y, and Z are three pairwise disjoint symbol sets. The symbols in the sets X, Y, and Z will be called the *symbols in column-1, column-2*, and *column-3*, respectively.

Definition 1. *Let p and q be any two indices in the index set $\{1, 2, 3\}$, and let S be a set of triples in U. A matching M in the set S is (p, q)-properly colored by a coloring f of $\mathrm{Val}^p(M) \cup \mathrm{Val}^q(M)$ if no two symbols in $\mathrm{Val}^p(M) \cup \mathrm{Val}^q(M)$ are colored with the same color under f.*

Theorem 2. *Let p and q be any two indices in the index set $\{1, 2, 3\}$. There is an algorithm of time $O^*(\sum_{j=0}^{k} \binom{g}{2j})$ that, on an integer k and a set S of triples in which the symbols in $\mathrm{Val}^p(S) \cup \mathrm{Val}^q(S)$ are colored by a g-coloring f, constructs a (p, q)-properly colored k-matching in S when such matchings exist in S.*

Proof. Consider the algorithm in Fig. 2. By steps 6.3-6.6, for every set C in the collection $\mathcal{Q}_{old}$, all symbols in C are from $\mathrm{Val}^p(S) \cup \mathrm{Val}^q(S)$, and no two symbols in C are of the same color. The algorithm **3DMatch**$(S, k, f, g; p, q)$ either

Algorithm 3DMatch$(S, k, f, g; p, q)$
input: A set S of triples, an integer k, a g-coloring f of the symbols in $\mathrm{Val}^p(S) \cup \mathrm{Val}^q(S)$
output: A (p, q)-properly colored k-matching in S if such a matching exists

1.   remove any triples in S in which any two symbols have the same color under f;
2.   let the set of remaining triples be S';
3. $r = \{1, 2, 3\} - \{p, q\}$;
4.   let the symbols in $\mathrm{Val}^r(S')$ be $x_1, x_2, \ldots, x_m$;
5. $\mathcal{Q}_{old} = \{\emptyset\}$; $\mathcal{Q}_{new} = \{\emptyset\}$;
6.   **for** $i = 1$ **to** m **do**
6.1.   **for** each set C of symbol pairs in $\mathcal{Q}_{old}$ **do**
6.2.   **for** each $t \in S'$ with $\mathrm{Val}^r(t) = x_i$ **do**
6.3.   **if** no symbol in C is of the same color as a symbol in $\mathrm{Val}^p(t) \cup \mathrm{Val}^q(t)$
6.4.   **then** $C' = C \cup \{(\mathrm{Val}^p(t), \mathrm{Val}^q(t))\}$;
6.5.   **if** C' contains no more than k symbol pairs and $\mathcal{Q}_{new}$ contains no set
                       of symbol pairs that uses exactly the same colors as that used by C'
6.6.   **then** add C' to $\mathcal{Q}_{new}$;
6.7. $\mathcal{Q}_{old} = \mathcal{Q}_{new}$;
7.   return a set C of k symbol pairs in $\mathcal{Q}_{old}$ if such a set exists.

Fig. 2. Dynamic programming for 3-D MATCHING

outputs a set of k symbol pairs in the collection $\mathcal{Q}_{old}$ or reports that no (p, q)-properly colored k-matchings exist in S. We say that a set $C = \{w_1, \ldots, w_i\}$ of i symbol pairs is *extendable to an i-matching* in S if there is an i-matching $M = \{t_1, \ldots, t_i\}$ in S such that for all j, the pair $(\mathrm{Val}^p(t_j), \mathrm{Val}^q(t_j))$ is identical to the symbol pair w_j. For each i, let S_i' be the set of triples in S' whose symbols in column-r are among $\{x_1, x_2, \ldots, x_i\}$. For a matching M, we will denote by $cl(M) = \{f(y) \mid y \in \mathrm{Val}^p(M) \cup \mathrm{Val}^q(M)\}$ the set of colors used by the symbols in $\mathrm{Val}^p(M) \cup \mathrm{Val}^q(M)$.

We prove the following claim by induction on i:

> **Claim.** For each i, $0 \leq i \leq m$, and for all $h \leq k$, there is a (p, q)-properly colored h-matching M_h in S_i' if and only if after the i-th execution of the loop 6.1-6.7 of algorithm **3DMatch**$(S, k, f, g; p, q)$, the collection $\mathcal{Q}_{old}$ contains a set C_h of h symbol pairs such that the set of colors used for the symbols in C_h is exactly $cl(M_h)$. Moreover, each set C_h of h symbol pairs in the collection $\mathcal{Q}_{old}$ after the i-th execution of the loop 6.1-6.7 is extendable to an h-matching in S_i'.

The case $i = 0$ is obvious because we initially set $\mathcal{Q}_{old}$ to $\{\emptyset\}$. Consider $i \geq 1$. First note that the claim is always true for $h = 0$ because the collection $\mathcal{Q}_{old}$ always contains the empty set $\emptyset$ while the set S_i' always contains a 0-matching (which by the definition is (p, q)-properly colored).

Suppose that after the i-th execution of the loop 6.1-6.7, the collection $\mathcal{Q}_{old}$ contains a set C_h of h symbol pairs, where $h \geq 1$. Suppose that the set C_h was created during the j-th execution of the loop 6.1-6.7, where $j \leq i$, by adding a symbol pair $(\mathrm{Val}^p(t), \mathrm{Val}^q(t))$ to a set C_{h-1} of $h - 1$ symbol pairs, where t

is a triple with $\text{Val}^r(t) = x_j$ and the set C_{h-1} is contained in $\mathcal{Q}_{old}$ after the $(j-1)$-st execution of the loop 6.1-6.7. By the inductive hypothesis, the set C_{h-1} is extendable to an $(h-1)$-matching M_{h-1} in S'_{j-1}, which is obviously (p,q)-properly colored. Since no symbol in C_{h-1} uses the same color as a symbol in $\text{Val}^p(t) \cup \text{Val}^q(t)$, and the matching M_{h-1} does not contain the symbol x_j, the set $M_h = M_{h-1} \cup \{t\}$ makes a (p,q)-properly colored h-matching in S'_j. Since $j \le i$ and $S'_j \subseteq S'_i$, we conclude that the set S'_i contains a (p,q)-properly colored h-matching M_h such that the symbols in the set C_h use exactly the color set $cl(M_h)$. Moreover, it is obvious that the symbol set C_h is extendable to the h-matching M_h.

To prove the other direction, suppose that the set S'_i contains a (p,q)-properly colored h-matching M_h.

Case 1. There is a (p,q)-properly colored h-matching M'_h in S'_j for some $j < i$ such that $cl(M'_h) = cl(M_h)$. By the inductive hypothesis, after the j-th execution of the loop 6.1-6.7, the collection $\mathcal{Q}_{old}$ contains a set C_h of h symbol pairs such that (1) the set of colors used for the symbols of C_h is exactly $cl(M'_h)$; and (2) C_h is extendable to an h-matching in S'_j. Since $j < i$, $S'_j \subseteq S'_i$, and we never remove symbol pairs from $\mathcal{Q}_{old}$, we conclude that in this case, after the i-th execution of the loop 6.1-6.7, the set C_h is still contained in the collection $\mathcal{Q}_{old}$ such that (1) the set of colors used for the symbols of C_h is exactly $cl(M'_h) = cl(M_h)$; and (2) C_h is extendable to an h-matching in $S'_j \subseteq S'_i$.

Case 2. There is no (p,q)-properly colored h-matching M'_h in S'_j for any $j < i$ such that $cl(M'_h) = cl(M_h)$. Then by the inductive hypothesis, after the j-th execution of the loop 6.1-6.7 for any $j < i$, the collection $\mathcal{Q}_{old}$ contains no set C of symbol pairs such that the symbols in C use exactly the color set $cl(M_h)$. Let the (p,q)-properly colored h-matching M_h be $M_h = \{t_1, \cdots, t_h\}$, where for each j, $\text{Val}^r(t_j) = x_{d_j}$, with $d_1 < \cdots < d_{h-1} < d_h$. In this case, we must have $d_h = i$ and $\text{Val}^r(t_h) = x_i$. Let $y = \text{Val}^p(t_h)$ and $z = \text{Val}^q(t_h)$. Since $d_{h-1} < d_h = i$, the triple set $M_{h-1} = M_h - \{t_h\}$ is a (p,q)-properly colored $(h-1)$-matching in S'_{i-1}. By the inductive hypothesis, after the $(i-1)$-st execution of the loop 6.1-6.7 in the algorithm, the collection $\mathcal{Q}_{old}$ contains a set C_{h-1} of $h-1$ symbol pairs such that the set of colors used for the symbols in C_{h-1} is exactly $cl(M_{h-1})$. Now in the i-th execution of the loop 6.1-6.7 when the set C_{h-1} and the triple t_h are examined in step 6.3, a set C of symbol pairs using the color set $cl(M_{h-1}) \cup \{f(y), f(z)\} = cl(M_h)$ will be created. Therefore, after the i-th execution of the loop 6.1-6.7, a set C_h of h symbol pairs using the color set $cl(M_h)$ must be contained in the collection $\mathcal{Q}_{old}$. Suppose that the set C_h was created during the i-th execution by adding a symbol pair $(\text{Val}^p(t'_h), \text{Val}^q(t'_h))$ to a set C'_{h-1} of $h-1$ symbol pairs, where t'_h satisfies $\text{Val}^r(t'_h) = x_i$ and C'_{h-1} is contained in the collection $\mathcal{Q}_{old}$ after the $(i-1)$-st execution of the loop 6.1-6.7 (note that t'_h and C'_{h-1} are not necessarily t_h and C_{h-1}, respectively). By the inductive hypothesis, the set C'_{h-1} is extendable to a (p,q)-properly colored $(h-1)$-matching M'_{h-1} in S'_{i-1}. In consequence, the set C_h is extendable to the (p,q)-properly colored h-matching $M'_{h-1} \cup \{t'_h\}$ in S'_i. This completes the proof of the claim.

By the claim and let $i = m$, the algorithm **3DMatch**$(S, k, f, g; p, q)$ returns a set C_k of k symbol pairs if and only if the triple set S contains a (p, q)-properly colored k-matching, and the set C_k is extendable to a k-matching in S. To construct such a k-matching from C_k, we can use the graph matching technique suggested in [3]. Formally, from the set C_k of symbol pairs, we construct a bipartite graph $B_k = (V_L \cup V_R, E)$, where V_L contains k vertices, corresponding to the k symbol pairs in C_k, and V_R is the set of all symbols in $\mathrm{Val}^r(S)$. There is an edge in B_k from a vertex (y, z) in V_L to a vertex x in V_R if and only if the symbols y, z, and x form a triple in S. It is easy to see that a (p, q)-properly colored k-matching M_k in S can be obtained by constructing a graph matching of k edges in the bipartite graph B_k, which takes polynomial time [5].

In terms of the time complexity of the above algorithm, note that since for each set of $2j$ colors, we record at most one set of symbol pairs that uses exactly these $2j$ colors, the collection $\mathcal{Q}_{old}$ contains at most $\sum_{j=0}^{k} \binom{g}{2j}$ sets of symbol pairs. For each set C of symbol pairs in $\mathcal{Q}_{old}$, steps 6.2-6.6 of the algorithm take time polynomial in n and k. Therefore, each execution of the loop 6.1-6.7 of the algorithm runs in time $O^*(\sum_{j=0}^{k} \binom{g}{2j})$. In consequence, the running time of the algorithm **3DMatch**$(S, k, f, g; p, q)$ is bounded by $O^*(\sum_{j=0}^{k} \binom{g}{2j})$.   □

To solve the 3-D MATCHING AUGMENTATION problem (S, M_k), we only color two columns of a $(k + 1)$-matching properly if it exists. In this case, by Lemma 2, there is a $(k + 1)$-matching M_{k+1} such that M_{k+1} has two columns that contain at least $4k/3$ symbols in M_k. Thus at most $2k/3 + 2$ symbols in these two columns in M_{k+1} are missing in M_k. By introducing $2k$ new colors for each symbol in these two columns in M_k and by Proposition 1, in time $O^*(6.1^{2k/3})$, the two columns of M_{k+1} can be colored properly into $8k/3 + 2$ colors. By Theorem 2, the 3-D MATCHING AUGMENTATION problem (S, M_k) can be solved in time $O^*(6.1^{2k/3}2^{8k/3}) = O^*(2.77^{3k})$.

Theorem 3. *The* 3-D MATCHING *problem can be solved in time* $O^*(2.77^{3k})$.

If we use a randomized color-coding scheme that properly colors a subset of size k into k colors with high probability in time $O^*(e^k)$ [2], then the time complexity to solve 3-D MATCHING problem can be improved to $O^*(e^{2k/3}2^{8k/3}) = O^*(2.32^{3k})$.

Theorem 4. *The* 3-D MATCHING *problem can be solved by a randomized algorithm of time* $O^*(2.32^{3k})$.

5   Final Remarks

Recently there has been much interest in parameterized algorithms for *graph packing* problems, i.e., algorithms for constructing k disjoint isomorphic subgraphs in a given graph [8,12,14,15]. In particular, Fellows et al. [8] presented a parameterized algorithm of time $O^*(2^{2k \log k + 1.869k})$ for packing k vertex-disjoint triangles in a given graph, and Mathieson, Prieto, and Shaw [14] proposed a parameterized algorithm of time $O^*(2^{4.5k \log k + 4.5k})$ for packing k edge-disjoint triangles in a given graph. Since these problems can be trivially reduced to the 3-SET PACKING problem, by Corollary 1, they can be solved in time $O^*(4.61^{3k})$. This again gives significant improvements over the previous algorithms.

References

1. Alon, N., Goldreich, O., Håstad, J., Peralta, R.: Simple constructions of almost k-wise independent random variables. Random Struct. Alg. **3** (1992) 289–304
2. Alon, N., Yuster, R., Zwick, U.: Color-coding. J. ACM **42** (1995) 844–856
3. Chen, J., Friesen, D.K., Jia, W., Kanj, I.A.: Using nondeterminism to design efficient deterministic algorithms. Algorithmica **40** (2004) 83–97
4. Chen, J., Lu, S., Sze, S.-H., Zhang, F.: Improved algorithms for path, matching, and packing problems. Manuscript (2006)
5. Cormen, T.H., Leiserson, C.E., Rivest, R.L., Stein, C.: Introduction to Algorithms, 2nd ed. MIT Press, Cambridge, MA (2001)
6. Dehne, F., Fellows, M.R., Rosamond, F.A., Shaw, P.: Greedy localization, iterative compression, and modeled crown reductions: new FPT techniques, an improved algorithm for set splitting, and a novel $2k$ kernelization for vertex cover. Lect. Notes Comp. Sci. **3162** (IWPEC 2004) 271–280
7. Downey, R.G., Fellows, M.R.: Parameterized Complexity. Springer, New York (1999)
8. Fellows, M.R., Heggernes, P., Rosamond, F.A., Sloper, C., Telle, J.A.: Finding k disjoint triangles in an arbitrary graph. Lect. Notes Comp. Sci. **3353** (WG 2004) 235–244
9. Fellows, M.R., Knauer, C., Nishimura, N., Ragde, P., Rosamond, F., Stege, U., Thilikos, D.M., Whitesides, S.: Faster fixed-parameter tractable algorithms for matching and packing problems. Lect. Notes Comp. Sci. **3221** (ESA 2004) 311–322
10. Garey, M.R., Johnson, D.S.: Computers and Intractability: A Guide to the Theory of NP-Completeness. Freeman, San Francisco (1979)
11. Jia, W., Zhang, C., Chen, J.: An efficient parameterized algorithm for m-set packing. J. Alg. **50** (2004) 106–117
12. Kneis, J., Mölle, D., Richter, S., Rossmanith, P.: Divide-and-color. Lect. Notes Comp. Sci. (WG 2006) to appear
13. Koutis, I.: A faster parameterized algorithm for set packing. Inf. Process. Lett. **94** (2005) 7–9
14. Mathieson, L., Prieto, E., Shaw, P.: Packing edge disjoint triangles: a parameterized view. Lect. Notes Comp. Sci. **3162** (IWPEC 2004) 127–137
15. Prieto, E., Sloper, C.: Looking at the stars. Theor. Comp. Sci. **351** (2006) 437–445
16. Schmidt, J.P., Siegel, A.: The spatial complexity of oblivious k-probe hash functions. SIAM J. Comp. **19** (1990) 775–786

Fixed-Parameter Approximation: Conceptual Framework and Approximability Results[*]

Liming Cai[1] and Xiuzhen Huang[2]

[1] Department of Computer Science, The University of Georgia, Athens,
Georgia 30605 USA
cai@cs.uga.edu
[2] Department of Computer Science, Arkansas State University, State University,
Arkansas 72467 USA
xzhuang@csm.astate.edu

Abstract. The notion of fixed-parameter approximation is introduced to investigate the approximability of optimization problems within the framework of fixed-parameter computation. This work partially aims at enhancing the world of fixed-parameter computation in parallel with the conventional theory of computation that includes both exact and approximate computations. In particular, it is proved that fixed-parameter approximability is closely related to the approximation of small-cost solutions in polynomial time. It is also demonstrated that many fixed-parameter intractable problems are not fixed-parameter approximable. On the other hand, fixed-parameter approximation appears to be a viable approach to solving some inapproximable yet important optimization problems. For instance, all problems in the class MAX SNP admit fixed-parameter approximation schemes in time $O(2^{O((1-\epsilon/O(1))k)}p(n))$ for any small $\epsilon > 0$.

1 Introduction

The theory of fixed-parameter complexity was initiated by Downey and Fellows [14,16] to study the exact computation of important computational problems whose input contains a significant numerical parameter. The complexity of such fixed-parameter problems is measured in the value of the parameter, as well as the size of the input. For example, the problem of determining if a given graph G has a vertex cover of size k can be accomplished in time $O(1.2852^k + n)$ [12]. Thus the problem is called fixed-parameter tractable since it can be solved feasibly for small or fixed parameters. In contrast, other problems, such as to determine if a given graph has a dominating set of size k, seem not to behave well enough from the fixed-parameter tractability perspective. All known algorithms for such problems run in time $\Omega(n^{k+1})$, formidable even for small values of k. Studies [16] have shown that the fixed-parameter tractability of many computational problems hinges upon the answers to some long-standing open

---

[*] The authors would like to thank Rod Downey and Mike Fellows regarding the definition of the fixed-parameter approximability for the problem DOMINATING SET.

H.L. Bodlaender and M.A. Langston (Eds.): IWPEC 2006, LNCS 4169, pp. 96–108, 2006.

questions in conventional complexity theory [14,15]. In particular, problems, such as determining the existence of a dominating set of size k, are not fixed-parameter tractable unless SAT $\in DTIME(2^{o(n)})$. Given the phenomenon unfavorable to such problems, it is natural to ask whether fixed-parameter feasible algorithms may exist that can give "approximate" answers to decision questions involving these computational problems [19].

Another issue motivated this research is the "non-parity" phenomenon of fixed-parameter algorithms when they are used to find the optimal solution of optimization problems. For instance, consider the algorithm of running time $O(1.2852^k + n)$ that can determine if a given graph has a vertex cover of (the given) size k. It can be used to determine the size k_0 of the minimum vertex cover and find the cover when $k_0 \leq k$. However, it does not guarantee, with the same running time in k, to determine the size k_0 of the minimum cover for $k_0 > k$. Therefore, one fundamental issue is how much time is needed in estimating value of k_0 with respect to the value of k. Such issue has been addressed for a few individual fixed-parameter problems in the past. A typical example is the algorithm developed for the problem of graph tree decomposition. In a fixed-parameter feasible time, it can either determine if a given graph has a tree width larger than parameter value k or produces a tree decomposition of width $\leq 3k$. The yielded tree decomposition is an approximate solution with a "guaranteed ratio" of 3 with respect to the given parameter k but not to the optimum. (Note that later there is an improved algorithm [5], which, for a given graph G and a parameter value k, could either yield a tree decomposition of k if such a tree decomposition of G exists, or answer "no" otherwise.) Like the conventional approximation, this new type of approximation produces a range of values as an estimation to the optimum.

We develop this concept into the notion of fixed-parameter (FP-)approximation to study the existence of fixed-parameter feasible algorithms that may give approximate answers for fixed-parameter problems. Intuitively, an FP-approximation algorithm for a minimization problem either give a negative answer to the question of "$OPT(I) \leq k$?" or produces a solution with value bounded by $g(k)$ for some fixed function g. In addition, the algorithm runs in time $O(f(k)n^c)$ for some fixed function f and a constant $c > 0$. In section 2 we give the formal definition for FP-approximation. The FP-approximation is well defined because we can show that either fixed-parameter tractability or polynomial-time approximation would imply FP-approximation. We believe the new notion enhances the parameterization framework to be in parallel with the conventional theory of computing that includes both exact and approximation computation.

We show that fixed-parameter approximability is equivalent to approximating solutions of a small (but unbounded) cost in polynomial time. This complements the known results that fixed-parameter tractability is equivalent to determining solutions of a small cost in polynomial time [8,9]. In general, we prove that an optimization problem is not fixed-parameter approximable to ck unless the problem of finding solutions of cost bounded by $s(n)$ is approximable to the ratio c in polynomial time, for some unbounded and nondecreasing function $s(n)$. According to the results in [18,11], for many optimization problems, approximating solutions of a small cost cannot be done in polynomial time. Therefore, our research essentially shows the fixed-parameter inapproximability for many optimization problems. In section 3, we show that a number of fixed-parameter problems, including finding the minimum dominating set, are not FP-approximable.

The FP-approximation is also considered an alternative to the polynomial-time approximation. In general, we expect the approximability of certain optimization problems to be improved under the fixed-parameter setting. Indeed, in section 4 we show that all optimization problems in the optimization class MAX SNP admit FP-approximation schemes, i.e., they all have algorithms that run in time $O(2^{O((1-\epsilon/O(1))k)}p(n))$ to produce a solution of value bounded by $(1+\epsilon)k$ for any $\epsilon > 0$ for a minimization problem (or, a solution of value bounded by $(1 - \epsilon)k$ for any $\epsilon > 0$ for a maximization problem). This complements the known results that MAX SNP-complete problems do not admit polynomial time approximation schemes unless P=NP.

In section 5, we discuss some further results in the approximability improvements with the FP-approximation for some other problems. In fact, Bodlaender and Fellows [6] are able to show the fixed-parameter approximability for the problem of BANDWIDTH, which is W[1]-hard under the uniform reduction due to Bodlaender [4]. Moreover, we suspect that for many polynomial-time approximable optimization problems, the approximation ratios can be improved when fixed-parameter feasibility is the only concern regarding the running time of the algorithm. In particular, we show that problem BIN PACKING, which is MAX SNP-hard and fixed-parameter intractable unless P=NP [7], admits an (asymptotic) FP-approximation scheme. Our result improves the one devised by Karmarkar and Karp [23] in the sense that the ϵ we obtain in the ratio $1 + \epsilon$ decreases much faster with respect to $OPT(I)$. We point out that the techniques used in the proof may be applied to improving the approximation performance of other fixed-parameter problems.

2   Definitions

We first briefly introduce some basic concepts in the theory of fixed-parameter complexity. We refer the reader to Downey and Fellows [16] for further details. A fixed-parameter problem is defined over $\Sigma^* \times N$, where Σ is a finite alphabet and N is the set of natural numbers. Each instance of the fixed-parameter problem Π is a pair (I, k), where k is called the *parameter*. A fixed-parameter problem is *fixed-parameter tractable* if there is an algorithm to decide the membership of the problem in time $O(f(k)n^c)$, where $f(k)$ is a recursive function and c is a constant. Given an optimization problem Π, the fixed-parameter (decision) version of the problem is defined by the question whether a given input has the optimal cost $OPT(I)Rk$, where R is "$\geq$" or "$\leq$" depending on whether Π is a maximization or minimization problem [7]. In this paper whenever there is no confusion, we do not distinguish between an optimization problem and its fixed-parameter decision version.

We define the fixed-parameter approximation in the following. This notion was originally conceived by Downey and Fellows [19].

Definition 1. Let Π be the fixed-parameter version of a minimization problem. Let f be a (recursive) function, p be a polynomial independent of f, and k be a constant. Then Π is *fixed-parameter approximable to $g(k)$ in time $O(f(k)n^c)$* for some fixed function g if there is an algorithm such that given any instance I with parameter k, and question $OPT(I) \leq k$, the algorithm which runs in $O(f(k)n^c)$ steps, where $n = |I|$, (1) either outputs "no" (asserting the optimal cost is larger than k), or (2) produces a solution of cost at most $g(k)$.

Fixed-parameter approximation for a maximization problem can be defined similarly.

The problem Π is called *FP-approximable to $g(k)$* or simply *FP-approximable* when g is linear in k. The algorithm is called an *FP approximation algorithm* for the problem. We define *FPA* to be the class of fixed-parameter optimization problems that can be FP-approximation to $g(k)$ for some recursive function g.

Definition 2. A minimization (or, maximization) problem Π is said to have a *fixed-parameter approximation scheme*, if the minimization problem Π can be FP-approximable to $(1 + \epsilon)k$ for any given small constant $\epsilon > 0$ (or, if the maximization problem Π can be FP-approximable to $(1-\epsilon)k$ for any given small constant $\epsilon > 0$).

In the above definition, the time function of the approximation algorithm may depend on the given value of ϵ. The fixed-parameter approximation scheme is *efficient* if the polynomial p in the time function $O(f(k)p(n))$ does not depend on ϵ. In addition, when Π is fixed-parameter approximable under the situation for sufficiently large k, it is called *asymptotic* fixed-parameter approximable.

We show that the fixed-parameter approximability is well-defined under the parameterization framework. Essentially, we prove that fixed-parameter tractability implies fixed-parameter approximability. As shown in the following, this relationship holds for many optimization problems that are *solution-constructible*, a property first introduced by Cai and Chen [7]. (Note that in [7] the term *fixed-parameter tractability with witness* is used instead of solution constructible. Interested readers are referred to [25] for another related notion of self-reducibility.)

Definition 3. Assume the existence of a fixed algorithm with running time $T(|I|, k)$ that can determine $OPT(I) \leq k$ (or, $OPT(I) \geq k$) for each input (I, k). A minimization (or, maximization) problem is *solution-constructible* if for any input (I, k), a solution of cost at most (or, at least) k (if it does exist) can be constructed in time polynomial in both $|I|$ and $T(|I|, k)$.

Almost all NP-hard optimization problems studied in the literature [2,20,22] are solution-constructible. In fact, solution-constructibility is one of the necessities that the decision problem formulation characterizes the difficulty of the original optimization problem.

Theorem 1. *Let Π be a fixed-parameter version of some solution-constructible optimization problem. If Π is fixed-parameter tractable, then Π is fixed-parameter approximable to k.*

Proof. We briefly verify this for minimization problems. The proof for maximization problems is similar. We describe a fixed-parameter feasible approximation algorithm for the problem Π as follows.

Let B be an algorithm that solves the decision problem Π. For each input of size n and parameter k, it runs in time $O(h(k)n^c)$ for some (recursive) function $h(k)$ and some constant $c > 0$. Given an input (I, k), B is called to determine $OPT(I) \leq k$. If $k < OPT(I)$, output "no" and halt. Otherwise, because Π is solution-constructible, there is an algorithm that constructs a solution of cost at most k. The time used for the construction is $O(T^d p(n))$ for some constant d and polynomial p, where T is the running time of B. So the total running time is bounded by $O(h(k)^d q(n))$ for some polynomial q. Therefore, Π is fixed-parameter approximable to k in time $O(h(k)^d q(n))$. □

We establish a tight relationship between polynomial-time approximation and fixed-parameter approximability. We show the simple fact that if an optimization problem admits a polynomial-time approximation algorithm then the problem is fixed-parameter approximable. We consider minimization problems only. Similar results hold for maximization problems as well.

Theorem 2. *Let Π be a minimization problem that is approximable in polynomial time to the ratio $r \geq 1$ for some constant r. Then Π is fixed-parameter approximable to rk.*

Proof. Let A be a polynomial-time approximation algorithm for Π achieving the ratio r. Let $A(I)$ be the cost of a solution S obtained by the algorithm A such that $A(I)/OPT(I) \leq r$. Given an input I and a value for the parameter k, one can approximately answer the question "is $OPT(I) \leq k$?" by utilizing the algorithm A in the following way. First, answer "no" if $k < A(I)/r$. According to the definition of fixed-parameter approximability, this is the only correct answer since $k < A(I)/r \leq OPT(I)$. Second, if $k > A(I)$, output the solution S. Finally, if $A(I)/r \leq k \leq A(I)$, then output the solution S. Note that in the second and third cases, the algorithm returns a solution with cost $A(I) \leq rk$. □

3   Fixed-Parameter Inapproximability

Many optimization problems including CLIQUE, DOMINATING SET, and LONGEST COMMON SUBSEQUENCE are provably fixed-parameter intractable, i.e., they are hard for various levels of the W-hierarchy. Since the set of all fixed-parameter tractable problems forms the lowest level of the W-hierarchy, the above problems are not fixed-parameter tractable unless the W-hierarchy collapses. Since it is likely that there are no fixed-parameter feasible algorithms for the decision question for any of these problems, it is desirable to know whether there are fixed-parameter feasible algorithms that give approximate answers. In this section, we show that for many optimization problems such approximation algorithms do not exist either. First, we relate fixed-parameter approximability to the approximation of solutions with a small cost in polynomial time.

Lemma 1. *Let Π be an optimization problem and r be a constant. Then Π is not fixed-parameter approximable to rk unless Π is approximable in time $t(OPT(I))n^c$ to the ratio r for some (recursive) function t and some constant c.*

Proof. We give the proof for minimization problems. The proof for maximization problems is similar.

Let Π be a minimization problem. Assume that Π is fixed-parameter approximable to rk. Then there is an algorithm A for Π that runs in time $O(f(k)n^c)$, for some nondecreasing function f and constant c. Furthermore, given an input (I, k), the algorithm gives an approximate answer to the question whether $OPT(I) \leq k$, for each value of the parameter k. Fix an input I. Then for the parameter $k = 0, 1, ..., k_0$, where $k_0 = (OPT(I) - 1)/r$, the algorithm will output "no" because producing a solution of cost $A(I) \leq rk \leq r(OPT(I) - 1)/r$ implies that $A(I) \leq OPT(I) - 1$. Let k_1 be the largest value of the parameter k that allows the algorithm to output "no". Then $k_1 + 1$ allows the algorithm to produce a solution with cost $\leq r(k_1 + 1) \leq rOPT(I)$ because the answer of the algorithm on the parameter $k = OPT(I)$ cannot be "no".

We construct an algorithm B that calls the fixed-parameter approximation algorithm A for $k = 0, 1, ...,$ until A produces a solution. According to the analysis above, the solution has the cost $rk \leq r(k_1 + 1) \leq rOPT(I)$. The solution achieves the ratio $rOPT(I)/OPT(I) = r$. The total running time for algorithm B is bounded by $(k_1 + 1)f(k_1 + 1)n^c \leq OPT(I)f(OPT(I))n^c = t(OPT(I))n^c$, where $t(OPT(I)) = OPT(I)f(OPT(I))$. $\qquad\square$

Lemma 1 gives some surprisingly nice implications.

Theorem 3. *For any constant $r > 1$, a minimization problem Π is not fixed-parameter approximable to rk, unless the problem of finding solutions of cost bounded by $s(n)$ can be approximated to the ratio r in polynomial time, for some unbounded non-decreasing function $s(n)$.*

Proof. Assume that Π is fixed-parameter approximable to rk. By Lemma 1 and its proof, Π can be approximated to the ratio r by an algorithm A of time $t(OPT(I))n^d$ for some recursive function t and some constant d. When the input instance I is subject to the restriction that $OPT(I) \leq s(n)$, where s is the inverse function of t, the algorithm A runs in time $t(s(n))n^d = n^{d+1}$ to approximate solutions of cost at most $s(n)$ (see [7,8] for techniques for constructing and computing inverse functions). $\qquad\square$

Theorem 3 has a direct impact on the fixed-parameter approximability of the problem DOMINATING SET. We have the following result. Note that LOG DOMINATING SET is the problem of finding the minimum dominating set whose size is $\leq \log n$ in a given graph of n vertices.

Corollary 1. *For any* $r \geq 1$, DOMINATING SET *is not fixed-parameter approximable to* rk *in time* $O(2^k n^d)$, *where* d *is a constant, unless* LOG DOMINATING SET *is approximable to the ratio* r *in polynomial time.*

Theorem 3 shows the close relationship between fixed-parameter approximability and the ability to approximate small-value instances in polynomial time. The approximability of optimization problems with small value constrained objective functions have been studied in the literature (see for example, [11,18]). In particiular, it has been shown that, for a number of important problems, including CLIQUE, DOMINATING SET that are hard for the W-hierarchy, the approximation for small value instances, such as LOG CLIQUE, LOG DOMINATING SET, is as hard as subexponential-time (randomized) simulation of NP-computation. Note that subexponential-time (randomized) simulation of NP-computation is a seemingly weaker assumption than P=NP but still highly unlikely. Since the collapsing of the W-hierarchy would also imply subexponential-time simulation of NP-computation, the fixed-parameter inapproximability results developed in this section complement nicely the fixed-parameter intractability results developed in [16,7,9].

4   FP-Approximation Scheme for Problems in MAX SNP

In this section, we show that the notion of FP-approximation provides a viable alternative for approximation. In particular, we prove that FP-approximation schemes exist for all problems in the class MAX SNP. This is somewhat surprising since it is well-known result that MAX SNP-complete problems do not admit polynomial-time approximation scheme unless P=NP.

The class MAX SNP was introduced by Papadimitriou and Yannakakis [24] to capture a collection of optimization problems. For the purpose of investigating the approximability of these problems, the following approximation-preserving reduction was introduced.

Definition 4 ([24]). *Let* Π_1 *and* Π_2 *be two optimization problems with cost functions* f_1 *and* f_2. Π_1 *L-reduces to* Π_2 *if there are two polynomial time algorithms* A *and* B *and two constants* α, $\beta > 0$ *such that for each instance* I_1 *of* Π_1, (1) *the algorithm A produces an instance* $I_2 = A(I_1)$ *such that* $OPT_{\Pi_2}(I_2) \leq \alpha OPT_{\Pi_1}(I_1)$, *and* (2) *given any solution* S_2 *for* I_2 *with cost* $f_2(I_2, S_2)$, *algorithm* B *produces a solution* S_1 *for* I_1 *with cost* $f_1(I_1, S_1)$ *such that* $|OPT_{\Pi_1}(I_1) - f_1(I_1, S_1)| \leq \beta |OPT_{\Pi_2}(I_2) - f_2(I_2, S_2)|$.

It is known from the work of Cai and Chen [7] that the standard parameterized versions of all maximization problems in the MAX SNP are parameterized tractable. The proof of this earlier result was later refined by Cai and Juedes [10] to show that L-reductions actually preserve subexponential-time computability.

Lemma 2. *Let* Π_1 *and* Π_2 *are two minimization problems and* Π_1 *is L-reducible to* Π_2. *Then* Π_1 *has an FP-approximation scheme if* Π_2 *has one.*

Proof. Let algorithm A and B be the two algorithms associated with the L-reduction of running time $q(n)$ and $r(n)$ respectively. Also let α, β be the two associated constants. We assume that Π_2 admits a FP-approximation scheme M with running time $f(k, \epsilon)p(n, \epsilon)$. We give in the following an FP-approximation scheme for problem Π_1.

FP-APPROXIMATION SCHEME FOR Π_1
1. Given an input instance I_1 for Π_1, parameter k, and $\epsilon > 0$, call algorithm A to produce an instance $I_2 = A(I_1)$ for Π_2.
2. For $h = 0, 1, 2, ..., \alpha k$, and $\epsilon' = \epsilon/\alpha\beta$, run M on the instance I_2 to decide if $OPT_{\Pi_2}(I_2) \leq h$. Then there must exist an h_0, $0 \leq h_0 \leq \alpha k$, such that M answers "no" on all $h = 0, 1, ..., h_0$ but yields on $h_0 + 1$ an solution S_2 with value $f_2(I_2, S_2) \leq (1 + \epsilon')(h_0 + 1)$.
3. If $h_0 = \alpha k$, it is clear that $OPT_{\Pi_2}(I_2) > \alpha \cdot k$. By the L-reduction, $OPT_{\Pi_2}(I_2) \leq \alpha OPT_{\Pi_1}(I_1)$, it can be concluded that $OPT_{\Pi_1}(I_1) > k$. So output "no" and stop.
4. Otherwise, the solution S_2 satisfies $f_2(I_2, S_2) \leq (1 + \epsilon')(h_0 + 1) \leq (1 + \epsilon')\alpha k$. Because $OPT_{\Pi_2}(I_2) > h_0$, $OPT_{\Pi_2}(I_2) \geq h_0 + 1$. This means the ratio $f_2(I_2, S_2)/OPT_{\Pi_2}(I_2) \leq 1 + \epsilon'$.
5. Call algorithm B on S_2 and I_2 to produce a solution S_1 of value $f_1(I_1, S_1)$. This solution S_1 satisfies ratio $f_1(I_1, S_1)/OPT_{\Pi_1}(I_1) \leq 1 + \alpha\beta\epsilon' = 1 + \epsilon$ (refer to [24], Proposition 2).
6. If $f_1(I_1, S_1) \leq (1 + \epsilon)k$, output solution S_1 and stop.
7. Otherwise, $f_1(I_1, S_1) > (1+\epsilon)k$. Because $f_1(I_1, S_1) \leq (1+\epsilon)OPT_{\Pi_1}(I_1)$, this implies $OPT_{\Pi_1}(I_1) > k$. Output "no" and stop.

On the question "$OPT_{\Pi_1}(I_1) \leq k$?", the above algorithm outputs either "no" or a solution S_1 of value $f_1(I_1, S_1) \leq (1+\epsilon)k$. The total time is bounded by $q(n) + r(n) + \alpha k f(k, \epsilon')p(n, \epsilon')$. This bound remains the format of $O(f'(k, \epsilon)p'(n, \epsilon))$ for some polynomial p' and recursive function f'. □

By carefully examining the time complexity analysis given in the proof of Lemma 2, we have the following technical lemma

Lemma 3. *Let Π_1 and Π_2 are two minimization problems and Π_1 is L-reducible to Π_2 associated with two constants α and β. If Π_2 has an FP-approximation scheme of time $O(2^{O((1-\epsilon)k)}p(n))$ for some polynomial p, then Π_1 has an FP-approximation scheme of time $O(2^{O((1-\epsilon/\alpha\beta)k)}q(n))$ for some polynomial q.*

Before we establish the main result of this section, we prove the following

Lemma 4. *The MAX SNP-complete problem* VERTEX COVER *with bounded degree 3 (abbreviated as VC-3) admits an FP-approximation scheme of running time $O(2^{O((1-\epsilon/O(1))k)}p(n))$, where p is a polynomial.*

Proof. Let $G = (V, E)$ be a connected graph with bounded degree 3. Suppose $n = |V| > 4$, k is the parameter, and $0 < \epsilon < 1$. We first partition the vertex set V into two subsets V_1 and V_2 such that $|V_1| = \delta n$ and $|V_2| = (1 - \delta)n$, where $\delta = \epsilon/4$. Let $G_1 = (V_1, E_1)$ and $G_2 = (V_2, E_2)$ be the two subgraphs of G induced by V_1 and V_2 respectively. Then we apply the fixed-parameter algorithm in [12] to the subgraph G_2.

Suppose the minimum vertex cover found by the fixed-parameter algorithm for the subgraph G_2 be C_2. It is clear that the vertices in C_2 plus the vertices in V_1 is a vertex cover of G. Let $OPT(G)$ be the size of the minimum vertex cover of G. We claim that $(|C_2| + |V_1|)/OPT(G) \le 1 + \epsilon$.

Since $G_2 = (V_2, E_2)$ is a subgraph of the graph G, the size of its minimum vertex cover is not larger than the size of the minimum vertex cover of G. That is,

$$|C_2| \le OPT(G),$$

For the graph G with n vertices, the number of edges $|E| \ge n - 1$. Since G has bounded degree 3, the size of the minimum vertex cover $|OPT|$ satisfies:

$$|OPT| \ge (n - 1)/3 \ge n/4,$$

since $n > 4$.

Therefore,

$$(|C_2| + |V_1|)/OPT(G) \le (OPT(G) + |V_1|)/OPT(G)$$
$$= 1 + \delta n/OPT(G) \le 1 + \delta n/(n/4) \le 1 + 4\delta = 1 + \epsilon.$$

The running time of the FP-approximation algorithm is bounded by the time for finding the minimum vertex cover for G_2, which is $O(1.2852^{O((1-\delta)n)}p(n))$ [12]. Since $k \ge OPT(G) \ge n/4$, and $\delta = \epsilon/4$, we have the time complexity $O(1.2852^{O((1-\delta)k)}p(n)) = O(1.2852^{O((1-\epsilon/4)k)}p(n))$. The lemma is proved. $\square$

By Lemma 3 and Lemma 4, we have

Theorem 4. *All minimization problems in the class MAX SNP admit FP-approximation schemes of time $O(2^{O((1-\epsilon/O(1))k)}p(n))$, where p is a polynomial.*

Note that for maximization problems in the class MAX SNP, it is proved by Cai and Chen (Theorem 3.4, [7]) that they are in FPT and can be solved in time $O(2^k p(n))$. These exact algorithms can determine if $OPT \ge (1 - \epsilon)k$ in time $O(2^{O((1-\epsilon)k)}p(n))$. Therefore,

Theorem 5. *All maximization problems in the class MAX SNP admit FP-approximation schemes of time $O(2^{O((1-\epsilon)k)}p(n))$, where p is a polynomial.*

In conclusion,

Corollary 2. *All problems in the class MAX SNP admit FP-approximation schemes of time $O(2^{O((1-\epsilon/O(1))k)}p(n))$, where p is a polynomial.*

Theorem 4.1 shows that, with an amount of time less than what is needed by exact algorithms, the optimization problems in MAX SNP can be approximated to an arbitrary accuracy. On the other hand, any improvement in the running time of the parameterized algorithms for the parameterized decision problems would imply more efficient FP-approximation schemes for the problems. For example, by applying the parameterized algorithm of time $O(1.194^k k^2 + n)$ for the problem VC-3 in [13], the time complexity of the FP-approximation scheme for VC-3 could be improved from the aforementioned $O(1.2852^{O((1-\epsilon/4)k)} p(n))$ to $O(1.194^{O((1-\epsilon/4)k)} q(n))$, where p and q are polynomials.

5   Improving Approximability for BIN PACKING

In this section, we show that the notion of FP-approximation can be used to improve the approximation of parameterized intractable problems as well.

Consider the BIN PACKING problem [20]: Given a finite set $U = \{u_1, ..., u_n\}$ of items and a rational size $s(u) \in [0, 1]$ for each item $u \in U$, find a partition of U into disjoint subsets $U_1, ..., U_k$ such that the sum of the sizes of the items in each U_i is no more than 1 and k is minimized. Note that for the decision problem for $k = 3$ is NP-hard [20] and that it is not fixed-parameter tractable unless P=NP [7]. So, we cannot claim fixed-parameter approximability simply from the results shown in section 2. However, based on Theorem 2 and the fact that this problem can be approximated to the ratio $3/2$ [26], we obtain the following result. (Note that better asymptotic ratios for BIN PACKING are known [2].)

Corollary 3. BIN PACKING *is fixed-parameter approximable to* $3/2k$.

We show in the following that this approximation ratio can be significantly improved when fixed-parameter feasibility is the only concern regarding the running time of the algorithm.

Let $0 < \epsilon \leq 1/2$ be a rational number such that $\epsilon \leq 1/OPT(U)$, where U is the set of items for an instance of BIN PACKING. Let $V \subseteq U$ such that $u \in V$ if and only if $s(u) \geq \epsilon$. We have the following lemma.

Lemma 5. *Let B be an optimal packing for the set V. There is an algorithm A that uses B to construct an approximate packing for U such that $A(U)/OPT(U) \leq 1 + 2\epsilon + 2/OPT(U)$.*

Proof. (Sketch due to page limit) Assume $B = U_1, ..., U_p$ is an optimal packing for V. Then $B(V) = OPT(V) = p$. Algorithm constructs another packing A for U as follows. Pick items from $U - V$ and pack them into bins $U_1, ..., U_p$ using the well-known FFD (first fit decreasing) heuristic [20]. Notice that FFD may need to use some extra bins $U_{p+1}, ..., U_q$ to complete the packing.

The size of the space left in each bin U_i, $i = p+1, ..., q-1$ must be smaller than ϵ since each element in $U - V$ has size $< \epsilon$. Moreover, we observe that (1) if $q > p$, the size of the space left in each U_i, $i = 1, ..., p$ is also smaller than ϵ; and (2) if $p = q$, $A(U) = OPT(U)$. Based on these observations, $A(U)/OPT(U) = 1$

in the case when $p = q$. When $q > p$, we have two cases. When $\Sigma_{u \in U} s(u) \geq 1 - \epsilon$, we have $OPT(U) \geq \Sigma_{u \in U} s(u) = \Sigma_{i=1}^{q} \Sigma_{s \in U_i} \geq q(1 - \epsilon) \geq A(U)(1 - \epsilon)$. Since $\epsilon \leq 1/2$, we have that $A(U)/OPT(U) \leq 1 + 2\epsilon$. When $\Sigma_{u \in U_q} s(u) < 1 - \epsilon$, we have $OPT(U) \geq \Sigma_{u \in U} s(u) = \Sigma_{i=1}^{q} \Sigma_{s \in U_i} s(u) \geq (q - 1)(1 - \epsilon) + \Sigma_{u \in U_q} s(u) \geq A(U)(1 - \epsilon) - \alpha$, where $\alpha = -(\epsilon - 1 + \Sigma_{u \in U_q} s(u))$. It is clear that $\alpha \leq 1$. So, $A(U)/OPT(U) \leq 1 + 2\epsilon + 2\alpha/OPT(U) \leq 1 + 2\epsilon + 2/OPT(U)$. $\qquad\square$

Now we are ready to prove the main theorem of this section.

Theorem 6. *For any $\delta > 0$, problem* BIN PACKING *is (asymptotic) fixed-parameter approximable to $(1 + \delta)k$.*

Proof. (Sketch due to page limit) We describe a fixed-parameter feasible process for BIN PACKING in the following. For a given δ, let $\epsilon = \delta/4$. Given a set U of items and an integer k for the problem, first run the approximation algorithm FFD on U. Let the number of bins in the packing obtained by the algorithm be m and the approximation ratio be r. Output "no" and halt if $k < m/r$. Otherwise, let $V \subseteq U$ such that $u \in V$ if and only if $s(u) \geq \epsilon$. Then find an optimal packing B for set V by a brute force approach. According to Lemma 5, there is a packing algorithm A for the original set U of items achieving the ratio bounded by $1 + 4\epsilon = 1 + \delta$, because the ratio $1 + 2\epsilon + 2/OPT(U) \leq 1 + 4\epsilon$ asymptotically. Now output "no" and halt if $k < A(U)/(1 + \delta)$. If $k > A(U)/(1 + \delta)$, output the packing with cost $A(U)$ bounded by $(1 + \delta)k$. We can show that the total time for the above process is $O((ck)^{ck} + n^d)$ for some constant d. $\qquad\square$

According to the proof of Theorem 6, we obtain the following corollary regarding approximation of a small number of bins for the problem BIN PACKING.

Corollary 4. *There is an asymptotic polynomial-time approximation scheme for* BIN PACKING *when the number of bins is bounded by $\log n / \log \log n$.*

Note that BIN PACKING cannot be approximated to the ratio $3/2 - \epsilon$ in polynomial time for any $\epsilon > 0$ unless P=NP [20]. Therefore, it is not possible for the problem to have a polynomial-time approximation scheme. However, Karmarkar and Karp [23] devised an efficient approximation scheme for the problem with asymptotic ratio $1 + \beta$, where $\beta = O(\log^2 OPT(U)/OPT(U))$. Our proof in Lemma 5 shows an improvement in that the ratio $1 + \beta$ is determined by $\beta = O(1/OPT(U))$.

6   Further Research Work

This is a preliminary work for the new framework of fixed-parameter approximation. Further research may derive more valuable positive and negative results. For example, the work on the fixed-parameter approximability of planar graph problems, such as PLANAR VERTEX COVER, PLANAR DOMINATING SET, and PLANAR INDEPENDENT SET, will be very interesting, compared with the well-known current results of Alber et al. [1], the $O(2^{\sqrt{k}} n^c)$-time parameterized algorithms, and of Baker [3], the $O(2^{1/\epsilon} n^c)$-time PTAS algorithm for these problems. Also we believe the inapproximability results derived in the current paper may shed light on the study of the FP-approximability of the problems in the newly-proposed parameterized class MINI[1] (refer to [17]).

References

1. J. ALBER, H. FERNAU, AND R. NIEDERMEIER, Parameterized complexity: exponential speed-up for planar graph problems, *J. Algorithms 52*, pp. 26-56, (2004).
2. G. AUSIELLO, P. CRESCENZI, G. GAMBOSI, V. KANN, A. MARCHETTI SPACCAMELA, AND M. PROTASI, *Complexity and Approximation: Combinatorial Optimization Problems and their Approximability Properties*, Springer-Verlag, (1999).
3. B. S. BAKER, Approximation algorithms for NP-complete problems on planar graphs, *Journal of the ACM 41*, pp. 153-180, (1994).
4. H. BODLAENDER, unpublished manuscipt.
5. H. L. BODLAENDER, A Linear-Time Algorithm for Finding Tree-Decompositions of Small Treewidth. *SIAM J. Comput. 25*, pp. 1305-1317, (1996).
6. H. BODLAENDER AND M. R. FELLOWS, unpublished manuscipt.
7. L. CAI AND J. CHEN, On fixed-parameter tractability and approximability of NP optimization problems, *J. of Comp. and Sys. Sciences 54*, pp. 465-474, (1997).
8. L. CAI AND J. CHEN, On the amount of nondeterminism and the power of verifying, *SIAM Journal of Computing 26*, pp. 733-750, (1997).
9. L. CAI, J. CHEN, R. DOWNEY, AND M. FELLOWS, On the structure of parameterized problems in NP, *Information and Computation 123*, pp. 38-49, (1995).
10. L. CAI AND D.W. JUEDES, On the existence of sub-exponential time parameterized algorithms, *Journal of Computer and System Sciences 67*, pp. 789-807, 2003.
11. L. CAI, D. JUEDES, AND I. KANJ, The inapproximability of non NP-hard optimization problems, *Theoretical Computer Science 289*, pp. 553-571, (2002).
12. J. CHEN, I. KANJ, AND W. JIA, Vertex cover: further observations and further improvements, *Journal of Algorithms 41*, pp. 280-301, (2001).
13. J. CHEN, I. KANJ, G. XIA, Labeled Search Trees and Amortized Analysis: Improved Upper Bounds for NP-Hard Problems, *Algorithmica 43*, pp. 245-273, (2005).
14. R. G. DOWNEY AND M. R. FELLOWS, Fixed-parameter intractability, *Proc. of the 7th Annual Conference on Structure in Complexity Theory*, pp. 36-49, (1992).
15. R. G. DOWNEY AND M. R. FELLOWS, Fixed-parameter tractability and completeness I: Basic results, *SIAM Journal on Computing 24*, pp. 873-921, (1995).
16. R. G. DOWNEY AND M. R. FELLOWS, *Parameterized Complexity*, Monographs in Computer Science, Springer, New York, (1999).
17. R. G. DOWNEY, V. ESTIVILL-CASTRO, M. R. FELLOWS, E. PRIETO, AND F. A. ROSAMOND, Cutting Up is Hard to Do: the Parameterized Complexity of k-Cut and Related Problems, *Electr. Notes Theor. Comput. Sci. 78*, (2003).
18. U. FEIGE AND J. KILIAN, On limited versus polynomial nondeterminism, *CJTCS: Chicago Journal of Theoretical Computer Science*, (1997).
19. M. R. FELLOWS, Personal communications, (1999), (2005).
20. M. R. GAREY AND D. S. JOHNSON, *Computers and Intractability: A Guide to the Theory of NP-completeness*, W.H. Freeman and Company, San Francisco, (1979).
21. D. HOCHBAUM, *Approximation Algorithms for NP-hard Problems*, PWS Publishing, (1997).

22. V. KANN, *On the Approximability of NP-complete Optimization Problems*, PhD thesis, Royal Institute of Technology, Sweden, (1992).
23. N. KARMARKAR AND R. M. KARP, An efficient approximation scheme for the one-dimensional bin packing problem, *Proc. of FOCS*, pp. 312-320, (1982).
24. C. PAPADIMITRIOU AND M. YANNAKAKIS, Optimization, approximation, and complexity classes, *Journal Of Computer and System Sciences 43*, pp. 425-440, (1991).
25. C. SCHNORR, Optimal algorithms for self-reducible problems, *Proc. of ICALP*, pp. 322-337, (1976).
26. D. SIMCHI-LEVI, New worst-case results for the bin-packing problem, *Naval Research Logistics 41*, pp. 579-585, (1994).

On Parameterized Approximability

Yijia Chen[1], Martin Grohe[2], and Magdalena Grüber[2]

[1] BASICS, Department of Computer Science, Shanghai Jiaotong University,
Shanghai 200030, China
yijia.chen@cs.sjtu.edu.cn
[2] Institut für Informatik, Humboldt-Universität, Unter den Linden 6,
10099 Berlin, Germany
grohe@informatik.hu-berlin.de, grueber@informatik.hu-berlin.de

Abstract. Combining classical approximability questions with parameterized complexity, we introduce a theory of *parameterized approximability*. The main intention of this theory is to deal with the efficient approximation of small cost solutions for optimisation problems.

1 Introduction

Fixed-parameter tractability and approximability are two complementary approaches to dealing with intractability: Approximability relaxes the goal of finding exact or optimal solutions, but usually insists on polynomial time algorithms, whereas fixed-parameter tractable (fpt) algorithms are exact, but may have a super-polynomial running time that is controlled by a parameter associated with the problem instances in such a way that for small parameter values the running time can still be considered efficient. Obviously, the two approaches can be combined, which is what we do in this paper. Optimisation problems are often parameterized by the cost of the solution that is to be found. That is, together with an instance we are given a parameter k, and the goal is to find a solution of size at least k (for maximisation problems) or at most k (for minimisation problems). For small k, an fpt algorithm with a running time like $O(2^k \cdot n)$ can be quite efficient. For some problems, for example minimum vertex cover, such an algorithm exists, but for many other problems it does not, under plausible complexity theoretic assumptions. For such problems, can we at least find small solutions that approximately have the desired cost k? A slightly different, but closely related question can be asked starting from approximability: Suppose we have a problem that is hard to approximate. Can we at least approximate it efficiently for instances for which the optimum is small? The classical theory of inapproximability does not seem to help answering this question, because usually the hardness proofs require fairly large solutions.

Let us illustrate this with an example: The maximum clique problem is known to be hard to approximate — unless ZPP = NP not approximable with ratio $n^{1-\varepsilon}$ for any $\epsilon > 0$ [15] — and, most likely, not fixed-parameter tractable — the problem is W[1]-complete [9], and unless the exponential time hypothesis fails, it is not even solvable in time $n^{o(k)}$ [6]. Here, and in the following, k denotes the size of the desired clique and n the number of vertices of the input graph. Now we ask:

H.L. Bodlaender and M.A. Langston (Eds.): IWPEC 2006, LNCS 4169, pp. 109–120, 2006.
© Springer-Verlag Berlin Heidelberg 2006

Is there an fpt algorithm that, given a graph $\mathcal{G}$ and a $k \in \mathbb{N}$, finds a clique of size $k/2$ in $\mathcal{G}$ provided $\mathcal{G}$ has a clique of size at least k. (If $\mathcal{G}$ does not have a clique of size k, the algorithm may still find a clique of size at least $k/2$, or it may reject the input.) We would call such an algorithm an *fpt approximation algorithm with approximation ratio* 2 for the clique problem. If no such algorithm exists, we may still ask if there is an algorithm that finds a clique of size $\sqrt{k}$ or even $\log k$, provided the input graph $\mathcal{G}$ has a clique of size k. As a matter of fact, it would be interesting to have an fpt approximation algorithm with approximation ratio ρ for any function ρ on the positive integers such that $k/\rho(k)$ is unbounded (for technical reasons, we also require ρ to be computable and $k/\rho(k)$ to be nondecreasing). If such an algorithm existed, then the maximum clique problem would be *fpt approximable*. It is an open problem whether the clique problem is fpt approximable; unfortunately the strong known inapproximability results for the clique problem do not shed any light on this question. Note that when we go beyond a constant approximation ratio we express the ratio as a function of the cost of the solution rather than the size of the instance, as it is usually done in the theory of approximation algorithms. This is reasonable because in our parameterized setting we are mainly interested in solutions that are very small compared to the size of the instance.

Our main contribution is a framework for studying such questions. We define fpt approximability for maximisation and minimisation problems and show that our notions are fairly robust. We also consider a decision version of the fpt approximability problem that we call *fpt cost approximability*, where instead of computing a solution of cost approximately k, an algorithm only has to decide if such a solution exists. We observe that a few known results yield fpt approximation algorithms: Oum and Seymour [17] showed that the problem of finding a clique decomposition of minimum width is fpt approximable. Based on results due to Seymour [19], Even et al. [12] showed that the directed feedback vertex set problem is fpt approximable. Then it follows from a result due to Reed et al. [18] that the linear programming dual of the feedback vertex set problem, the vertex disjoint cycle problem, is fpt cost approximable. This is interesting because the standard parameterization of this maximisation problem is W[1]-hard.

The classes of the fundamental W-hierarchy of parameterized complexity theory are defined as closures of so called weighted satisfiability problems under fpt reductions. We prove that for all levels of the W-hierarchy, natural optimisation versions of the defining weighted satisfiability problems are not fpt approximable, not even fpt cost approximable, unless the corresponding level of the hierarchy collapses to FPT, the class of fixed-parameter tractable problems. Furthermore, we prove that the short halting problem, which is known to be W[1]-complete for single tape machines and W[2]-complete in general, is not fpt cost approximable unless W[2] = FPT.

As a final result, we show that every parameterized problem in NP is both equivalent to the standard parameterization of an optimisation problem that is fpt approximable and equivalent to the standard parameterization of an optimisation problem that is not fpt cost approximable; in other words: every parameterized complexity class above FPT that contains a problem in NP contains problems that are approximable and problems that are inapproximable.

Independently, Cai and Huang [4] and Downey, Fellows and McCartin [11] introduced similar frameworks of parameterized approximability.

Due to space limitations, proofs are omitted here, but can be found in the full version of this paper.

2 Preliminaries

$\mathbb{N}$ denotes the natural numbers (positive integers), $\mathbb{R}$ the real numbers, and $\mathbb{R}_{\geq 1}$ the real numbers greater than or equal to 1. We recall a few basic definitions on optimisation problems and parameterized complexity. For further background, we refer the reader to [2] and [10,14].

2.1 Optimisation Problems

In this paper we consider NP-optimisation problems O over a finite alphabet Σ consisting of triples $(\mathrm{sol}_O, \mathrm{cost}_O, \mathrm{goal}_O)$ where

1. sol_O is a function that associates to any input instance $x \in \Sigma^*$ the set of feasible solutions of x such that the relation $\{(x, y) \mid x \in \Sigma^* \text{ and } y \in \mathrm{sol}_O(x)\}$ is polynomially balanced and decidable in polynomial time;
2. cost_O is the measure function and is defined on the class $\{(x, y) \mid x \in \Sigma^*, \text{ and } y \in \mathrm{sol}_O(x)\}$; the values of cost_O are positive natural numbers and cost_O is polynomial time computable;
3. $\mathrm{goal}_O \in \{\max, \min\}$

The objective of an optimisation problem O is to find an optimal solution z for a given instance x, that is a solution z with $\mathrm{cost}_O(x, z) = \mathrm{opt}_O(x) := \mathrm{goal}_O\{\mathrm{cost}_O(x, y) \mid y \in \mathrm{sol}_O(x)\}$. If O is clear from the context, we omit the subscript and just write opt, sol, cost and goal.

2.2 Parameterized Problems

We represent decision problems over a finite alphabet Σ as sets $Q \subseteq \Sigma^*$ of strings. Let us briefly recall the basic definitions of parameterized problems that we will need:

1. A *parameterization* of Σ^* is a polynomial time computable mapping $\kappa : \Sigma^* \to \mathbb{N}$.
2. A *parameterized decision problem* is a pair (Q, κ) consisting of a set $Q \subseteq \Sigma^*$ and a parameterization κ of Σ^*.
3. An algorithm $\mathbb{A}$ with input alphabet Σ is an *fpt-algorithm with respect to κ* if there is a computable function $f : \mathbb{N} \to \mathbb{N}$ such that for every instance x the running time of $\mathbb{A}$ on this input x is at most $f(\kappa(x)) \cdot |x|^{O(1)}$.
4. A parameterized decision problem (Q, κ) is *fixed-parameter tractable* if there is an fpt-algorithm with respect to κ that decides Q. FPT denotes the class of all fixed-parameter tractable decision problems. In parameterized complexity theory the analogue to polynomial time reductions are *fpt-reductions*.

5. A parameterized decision problem (Q, κ) belongs to the class XP if there is a computable function $f : \mathbb{N} \to \mathbb{N}$ and an algorithm that decides if $x \in Q$ for a given $x \in \Sigma^*$ in at most $O(|x|^{f(\kappa(x))})$ steps.

An important class of parameterized problems is the class of weighted satisfiability problems. We look at weighted satisfiability problems for propositional formulas and circuits: A formula α is k-satisfiable if there exists a satisfying assignment that sets exactly k many variables to TRUE (this assignment has weight k). A circuit γ is k-satisfiable if there is a possibility of setting exactly k many input nodes to TRUE and getting the value TRUE at the output node (γ is satisfied by an input tuple of weight k). We are interested in special classes of propositional formulas, $\Gamma_{t,d}$ and $\Delta_{t,d}$, defined inductively for $t \geq 0$, $d \geq 1$ as follows:

$$\Gamma_{0,d} := \{\lambda_1 \wedge \ldots \wedge \lambda_c \mid c \in [d], \lambda_1, \ldots, \lambda_c \text{ literals}\},$$
$$\Delta_{0,d} := \{\lambda_1 \vee \ldots \vee \lambda_c \mid c \in [d], \lambda_1, \ldots, \lambda_c \text{ literals}\},$$
$$\Gamma_{t+1,d} := \{\bigwedge_{i \in I} \delta_i \mid I \text{ finite and nonempty}, \delta_i \in \Delta_{t,d} \text{ for all } i \in I\},$$
$$\Delta_{t+1,d} := \{\bigvee_{i \in I} \gamma_i \mid I \text{ finite and nonempty}, \gamma_i \in \Gamma_{t,d} \text{ for all } i \in I\}.$$

For a class Γ of propositional formulas or Boolean circuits, the parameterized weighted satisfiability problem for Γ is

---
p-WSAT(Γ)

        Input: $\gamma \in \Gamma$ and $k \in \mathbb{N}$.

    Parameter: k.

      Problem: Decide whether γ is k-satisfiable.

The problems p-WSAT($\Gamma_{t,d}$) with $t, d \geq 1$ are used to define the classes W[t] of the W-hierarchy: A parameterized problem (Q, κ) belongs to the class W[t] if there is a $d \geq 1$ such that (Q, κ) is fpt-reducible to p-WSAT($\Gamma_{t,d}$). In the same way, the weighted satisfiability problem p-WSAT(CIRC) for the class CIRC of all Boolean circuits defines the parameterized complexity class W[P]. It holds

$$\text{FPT} \subseteq \text{W}[1] \subseteq \text{W}[2] \subseteq \cdots \subseteq \text{W}[\text{P}] \subseteq \text{XP}$$

where FPT is known to be strictly contained in XP and all other inclusions are believed to be strict as well.

3 Parameterized Approximability

Definition 1. *Let O be an NP-optimisation problem over the alphabet Σ, and let $\rho : \mathbb{N} \to \mathbb{R}_{\geq 1}$ be a computable function. Let $\mathbb{A}$ be an algorithm that expects inputs $(x, k) \in \Sigma^* \times \mathbb{N}$.*

1. $\mathbb{A}$ *is a* parameterized approximation algorithm *for O with* approximation ratio ρ *if for every input* $(x, k) \in \Sigma^* \times \mathbb{N}$ *with* $\text{sol}(x) \neq \emptyset$ *that satisfies*

$$\begin{cases} \text{opt}(x) \geq k & \text{if goal} = \max, \\ \text{opt}(x) \leq k & \text{if goal} = \min, \end{cases} \qquad (\star)$$

$\mathbb{A}$ *computes a* $y \in \text{sol}(x)$ *such that*

$$\begin{cases} \text{cost}(x, y) \geq \dfrac{k}{\rho(k)} & \text{if goal} = \max, \\ \text{cost}(x, y) \leq k \cdot \rho(k) & \text{if goal} = \min. \end{cases} \qquad (\star\star)$$

For inputs $(x, k) \in \Sigma^* \times \mathbb{N}$ *not satisfying condition* $(\star)$, *the output of* $\mathbb{A}$ *can be arbitrary.*

2. $\mathbb{A}$ *is an* fpt approximation algorithm *for O with* approximation ratio ρ *if it is a parameterized approximation algorithm for O with approximation ratio ρ and an fpt-algorithm with respect to the parameterization $(x, k) \mapsto k$ of its input space (that is, the running time of $\mathbb{A}$ is $f(k) \cdot |x|^{O(1)}$ for some computable function f).*
   $\mathbb{A}$ *is a* constant fpt approximation algorithm *for O if there is a constant $c \geq 1$ such that $\mathbb{A}$ is an fpt approximation algorithm for O with approximation ratio $k \mapsto c$ (the constant function with value c).*

3. *The problem O is* fpt approximable with approximation ratio ρ *if there is an fpt approximation algorithm for O with approximation ratio ρ. The problem O is* fpt approximable *if it is fpt approximable with approximation ratio ρ for some computable function $\rho : \mathbb{N} \to \mathbb{R}_{\geq 1}$ such that*

$$\begin{cases} \dfrac{k}{\rho(k)} \text{ is unbounded and nondecreasing} & \text{if O is a maximisation problem,} \\ k \cdot \rho(k) \text{ is nondecreasing} & \text{if O is a minimisation problem} \end{cases}$$

O *is* constant fpt approximable *if there is a constant fpt approximation algorithm for O.*

Remark 2. Since it is decidable by an fpt-algorithm whether an output y is an element of $\text{sol}(x)$ that satisfies $(\star\star)$, we can assume that an fpt approximation algorithm always (that is, even if the input does not satisfy $(\star)$) either outputs a $y \in \text{sol}(x)$ that satisfies $(\star\star)$ or outputs a default value, say "reject". Let us call an fpt approximation algorithm that has this property *normalised.*

Remark 3. We have decided to let the approximation ratio ρ be a function of the parameter k, because this is what we are interested in here. One could easily extend the definition to approximation ratios ρ depending on the input size as well, or even to arbitrary functions $\rho : \Sigma^* \times \mathbb{N} \to \mathbb{R}$. A technical condition that should be imposed then is that ρ be computable by an fpt-algorithm with respect to the parameterization $(x, k) \mapsto k$.

An alternative approach to parameterized approximability could be to parameterize optimisation problems by the optimum, with the goal of designing efficient approximation algorithms for instances with a small optimum. Interestingly, for minimisation problems, this yields exactly the same notion of parameterized approximability, as the following proposition shows.

Proposition 4. *Let O be an NP-minimisation problem over the alphabet Σ, and let $\rho : \mathbb{N} \to \mathbb{R}_{\geq 1}$ be a computable function such that $k \cdot \rho(k)$ is nondecreasing. Then the following two statements are equivalent:*

1. *O has an fpt approximation algorithm with approximation ratio ρ.*
2. *There exists a computable function g and an algorithm $\mathbb{B}$ that on input $x \in \Sigma^*$ computes a solution $y \in \mathrm{sol}(x)$ such that $\mathrm{cost}(x, y) \leq \mathrm{opt}(x) \cdot \rho(\mathrm{opt}(x))$ in time $g(\mathrm{opt}(x)) \cdot |x|^{O(1)}$.*

For maximisation problems, our definition of fpt approximation algorithm does not coincide with the analogue of Proposition 4(2). Yet we do have an analogue of the implication (1) $\Rightarrow$ (2) of Proposition 4 for maximisation problems:

Proposition 5. *Let O be an NP-maximisation problem over the alphabet Σ, and let $\rho : \mathbb{N} \to \mathbb{R}_{\geq 1}$ be a computable function such that $k/\rho(k)$ is nondecreasing and unbounded.*

Suppose that O has an fpt approximation algorithm with approximation ratio ρ. Then there exists a computable function g and an algorithm $\mathbb{B}$ that on input $x \in \Sigma^*$ computes a solution $y \in \mathrm{sol}(x)$ such that $\mathrm{cost}(x, y) \geq \frac{\mathrm{opt}(x)}{\rho(\mathrm{opt}(x))}$ in time $g(\mathrm{opt}(x)) \cdot |x|^{O(1)}$.

The problem with the converse direction is best illustrated for NP-optimisation problems where the optimal value is always large (say, of order $\Omega(|x|)$ for every instance x). An example of such a problem is the maximum independent set problem on planar graphs. Then an algorithm $\mathbb{B}$ as in Proposition 5 trivially exists even for $\rho = 1$, because all NP-optimisation problems can be solved exactly in exponential time. But this does not seem to help much for finding a solution of size approximately k for a given, small value of k.

3.1   Cost Approximability

Sometimes, instead of computing an optimal solution of an optimisation problem O, it can be sufficient to just compute the cost of an optimal solution (called *evaluation problem* in [2]). This is equivalent to solving the *standard decision problem* associated with O: Given an instance x and a natural number k, decide whether

$$\begin{cases} \mathrm{opt}(x) \geq k & \text{if } O \text{ is a maximisation problem,} \\ \mathrm{opt}(x) \leq k & \text{if } O \text{ is a minimisation problem.} \end{cases}$$

If we parameterize the standard decision problem by the input number k, we obtain the *standard parameterization* of O:

> *Input:* $x \in \Sigma^*$, $k \in \mathbb{N}$.
> *Parameter:* k.
> *Problem:* Decide whether $\text{opt}(x) \geq k$ (if goal = max) or
> $\text{opt}(x) \leq k$ (if goal = min).

To simplify the notation, for the rest of this section we only consider maximisation problems. All definitions and results can easily be adapted to minimisation problems.

What if we only want to compute the cost of the optimal solution approximately, say, with ratio ρ? On the level of the decision problem, this means that we allow an algorithm that is supposed to decide if $\text{opt}(x) \geq k$ to err if k is close to the optimum. The following definition makes this precise:

Definition 6. *Let O be an NP-maximisation problem over the alphabet Σ, and let $\rho : \mathbb{N} \to \mathbb{R}_{\geq 1}$ be a computable function.*

Then a decision algorithm $\mathbb{A}$ is a parameterized cost approximation algorithm *for O with* approximation ratio ρ *if it satisfies the following conditions for all inputs $(x,k) \in \Sigma^* \times \mathbb{N}$:*

- *If $k \leq \dfrac{\text{opt}(x)}{\rho(\text{opt}(x))}$, then $\mathbb{A}$ accepts (x,k).*
- *If $k > \text{opt}(x)$, then $\mathbb{A}$ rejects (x,k).*

The notions of an fpt cost approximation algorithm *and a* constant fpt cost approximation algorithm *and of a problem being* (constant) fpt cost approximable *are defined accordingly.*

A parameterized cost approximation algorithm may be thought of as deciding a parameterized problem that approximates the standard parameterization of an optimisation problem. This is made precise in the following simple proposition:

Proposition 7. *Let O be an NP-maximisation problem over the alphabet Σ, and let $\rho : \mathbb{N} \to \mathbb{R}_{\geq 1}$ be a computable function such that $k/\rho(k)$ is nondecreasing and unbounded. Then the following two statements are equivalent:*

1. *O has an fpt cost approximation algorithm with approximation ratio ρ.*
2. *There exists a parameterized problem $(Q', \kappa') \in \text{FPT}$ with $Q' \subseteq \Sigma^* \times \mathbb{N}$ and with $\kappa' : \Sigma^* \times \mathbb{N} \to \mathbb{N}$ defined by $\kappa'(x,k) := k$ that "approximates" the standard parameterization (Q_O, κ_O) of O with approximation ratio ρ: Given input $(x,k) \in \Sigma^* \times \mathbb{N}$, if $(x,k) \in Q_O$ then $(x, \lfloor k/\rho(k) \rfloor) \in Q'$ and if $(x,k) \notin Q_O$ then $(x,k) \notin Q'$.*

Mike Fellows (in a recent Dagstuhl Seminar) proposed a taxonomy of hard parameterized problems which is based on their approximability. In his terminology, the standard parameterization of an optimisation problem is *good* if it is fixed-parameter tractable; it is *bad* if it is not good, but constant fpt cost approximable; it is *ugly* if it is not bad, but fpt cost approximable; otherwise, it is *hideous*.

Proposition 8. *Let O be an NP-maximisation problem over the alphabet Σ, and let $\rho : \mathbb{N} \rightarrow \mathbb{R}_{\geq 1}$ be a computable function such that $k/\rho(k)$ is nondecreasing and unbounded.*

Suppose that O is fpt approximable with approximation ratio ρ. Then O is fpt cost approximable with approximation ratio ρ.

The following two propositions show that for cost approximability, we obtain a full analogue of Proposition 4 for maximisation problems and that the notion of fpt approximability is strictly stronger than that of fpt cost approximability:

Proposition 9. *Let O be an NP-maximisation problem over the alphabet Σ, and let $\rho : \mathbb{N} \rightarrow \mathbb{R}_{\geq 1}$ be a computable function such that $k/\rho(k)$ is nondecreasing and unbounded. Then the following two statements are equivalent:*

1. *O has an fpt cost approximation algorithm with approximation ratio ρ.*
2. *There exist a computable function g and an algorithm $\mathbb{B}$ that on input $x \in \Sigma^*$ computes an $\ell \in \mathbb{N}$ with $\mathrm{opt}(x) \geq \ell \geq \frac{\mathrm{opt}(x)}{\rho(\mathrm{opt}(x))}$ in time $g(\mathrm{opt}(x)) \cdot |x|^{O(1)}$.*

Proposition 10. *Assume that $\mathrm{NP} \cap \mathrm{co\text{-}NP} \neq P$. Then there exists an NP-optimisation problem that is fpt cost approximable but not fpt approximable.*

3.2   Examples

Example 11. MIN-CLIQUE-WIDTH is the problem of computing a decomposition of minimum clique-width for a given graph where Clique-width [7] is a graph parameter that is defined by a composition mechanism for vertex-labelled graphs and measures the complexity of a graph according to the difficulty of decomposing the graph into a kind of tree-structure. Given a decomposition of clique-width k (also called k-expression), many hard graph problems are solvable in polynomial time for graphs of bounded clique-width. Fellows et al. [13] recently proved that deciding whether the clique-width of $\mathcal{G}$ is at most k is NP-hard and that the minimisation problem MIN-CLIQUE-WIDTH cannot be absolutely approximated in polynomial time unless P = NP.

Oum and Seymour [17] defined the notion of rank-width to investigate clique-width and showed that $\mathrm{rwd}(\mathcal{G}) \leq \mathrm{cwd}(\mathcal{G}) \leq 2^{\mathrm{rwd}(\mathcal{G})+1} - 1$ for the clique-width $\mathrm{cwd}(\mathcal{G})$ and the rank-width $\mathrm{rwd}(\mathcal{G})$ of a given simple, undirected and finite graph $\mathcal{G}$. In [16], Oum presents two algorithms to compute rank-decompositions approximately: For a graph $\mathcal{G} = (V, E)$ and $k \in \mathbb{N}$ the algorithms either output a rank-decomposition of width at most $f(k)$ with $f(k) = 3k + 1$ or $f(k) = 24k$, respectively, or confirm that the rank-width is larger than k where the running time of these algorithms for fixed k is $O(|V|^4)$ for the first one and $O(|V|^3)$ for the second. Returning to clique-width there therefore exist algorithms that either output an $(2^{1+f(k)} - 1)$-expression or confirm that the clique-width is larger than k and that have the above running times for fixed k.

As both algorithms fulfil the properties of parameterized approximation algorithms with approximation ratio ρ defined by $\rho(k) := (2^{1+f(k)} - 1)/k$, we get that MIN-CLIQUE-WIDTH is fpt approximable.

Example 12. One of the major open problems in parameterized complexity is whether the following problem is fixed-parameter tractable.

---

p-DIRECTED-FEEDBACK-VERTEX-SET
  Input: A directed graph $\mathcal{G} = (V, E)$ and $k \in \mathbb{N}$.
Parameter: k.
  Problem: Decide whether there is a set $S \subseteq V$ with $|S| \leq k$ such
     that $\mathcal{G} \setminus S$ is acyclic.

Although still far from settling it, we note that the corresponding optimisation problem MIN-DIRECTED-FEEDBACK-VERTEX-SET is at least fpt approximable.

It is well-known that MIN-DIRECTED-FEEDBACK-VERTEX-SET can be described by the following integer linear program for a given directed graph $\mathcal{G} = (V, E)$, where x_v is a variable for each vertex $v \in V$:

$$\text{Minimise } \sum_{v \in V} x_v$$
$$\text{subject to } \sum_{v \in C} x_v \geq 1 \text{ for every cycle } C \text{ in } \mathcal{G}, \tag{1}$$
$$x_v \in \{0, 1\} \text{ for every vertex } v \in V.$$

We denote the minimum size of a feedback vertex set in a directed graph $\mathcal{G}$ by $\tau(\mathcal{G})$ and the size of a *fractional feedback vertex set* $(x_v)_{v \in V}$ with $0 \leq x_v \leq 1$ for every $v \in V$ by $\tau^*(\mathcal{G})$, where (1) without the integrality constraints can be solved in polynomial time (see e.g. [12]). Clearly we have $\tau^*(\mathcal{G}) \leq \tau(\mathcal{G})$ and Seymour [19] proved that the integrality gap of the feedback vertex set problem can be at most $O(\log \tau^* \cdot \log \log \tau^*)$. This proof can be modified to obtain a polynomial time approximation algorithm for MIN-DIRECTED-FEEDBACK-VERTEX-SET with an approximation ratio of $O(\log \tau^* \cdot \log \log \tau^*)$ [12]. Using Proposition 4 we conclude that MIN-DIRECTED-FEEDBACK-VERTEX-SET is fpt approximable.

Example 13. The linear programming dual of MIN-DIRECTED-FEEDBACK-VERTEX-SET is the optimisation problem MAX-DIRECTED-VERTEX-DISJOINT-CYCLES, whose standard parameterization is the following problem:

---

p-DIRECTED-VERTEX-DISJOINT-CYCLES
  Input: A directed graph $\mathcal{G}$ and $k \in \mathbb{N}$.
Parameter: k.
  Problem: Decide whether there are k vertex-disjoint cycles in $\mathcal{G}$.

It is implicit in [20] that p-DIRECTED-VERTEX-DISJOINT-CYCLES is W[1]-hard. For the maximum number $\nu(\mathcal{G})$ of vertex-disjoint cycles and for the minimum size $\tau(\mathcal{G})$ $(\tau^*(\mathcal{G}))$ of a (fractional) feedback vertex set in a given directed graph $\mathcal{G}$ it holds that $\nu(\mathcal{G}) \leq \tau^*(\mathcal{G}) \leq \tau(\mathcal{G})$. Furthermore, there is an upper bound of $\tau(\mathcal{G})$, in terms of $\nu(\mathcal{G})$ *only* as Reed et al. [18] proved the existence of a computable function $f : \mathbb{N} \cup \{0\} \to \mathbb{N}$, such that

$$\tau(\mathcal{G}) \leq f(\nu(\mathcal{G})) \tag{2}$$

for any directed graph $\mathcal{G}$. (The function f constructed in [18] is very large, a multiply iterated exponential, where the number of iterations is also a multiply iterated exponential; the best known lower bound is $f(x) \geq O(x \cdot \log x)$ for any $x \in \mathbb{N}$, a result attributed to Alon in [18].)

Together with the above inequalities, we can derive a very simple fpt cost approximation algorithm for MAX-VERTEX-DISJOINT-CYCLES: Let f be the function with property (2). Without loss of generality, we can assume f is increasing and time-constructible. Now let $\iota_f : \mathbb{N} \to \mathbb{N}$ be defined by $\iota_f(n) := \min\{i \in \mathbb{N} \mid f(i) \geq n\}$. Then ι_f is nondecreasing and unbounded, $\iota_f(n)$ is computable in time polynomial in n and $\iota_f(f(k)) \leq k$ for every $k \in \mathbb{N}$. Therefore we conclude

$$\iota_f(\nu(\mathcal{G})) \leq \iota_f(\lceil \tau^*(\mathcal{G}) \rceil) \leq \iota_f(\tau(\mathcal{G})) \leq \iota_f(f(\nu(\mathcal{G}))) \leq \nu(\mathcal{G}).$$

Thus the algorithm that given an input $(\mathcal{G}, k)$ computes $\tau^*(\mathcal{G})$ in time polynomial in the size of $\mathcal{G}$ and accepts if $k \leq \iota_f(\lceil \tau^*(\mathcal{G}) \rceil)$ and rejects otherwise is an fpt cost approximation algorithm with approximation ratio ρ where $\rho(k) = k/\iota_f(k)$.

4   Inapproximability Results

Under assumptions from parameterized complexity theory, the following theorem states the non-approximability of weighted satisfiability optimisation problems for the above defined classes of propositional formulas:

Theorem 14. MIN-WSAT($\Gamma_{t,d}$) with $t \geq 2$ and $d \geq 1$ is not fpt cost approximable unless W[t] = FPT, where the optimisation problem MIN-WSAT($\Gamma_{t,d}$) is defined as follows:

> Input: A propositional formula $\alpha \in \Gamma_{t,d}$.
> Solutions: All satisfying assignments for α.
> Cost: $\max\{1, \text{weight of a satisfying assignment}\}$.
> Goal: min.

Similarly as above we define the problem MIN-WSAT(CIRC) and maximisation versions of weighted satisfiability problems and get the following results:

Theorem 15. MIN-WSAT(CIRC) is not fpt cost approximable unless W[P] = FPT.

Theorem 16.

(1) MAX-WSAT($\Gamma_{t,d}$) with $t \geq 2$ and $d \geq 1$ is not fpt cost approximable unless W[t] = FPT.
(2) MAX-WSAT(CIRC) is not fpt cost approximable unless W[P] = FPT.

We now look at the following two versions MIN-SHORT-NTM-HALT and MIN-SHORT-NSTM-HALT of optimising halting problems:

> *Input:* A nondeterministic Turing machine $\mathbb{M}$.
> *Solutions:* All accepting runs of $\mathbb{M}$ on the empty string.
> *Cost:* The number of steps in such an accepting run.
> *Goal:* min.

> *Input:* A nondeterministic single-tape Turing machine $\mathbb{M}$.
> *Solutions:* All accepting runs of $\mathbb{M}$ on the empty string.
> *Cost:* The number of steps in such an accepting run.
> *Goal:* min.

The corresponding parameterized problems p-SHORT-NTM-HALT and p-SHORT-NSTM-HALT are to decide for a given nondeterministic (single-tape) Turing machine $\mathbb{M}$ and a given parameter $k \in \mathbb{N}$ whether $\mathbb{M}$ accepts the empty string in at most k steps, which is W[2]-complete [3] (W[1]-complete [5], respectively).

Theorem 17.   MIN-SHORT-NTM-HALT *is not fpt cost approximable unless* W[2] = FPT.

Theorem 18. MIN-SHORT-NSTM-HALT *is not fpt cost approximable unless* W[1] = FPT.

Recall the definition of the standard parameterization of an optimisation problem from page 114. The previous results show that each level of the W-hierarchy contains natural complete problems that are not fpt cost approximable. Our final result shows that artificial approximable and inapproximable problems of any given complexity can be constructed (as long as it is in NP, because we are dealing with NP-optimisation problems).

Theorem 19. *Let* (Q, κ) *be a parameterized problem not in* FPT *such that* $Q \in$ NP.

(1) (Q, κ) *is fpt equivalent to the standard parameterization of an NP-optimisation problem that is fpt approximable with approximation ratio 2.*

(2) (Q, κ) *is fpt equivalent to the standard parameterization of an NP-optimisation problem that is not fpt cost approximable.*

5   Further Research

Next to finding additional natural examples for problems with existing fpt approximation algorithms, the main open problem at the moment is to specify the parameterized approximability properties of basic problems like MIN-DOMINATING-SET or the MAX-CLIQUE problem already mentioned as an introductory example. Non-approximability results in the classical framework were proved for the CLIQUE problem using the PCP-theorem [1,8], so it might be necessary to obtain a parameterized version of the PCP-theorem to solve these questions.

References

1. S. Arora and S. Safra. Probabilistic checking of proofs: a new characterization of NP. *Journal of the ACM*, 45(1), pages 70–122, 1998.
2. G. Ausiello, P. Crescenzi, G. Gambosi, V. Kann, A. Marchetti-Spaccamela, and M. Protasi. *Complexity and Approximation*. Springer-Verlag, 2003.
3. L. Cai, J. Chen, R.G. Downey, and M.R. Fellows. On the parameterized complexity of short computation and factorization. *Archive for Mathematical Logic*, 36:321–337, 1997.
4. L. Cai and X. Huang. Fixed-Parameter Approximation: Conceptual Framework and Approximability Results. These proceedings.
5. M. Cesati and M. Di Ianni. Computation models for parameterized complexity. *Mathematical Logic Quarterly*, 43:179–202, 1997.
6. J. Chen, B. Chor, M. Fellows, X. Huang, D. Juedes, I. Kanj, and G. Xia. Tight lower bounds for certain parameterized NP-hard problems. *Proceedings of the 19th IEEE Conference on Computational Complexity*, pages 150–160, 2004.
7. B. Courcelle, J. Engelfriet, and G. Rozenberg. Context-free handle-rewriting hypergraph grammars. In H. Ehrig, H.-J. Kreowski, and G. Rozenberg, editors, *Graph-Grammars and their Application to Computer Science, 4th International Workshop*, Proceedings, volume 532 of Lecture Notes in Computer Science, pages 253–268, 1991.
8. I. Dinur. The PCP theorem by gap amplification. *Proceedings of STOC 2006*, 38th ACM Symposium on Theory of Computing, Seattle, Washington, USA, to appear.
9. R.G. Downey and M.R. Fellows. Fixed-parameter tractability and completeness II: On completeness for W[1]. *Journal of Theoretical Computer Science*, volume 141, pages 109–131, 1995.
10. R.G. Downey and M.R. Fellows. *Parameterized Complexity*. Springer-Verlag, 1999.
11. R.G. Downey, M.R. Fellows, and C. McCartin. Parameterized Approximation Algorithms. These proceedings.
12. G. Even, J. S. Naor, B. Schieber, and M. Sudan. Approximating minimum feedback sets and multicuts in directed graphs. *Algorithmica*, 20(2):151–174, 1998.
13. M. Fellows, F. Rosamond, U. Rotics, and S. Szeider. Clique-width minimization is NP-hard. *Proceedings of STOC 2006*, 38th ACM Symposium on Theory of Computing, Seattle, Washington, USA, to appear.
14. J. Flum and M. Grohe. *Parameterized Complexity Theory*. Springer-Verlag, 2006.
15. J. Håstad. Clique is hard to approximate within $n^{1-\varepsilon}$. *Electronic Colloquium on Computational Complexity*, Report TR97-038, 1997.
16. S. Oum. Approximating rank-width and clique-width quickly. *31th International Workshop on Graph-Theoretic Concepts in Computer Science, WG 2005*, volume 3787 of Lecture Notes in Computer Science, pages 49–58, 2005.
17. S. Oum and P. Seymour. Approximating clique-width and branch-width. *Journal of Combinatorial Theory*, Series B, to appear.
18. B. Reed, N. Robertson, P. Seymour, and R. Thomas. Packing directed circuits. *Combinatorica*, 16(4):535–554, 1996.
19. P. Seymour. Packing directed circuits fractionally. *Combinatorica*, 15(2):281–288, 1995.
20. A. Slivkins. Parameterized tractability of edge-disjoint paths on directed acyclic graphs. In G. D. Battista and U. Zwick editors, *Proceedings of 11th Annual European Symposium on Algorithms, ESA'03*, volume 2832 of Lecture Notes in Computer Science, pages 482–493, 2003.

Parameterized Approximation Problems

Rodney G. Downey[1], Michael R. Fellows[2], and Catherine McCartin[3]

[1] Victoria University, Wellington, New Zealand
Rod.Downey@mcs.vuw.ac.nz
[2] The University of Newcastle, Calaghan, Australia
mfellows@cs.newcastle.edu.au
[3] Massey University, Palmerston North, New Zealand
C.M.McCartin@massey.ac.nz

Abstract. Up to now, most work in the area of parameterized complexity has focussed on exact algorithms for decision problems. The goal of this paper is to apply parameterized ideas to approximation. We begin exploration of *parameterized approximation problems*, where the problem in question is a parameterized decision problem, and the required approximation factor is treated as a second parameter for the problem.

1 Introduction

Parameterized complexity is fast becoming accepted as an important strand in the mainstream of algorithm design and analysis, alongside approximation, randomization, and the like. It is fair to say that most of the work in the area has focussed on exact algorithms for decision problems. On the other hand it is clear that parameterized ideas have applications to many other questions of algorithmic design. For example, in [6] and [8] the ideas have been applied to counting problems and in [5], [8] the ideas were applied to online problems.

The goal of the present paper is to apply the ideas to approximation. Already we have seen that there are close ties between classical approximation and the theory of parameterized complexity.

For example, the following is now well-known. We can define a classical optimization problem to have an *efficient P-time approximation scheme* (EPTAS) if it can be approximated to a goodness of $(1 + \epsilon)$ of optimal in time $f(1/\epsilon)n^c$ where c is a constant. If we set $k = 1/\epsilon$ as the parameter, and then produce a reduction to the PTAS from some parametrically hard problem, we can, in essence, demonstrate that no such EPTAS exists [1], [3].

In this paper, we begin exploration of *parameterized approximation problems*, where the problem in question is a parameterized decision problem, and the required approximation factor is treated as a second parameter for the problem. Consider the following 'classic' parameterized problem:

k-INDEPENDENT SET
Input: A graph $G = (V, E)$
Parameter: k, a positive integer
Output: An independent set $V' \subseteq V$ for G of size at least k, or 'NO' if none such exists.

H.L. Bodlaender and M.A. Langston (Eds.): IWPEC 2006, LNCS 4169, pp. 121–129, 2006.

How might we define a problem that provides an 'approximate' solution to this problem? Here are two possibilities which we will consider in this paper. In both cases we relax our requirements by introducing a 'gap' between YES and NO solutions to the problem. In the first case the gap size is additive in the approximation parameter , in the second case the gap size is multiplicative in the approximation parameter.

ADD-APPROX k-INDEPENDENT SET
Input: A graph $G = (V, E)$
Parameters: k, c, positive integers
Output: 'NO', asserting that no independent set $V' \subseteq V$ of size $\geq k$ for G exists, or an independent set $V' \subseteq V$ for G of size at least $k - c$.

MULT-APPROX k-INDEPENDENT SET
Input: A graph $G = (V, E)$
Parameters: k, c, positive integers
Output: 'NO', asserting that no independent set $V' \subseteq V$ of size $\geq k$ for G exists, or an independent set $V' \subseteq V$ for G of size at least k/c.

The first parameterized approximation question is the the parameterized version of absolute approximability. The question is, are there parameterized algorithms to solve the above questions, in spite of our belief that there is no such algorithm for the exact problem? More generally, we will be considering the following class of questions. Our setting will be languages $L \subseteq \Sigma^* \times \Sigma^*$ We state the following for maximization problems, the analogous definition would work for minimization.

$g(k)$-APPROXIMATION
Input $\langle x, k \rangle$
Parameter k, g
Output 'NO' asserting $\langle x, k \rangle \notin L$ or $\langle x, k' \rangle \in L$ for some $k' \leq g(k)$.

As stated, the approximation problem above is for the version where we ask for the certificate $\langle x, k' \rangle$. There could also be a version where we simply ask for the 'YES' asserting that some such certificate exists. Since all the practical examples are self-reducible, we will get the certificates from the problem for free.

Notice that we can take a arbitrarily 'bad' language $L = \{\langle x, 2k \rangle : k \in \mathbb{N}\}$ and consider $L' = L \cup \{\langle x, 2k + 1 \rangle : x \in \Sigma^* \wedge k \in \mathbb{N}\}$. Then, in spite of the fact that $\langle x, m \rangle \in L'$ is as bad as you like, we can always have an approximation with $g(m) = m + 1$. The problem is that the odd parameters give *no* information. On the other hand, we believe that some natural problems are sufficiently well-structured so that, for certain functions g, the approximation schemes should give enough information so as to be able to solve the original problem.

In this paper we will consider different kinds of functions g. We first consider $g(k) = k + c$, absolute additive approximation. We demonstrate that for many of the basic $W[1]$-hard problems no such approximation scheme can exist unless $W[1] = FPT$. These problems include k-INDEPENDENT SET, k-CLIQUE and

k-STEP TURING MACHINE ACCEPTANCE. We also demonstrate that no such approximation scheme can exist for k-DOMINATING SET unless $W[2] = FPT$.

Next we consider multiplicative and other values values for g. Notice that for instance, BIN PACKING parameterized by the number of bins, has (by First Fit) a natural approximation with $g(k) = 2k$, say. (See Garey and Johnson [7].) Thus there are natural problems with such multiplicative parameterized approximation schemes.

On the other hand we show that there exist problems where there is no approximation scheme for *any* function $g(k)$ unless $W[2] = FPT$. One example of this phenomenon is k-INDEPENDENT DOMINATING SET . That is, for any computable function $g(k) \geq k$, there is no algorithm which either asserts that there is no independent dominating set of size $\leq k$ for a given graph G, or otherwise asserts that there is one of size $\leq g(k)$. We call such problems *completely inapproximable*.

Up to this time there is no literature on this kind of approximation. The idea was introduced by Downey and Fellows in [4] for DOMINATING SET. As we were about to submit this paper, we were sent a copy of a new paper by Cai and Huang [2] also studying the same kind of problems; their results being quite complementary to ours. Interestingly, the fundamental problem we wished to classify, DOMINATING SET, remains open.

2   Additive Parameterized Approximation

We show that the following parameterized approximation problems are, in each case, reducible to the corresponding original parameterized problem: ADD-APPROX k-INDEPENDENT SET, ADD-APPROX k-CLIQUE, ADD-APPROX k-DOMINATING SET, ADD-APPROX k-STEP TURING MACHINE ACCEPTANCE.

Theorem 1. ADD-APPROX k-INDEPENDENT SET *is* $W[1]$-*hard.*

Proof. We transform from k-INDEPENDENT SET.

Let $G = (V, E)$ be a graph and let k be the parameter. We produce $G' = (V', E')$ such that G' has a c-additive approximate solution for dk-INDEPENDENT SET, i.e. G' contains an independent set of size at least $dk - c$, iff G contains an independent set of size at least k.

To build G' we begin with the original graph G, and proceed as follows:

1. Find smallest d such that $\lceil \frac{dk-c}{d} \rceil \geq k$
2. G' consists of d separate copies of G

$\Leftarrow$ Suppose that G contains an independent set of size at least k, then, by the construction of G', there must be an independent set of size at least dk in G'.
$\Rightarrow$ Suppose that G' contains an independent set of size at least $dk - c$, then some copy of G in G' must contain an independent set of size at least k, by the choice of d. $\qquad \square$

This simple *amplification technique* can be used in parallel fashion to show:

Theorem 2. ADD-APPROX k-CLIQUE *is* $W[1]$-*hard.*

The case of ADD-APPROX k-DOMINATING SET also employs the amplification technique, except now we are looking to minimize, rather than maximize, the solution.

Theorem 3. ADD-APPROX k-DOMINATING SET *is* $W[2]$-*hard.*

Proof. We transform from k-DOMINATING SET.

Let $G = (V, E)$ be a graph and let k be the parameter. We produce $G' = (V', E')$ such that G' has a c-additive approximate solution for dk-DOMINATING SET, i.e. G' contains a dominating set of size at most $dk + c$, iff G contains a dominating set of size at most k.

To build G' we begin with the original graph G, and proceed as follows:

1. Find smallest d such that $\lfloor \frac{dk+c}{d} \rfloor \leq k$
2. G' consists of d separate copies of G

$\Leftarrow$ Suppose that G contains a dominating set of size at most k, then, by the construction of G', there must be a dominating set of size at most dk in G'.
$\Rightarrow$ Suppose that G' contains a dominating set of size at most $dk + c$, then some copy of G in G' must contain a dominating set of size at most k, by the choice of d. $\qquad\square$

We now consider additive approximation for k-TURING MACHINE ACCEPTANCE, the problem of deciding if a nondeterministic Turing machine with arbitrarily large fanout has a k-step accepting path on the empty input string.

ADD-APPROX k-TURING MACHINE ACCEPTANCE
Input: A Turing machine M
Parameters: k, c, positive integers
Output: 'NO' asserting that no k-step accepting path for M exists, or an accepting path of length at most $k + c$ for M.

Theorem 4. ADD-APPROX k-TURING MACHINE ACCEPTANCE *is* $W[1]$-*hard.*

Proof. We transform from k-TURING MACHINE ACCEPTANCE.

Let M be a Turing machine and let k be the parameter. We define M' such that M' has a c-additive approximate solution for $dk + 1$-TURING MACHINE ACCEPTANCE, iff M has an accepting path of length at most k.

Choose $d \gg c$. Choose an alphabet for M' sufficiently large such that all d-sets of symbols from the alphabet for M may be represented. On the empty input string M' runs d copies of M in parallel, repeating each step of the computation for M d times, before proceeding to the next. M' will halt and accept immediately that all copies of M have halted and accepted.

$\Leftarrow$ Suppose that M has an accepting path of length at most k, then, by the construction of M', there must be an accepting path of length at most $dk + 1$ for M'.

$\Rightarrow$ Suppose that M' has an accepting path of length at most $dk + 1 + c$, then some copy of M run by M' must have an accepting path of length at most k, by the choice of d. $\square$

3 A Completely Inapproximable Parameterized Problem

In this section we show that k-INDEPENDENT DOMINATING SET is *completely inapproximable*. Specifically, we show that there is no approximation scheme for k-INDEPENDENT DOMINATING SET for *any* function $g(k)$ unless $W[2] = FPT$.

The natural parameterized version of the DOMINATING SET problem is the following.

k-DOMINATING SET
Input: A graph G.
Parameter: A positive integer k.
Question: Does G have a dominating set of size k? (A dominating set for G is a set $X \subseteq V(G)$ such that for all $y \in V(G)$, there is an $x \in X$ with $\langle x, y \rangle \in E(G)$.)

In [4] Downey and Fellows show that k-DOMINATING SET is $W[2]$-hard via a transformation from WEIGHTED CNF SATISFIABILITY.

For X a Boolean expression in conjunctive normal form consisting of m clauses $C_1, ..., C_m$ over the set of n variables $x_0, ..., x_{n-1}$, they show how to produce in polynomial-time by local replacement, a graph $G = (V, E)$ that has a dominating set of size $2k$ if and only if X is satisfied by a truth assignment of weight k.

The size $2k$ dominating set in G corresponding to a weight k truth assignment for X, is in fact an independent set as well. Thus the same transformation shows that k-INDEPENDENT DOMINATING SET is $W[2]$-hard.

We outline the construction of the graph G used in the reduction here. There are k gadgets arranged in a vertical line. Each of the gadgets has 3 main parts. Taken from top to bottom, these are variable selection, gap selection and gap and order enforcement. The variable selection component $A(r)$ is a clique and the gap selection component $B(r)$ consists of n cliques which are called columns. The first action is to ensure that in *any* dominating set of $2k$ elements, we must pick one vertex from each of these two components. This goal is achieved by $2k$ sets of $2k + 1$ enforcers, vertices from V_4 and V_5. (The names refer to the sets below.) Take the V_4, for instance. For a fixed r, these $2k + 1$ vertices are connected to all of the variable selection vertices in the component $A(r)$, and nowhere else. Thus if they are to be dominated by a $2k$ dominating set, then we must choose *some* element in the set $A(r)$, and similarly we must choose an element in the set $B(r)$ by virtue of the V_5 enforcers. Since we will need exactly $2k$ (or even $\leq 2k$) dominating elements it follows that we must pick *exactly* one from each of the $A(r)$ and $B(r)$ for $r = 1, ..., k$.

Each of the k variable selection components consists of a clique of n vertices labelled $0, ..., n - 1$. The intention being that the vertex labelled i represents a choice of variable i being made true in the formula X. Correspondingly in the next $B(r)$ we have columns (cliques) $i = 0, ..., n-1$. The intention is that column i corresponds to the choice of variable i in the preceding $A(r)$. We join the vertex $a[r, i]$ corresponding to variable i, in $A(r)$, to all vertices in $B(r)$ *except* those in column i. This means that the choice of i in $A(r)$ will cover all vertices of $B(r)$ except those in this column. It follows that we *must* choose the dominating element from this column and nowhere else. (There are no connections from column to column.) The columns are meant to be the gap selection saying how many 0's there will be till the next positive choice for a variable. We finally need to ensure that (i) if we chose variable i in $A(r)$ and gap j in column i from $B(r)$ then we need to pick $i+j+1$ in $A(r+1)$ and (ii) that the selections are in order. This is the role of the gap and order enforcement component which consists of a set of n vertices (in V_6.)

Thus the above provides a selection gadget that chooses k true variables with the gaps representing false ones. We enforce that the selection is consistent with the clauses of X via the clause variables V_3. These are connected in the obvious ways. One connects a choice in $A[r]$ or $B[r]$ corresponding to making a clause C_q true to the vertex c_q. Then if we dominate all the clause vertices too, we must have either chosen in some $A[r]$ a positive occurrence of a variable in C_q or we must have chosen in $B[r]$ a gap corresponding to a negative occurrence of a variable in C_q, and conversely.

The vertex set V of G is the union of the following sets of vertices:
$V_1 = \{a[r, s] : 0 \leq r \leq k - 1, 0 \leq s \leq n - 1\}$
$V_2 = \{b[r, s, t] : 0 \leq r \leq k - 1, 0 \leq s \leq n - 1, 1 \leq t \leq n - k + 1\}$
$V_3 = \{c[j] : 1 \leq j \leq m\}$
$V_4 = \{a'[r, u] : 0 \leq r \leq k - 1, 1 \leq u \leq 2k + 1\}$
$V_5 = \{b'[r, u] : 0 \leq r \leq k - 1, 1 \leq u \leq 2k + 1\}$
$V_6 = \{d[r, s] : 0 \leq r \leq k - 1, 0 \leq s \leq n - 1\}$

For convenience, we introduce the following notation for important subsets of some of the vertex sets above.
$A(r) = \{a[r, s] : 0 \leq s \leq n - 1\}$
$B(r) = \{b[r, s, t] : 0 \leq s \leq n - 1, 1 \leq t \leq n - k + 1\}$
$B(r, s) = \{b[r, s, t] : 1 \leq t \leq n - k + 1\}$

The edge set E of G is the union of the following sets of edges. In these descriptions we implicitly quantify over all possible indices.
$E_1 = \{c[j]a[r, s] : x_s \in C_j\}$
$E_2 = \{a[r, s]a[r, s'] : s \neq s'\}$
$E_3 = \{b[r, s, t]b[r, s, t'] : t \neq t'\}$
$E_4 = \{a[r, s]b[r, s', t] : s \neq s'\}$
$E_5 = \{b[r, s, t]d[r, s'] : s + t + 1 \leq n \wedge s' \neq s + t\} \cup \{b[k - 1, s, t]d[k - 1, s'] : s' \neq s + t(\mathrm{mod}\,n)\}$
$E_6 = \{a[r, s]a'[r, u]\}$
$E_7 = \{b[r, s, t]b'[r, u]\}$
$E_8 = \{c[j]b[r, s, t] : \exists i \; \overline{x_i} \in C_j, s < i < s + t\}$
$E_9 = \{d[r, s]a[r', s] : r' = r + 1 \bmod k\}$

Suppose X has a satisfying truth assignment τ of weight k, with variables $x_{i_0}, x_{i_1}, ..., x_{i_{k-1}}$ assigned the value *true*. Suppose $i_0 < i_2 < ... < i_{k-1}$. Let $d_r = i_{r+1(\text{mod }k)} - i_r \pmod{n}$ for $r = 0, ..., k - 1$. It is straightforward to verify that the set of $2k$ vertices

$$D = \{a[r, i_r] : 0 \leq r \leq k - 1\} \cup \{b[r, i_r, d_r] : 0 \leq r \leq k - 1\}$$

is a dominating set in G.

Conversely, suppose D is a dominating set of $2k$ vertices in G. The closed neighbourhoods of the $2k$ vertices $a'[0, 1], ..., a'[k - 1, 1], b'[0, 1], ..., b'[k - 1, 1]$ are disjoint, so D must consist of exactly $2k$ vertices, one in each of these closed neighbourhoods. Also, none of the vertices of $V_4 \cup V_5$ are in D, since if $a'[r, u] \in D$ then necessarily $a'[r, u'] \in D$ for $1 < u' < 2k + 1$ (otherwise D fails to be dominating), which contradicts that D contains exactly $2k$ vertices. It follows that D contains exactly one vertex from each of the sets $A(r)$ and $B(r)$ for $0 \leq r \leq k - 1$.

The possibilities for D are further constrained by the edges of E_4, E_5 and E_9. The vertices of D in V_1 represent the variables set to *true* in a satisfying truth assignment for X, and the vertices of D in V_2 represent intervals of variables set to *false*. Since there are k variables to be set to *true* there are, considering the indices of the variables mod n, also k intervals of variables to be set to *false*. Furthermore the set E_5 forces the chosen variables to be chosen so that if $r < r'$ and we choose $a[r, q]$ and $a[r', q']$ then $q < q'$.

The edges of E_4, E_5 and E_9 enforce that the $2k$ vertices in D must represent such a choice consistently. It remains only to check that the fact that D is a dominating set ensures that the truth assignment represented by D satisfies X. This follows by the definition of the edge sets E_1 and E_8.

We modify the construction described above to show that the following parameterized approximation problem is $W[2]$-hard for *any* choice of $g(k)$.

$g(k)$-APPROX INDEPENDENT DOMINATING SET
Input: $G = (V, E)$
Parameter: k, g
Output: 'NO' asserting that no independent dominating set $V' \subseteq V$ of size $\leq k$ for G exists, or an independent dominating set $V' \subseteq V$ for G of size at most $g(k)$.

Theorem 5. $g(k)$-APPROX INDEPENDENT DOMINATING SET *is* $W[2]$-*hard.*

Proof. We transform from WEIGHTED CNF SATISFIABILITY.

Given X, a Boolean expression in conjunctive normal form we construct a graph G that has a $g(k)$ approximate solution for $2k$-INDEPENDENT DOMINATING SET if and only if X is satisfied by a truth assignment of weight k.

To build G we begin with the construction from [4] described in detail above. We single out *one* of the variable selection components $A(q)$ and add to the construction $k' = g(k) - 2k + 1$ new variable selection components, $G_1, G_2, \ldots, G_{k'}$, along with new edges between all vertices in each of these new components and all vertices in $A(q)$. Each of the new components is connected to $B(q)$, to the

$(q - 1 \bmod k)$ gap and order enforcement component, and to the clause vertices of $V_3 = \{c[j] : 1 \leq j \leq m\}$ in exactly the same way as is $A(q)$. We modify the edge sets E_1, E_4 and E_9 accordingly.

For each of the new variable selection components except the last, that is G_i, $1 \leq i < k'$, we connect the jth vertex, $0 \leq j \leq n - 1$, in G_i, by an edge to all vertices in G_{i+1} except for the jth vertex in G_{i+1}.

Finally, we blow up the size of the enforcement vertex sets V_4, V_5 and V_6 so that
$$V_4 = \{a'[r, u] : 0 \leq r \leq k - 1, 1 \leq u \leq g(k) + 1\}$$
$$V_5 = \{b'[r, u] : 0 \leq r \leq k - 1, 1 \leq u \leq g(k) + 1\}$$
$$V_6 = \{d[r, s, t] : 0 \leq r \leq k - 1, 0 \leq s \leq n - 1, 1 \leq t \leq g(k) + 1\}$$
and modify the edge sets E_5, E_6, E_7 and E_9 accordingly.

The construction is illustrated in Fig. 1.

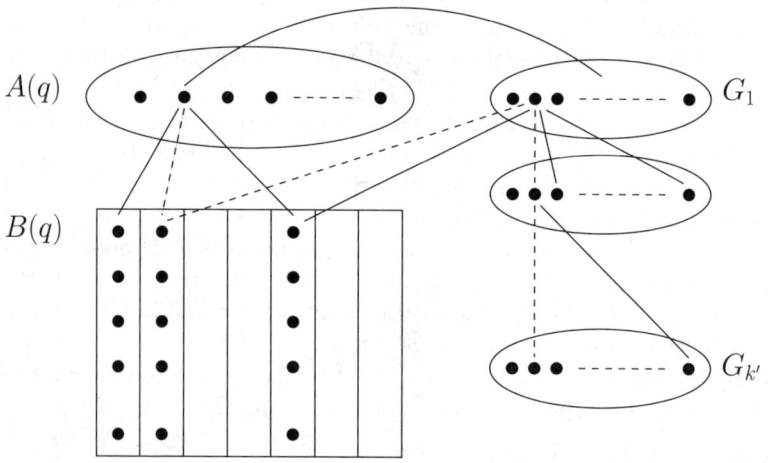

Fig. 1. Gadget for $g(k)$-Approx Independent Dominating Set

$\Leftarrow$ Suppose X has a satisfying truth assignment τ of weight k, with variables $x_{i_0}, x_{i_1}, ..., x_{i_{k-1}}$ assigned the value *true*. Suppose $i_0 < i_2 < ... < i_{k-1}$. Let $d_r = i_{r+1 (\bmod k)} - i_r \pmod{n}$ for $r = 0, ..., k - 1$. It is straightforward to verify that the set of $2k$ vertices

$$D = \{a[r, i_r] : 0 \leq r \leq k - 1\} \cup \{b[r, i_r, d_r] : 0 \leq r \leq k - 1\}$$

is an independent dominating set in G.

$\Rightarrow$ Suppose that G contains an independent dominating set D of size at most $g(k)$.

There are two possibilities here. In the first case, D has size $2k$ and contains exactly one vertex from each of the sets $A(r)$ and $B(r)$ for $0 \leq r \leq k - 1$. The closed neighbourhoods of the $2k$ vertices $a'[0, 1], ..., a'[k - 1, 1], b'[0, 1], ..., b'[k - 1, 1]$ are disjoint, so D must consist of at least $2k$ vertices, one in each of these closed neighbourhoods. Since D is independent we can choose at most one from each of the $A(r)$, $0 \leq r \leq k-1$, and at most one from each of the $B(r)$, $0 \leq r \leq k-1$ (in the correct non-dominated column.) Also, none of the vertices of $V_4 \cup V_5$ are in D, since if $a'[r,u] \in D$ then necessarily $a'[r,u'] \in D$ for $1 < u' < g(k)+1$ (otherwise D fails to be dominating), which contradicts that D contains at most $g(k)$ vertices. None of the vertices in V_6 are in D by a similar argument. Finally, if D contains a vertex from $A(q)$ then, since G is independent, none of the vertices in the new variable selection components, $G_i, 1 \leq i \leq k'$ are in D. It follows that D contains exactly one vertex from each of the sets $A(r)$ and $B(r)$ for $0 \leq r \leq k-1$.

In the second case, D has size $g(k)$ and contains exactly one vertex from each of the sets $A(r)$, $0 \leq r \leq k-1, r \neq q$, and $B(r)$, $0 \leq r \leq k-1$, and exactly one vertex from each of the sets $G_i, 1 \leq i \leq k'$. In the case of the G_i sets, if D contains the jth vertex of G_1, then D contains the jth vertex of each of the other $G_i, 2 \leq i \leq k'$. Note that, as in the first case, none of the vertices of $V_4 \cup V_5$ are in D, since if $a'[r,u] \in D$ then necessarily $a'[r,u'] \in D$ for $1 < u' < g(k)+1$ (otherwise D fails to be dominating), which contradicts that D contains at most $g(k)$ vertices. None of the vertices in V_6 are in D by a similar argument. Finally, if D contains a vertex from any of the G_i then, since G is independent, none of the vertices in $A(q)$ can be in D.

In either case, the truth assignment represented by D satisfies X. This follows by the definition of the edge sets E_1 (modified) and E_8. □

References

1. C. Bazgan: *Schemas d'approximation et complexite parametree*. Rapport de DEA, Universite Paris Sud, 1995.
2. L. Cai, X. Huang: *Fixed parameter Approximation: Conceptual Framework and Approximability Results*. Manuscript.
3. M. Cesati, L. Trevisan: *On the efficiency of polynomial approximation schemes*. Information Processing Letters, 64(4), pp 165-171, 1997.
4. R. G. Downey, M. R. Fellows: *Parameterized Complexity* Springer-Verlag, 1999.
5. R. G. Downey, C. M. McCartin: *Online Problems, Pathwidth, and Persistence*. Proceedings of IWPEC 2004, Springer-Verlag LNCS 3162, pp 13-24, 2004.
6. J. Flum, M. Grohe: *The Parameterized Complexity of Counting Problems*. SIAM Journal on Computing, 33(4), pp 892-922, 2004.
7. M. R. Garey and D. S. Johnson: *Computers and Intractability: A Guide to the Theory of NP-completeness* Freeman, New York, 1979.
8. C. M. McCartin: *Contributions to Parameterized Complexity* Ph.D. Thesis, Victoria University, Wellington, 2003.

An Exact Algorithm for the Minimum Dominating Clique Problem

Dieter Kratsch and Mathieu Liedloff

LITA, Université Paul Verlaine - Metz, 57045 Metz Cedex 01, France
{kratsch, liedloff}@univ-metz.fr

Abstract. A subset of vertices $D \subseteq V$ of a graph $G = (V, E)$ is a dominating clique if D is a dominating set and a clique of G. The existence problem 'Given a graph G, is there a dominating clique in G?' is NP-complete, and thus both the Minimum and the Maximum Dominating Clique problem are NP-hard. We present an $O(1.3390^n)$ time algorithm that for an input graph on n vertices either computes a minimum dominating clique or reports that the graph has no dominating clique. The algorithm uses the Branch & Reduce paradigm and its time analysis is based on the Measure & Conquer approach. We also establish a lower bound of $\Omega(1.2599^n)$ for the worst case running time of our algorithm.

1 Introduction

In the last decades of the 20th century the early research on exact exponential time algorithms for NP-hard problems had been concentrating on satisfiability problems (see e.g. [12,20]). Nevertheless, some interesting exponential time algorithms for graph problems had also been established during this period. In the last years the design and analysis of exact exponential time algorithms for NP-hard problems has gone through an exciting growth of interest. Various NP-hard graph problems have attracted attention. For some of them, such as Independent Set, Coloring and Hamiltonian Circuit, exact algorithms had been studied since a long time [19,21,15,11]. Other problems, such as Dominating Set, Treewidth and Feedback Vertex Set, have not been considered under this perspective until very recently [3,22,18].

For more details we refer to the following surveys on exact algorithms: [4,12,20,23,24]. In an important survey, Woeginger presents fundamental techniques to design and analyse exact exponential time algorithms [23]. In a survey by Fomin et al. various techniques for the design and analysis of Branch & Reduce algorithms are discussed, among them Measure & Conquer, Lower Bounds and Memorization [4].

In this paper we study the Minimum Dominating Clique problem (abbr. MinDC) and present an exact Branch & Reduce algorithm to solve it.

Basic Definitions. Let $G = (V, E)$ be an undirected and simple graph, i.e. without loops and multiple edges. We denote by n the number of vertices of G. The open neighborhood of a vertex v is denoted by $N(v) = \{u \in V : \{u, v\} \in E\}$, and the closed neighborhood of v is denoted by $N[v] = N(v) \cup \{v\}$. The degree

H.L. Bodlaender and M.A. Langston (Eds.): IWPEC 2006, LNCS 4169, pp. 130–141, 2006.

of a vertex v is $|N(v)|$. The subgraph of G induced by $S \subseteq V$ is denoted by $G[S]$. We will write $G - S$ short for $G[V - S]$. A set $S \subseteq V$ of vertices is a *clique*, if any two of its vertices are adjacent; S is *independent* if any two of its vertices are non adjacent.

For a vertex set $S \subseteq V$, we define $N[S] = \bigcup_{v \in S} N[v]$ and $N(S) = N[S] - S$. We also define $N_S(v) = N(v) \cap S$ and $N_S[v] = N[v] \cap S$. The S-degree of v, denoted by $d_S(v)$, is $|N_S(v)|$. Similarly, given two subsets of vertices $S \subseteq V$ and $X \subseteq V$, we define $N_S(X) = N(X) \cap S$. We will write $\overline{N_S[v]}$ for $S \setminus N_S[v]$, and $\overline{N_S(v)}$ for $\overline{N_S[v]} \cup \{v\}$.

Domination. Let $G = (V, E)$ be a graph. A set $D \subseteq V$ with $N[D] = V$ is called a *dominating set* of G; in other words, every vertex in G either belongs to D or has a neighbor in D. The Minimum Dominating Set problem asks to find a dominating set of minimum cardinality. It is one of the fundamental and well-studied NP-hard graph problems [6]. Various variants of the Dominating Set problem have been studied extensively. Dominating sets D of a graph G are often required to have particular additional properties, as e.g. to be an independent set or to be a clique of G. For such particular types of dominating sets the related problem asking to find such a set of minimum cardinality is often NP-hard. For a large and comprehensive survey on domination theory, we refer the reader to the books [9,10] by Haynes, Hedetniemi and Slater.

Dominating Cliques. A subset of vertices $D \subseteq V$ of a graph $G = (V, E)$ is a *dominating clique* if D is a dominating set and a clique. The study of dominating cliques was initiated in [1] with a motivation from social sciences. The existence problem 'Given a graph G, decide whether G has a dominating clique' (abbr. ExDC) is NP-complete even when restricted to weakly triangulated graphs or when restricted to cocomparability graphs [14]. This implies that both natural optimization versions of the problem are NP-hard. The Minimum Dominating Clique problem (abbr. MinDC) asks either to find a dominating clique of minimum cardinality of the input graph G, or to output that G has no dominating clique. The Maximum Dominating Clique problem (abbr. MaxDC) asks to find a dominating clique of maximum cardinality, or to output that there is none. The complexity of MinDC on various graph classes has been studied in the eighties and nineties (see [14]).

Related Results. Exact algorithms for the Minimum Dominating Set problem have been studied in a sequence of papers [5,17,8,3]. The fastest known algorithms today are due to Fomin et al. and they are based on a Branch & Reduce algorithm for the Minimum Set Cover problem. Their running time is $O(1.5263^n)$ using polynomial space and $O(1.5137^n)$ using exponential space [3]. The analysis is based on Measure & Conquer and the exponential space algorithm is obtained using memorization.

The above results and the power of the Branch & Reduce paradigm when combined with an analysis using Measure & Conquer, motivate studying exact algorithms for other domination problems. For example, Gaspers et al. established an $O(1.3575^n)$ time Branch & Reduce algorithm to compute a minimum independent dominating set [7].

Prior to our work, the only published exact algorithm solving a dominating clique problem is due to Culberson et al. and it solves the existence problem ExDC [2]. Their exact Branch & Reduce algorithm is a by-product and it is used for experimental studies and no analysis of the worst case running time is attempted.

There is also a simple $3^{n/3} n^{O(1)} = O(1.4423^n)$ time algorithm for ExDC or MaxDC enumerating all maximal cliques and verifying each for being a dominating set. It uses a polynomial delay algorithm to generate all maximal independent sets [13], and its running time follows from Moon and Moser's result [16] that the number of maximal independent sets of a graph is at most $3^{n/3}$.

Our Results. We present an $O(1.3390^n)$ time and polynomial space algorithm computing a minimum dominating clique of the input graph G, or reporting that G has no dominating clique. Thus our algorithm also solves the existence problem ExDC. It is designed using the Branch & Reduce paradigm aiming for simple reduction and branching rules. To analyse the worst case running time of the algorithm we use a non standard measure for the size of an input of a (sub)problem and we rely heavily on the Measure & Conquer technique. The theoretical analysis will be described in full detail. The numerical solution of a system of about 400 linear recurrences each one depending on up to 7 parameters is impossible without a computer. We use a program based on random local search and obtain an upper bound of $O(1.3390^n)$ for the worst case running time.[1]

Since current tools for the time analysis of Branch & Reduce algorithms (including Measure & Conquer) seem to overestimate the running time of the algorithm, lower bounds for the worst case running time are of interest. We show that the worst case running time of our algorithm is $\Omega(1.2599^n)$.

2   The Algorithm of Culberson et al.

In [2] Culberson et al. propose the Branch & Reduce algorithm DomClq to decide whether a given graph has a dominating clique or not. However the main interest of their work is in determining the phase transition of the existence problem ExDC on random graphs in the classical $G_{n,p}$ model. The algorithm DomClq has been developed for experimental studies. Culberson et al. report on experiments for random graphs with up to 1000 vertices (and p close to the threshold $\frac{3-\sqrt{5}}{2}$). A theoretical analysis of the worst case running time had not been attempted.

It is natural to compare the algorithm DomClq and our algorithm. However as Fomin et al. point out in [3], upper bounds of Branch & Reduce algorithms

---

[1] While determining optimal values of our parameters needs computing power, given the recurrences and the optimal values of the parameters, verification of correctness is relatively easy.

are likely to overestimate the worst case running time. Thus comparing upper bounds of the worst case running time of both algorithms might lead to wrong conclusions. Here lower bounds will be of great help.

Theorem 1. *The worst case running time of algorithm DomClq is* $\Omega(1.4142^n)$.

Proof. The graphs used to demonstrate the lower bound are denoted by $G_{k,\ell} = (V_{k,\ell}, E_{k,\ell})$, $k, \ell \geq 1$. Their vertex set is $V_{k,\ell} = \{w, v, a, b\} \cup \{x_i : 0 \leq i \leq \ell\} \cup \{y_i : 1 \leq i \leq \ell\} \cup \{u_i : 1 \leq i \leq k\} \cup \{s_i : 1 \leq i \leq k + \ell - 1\}$. Thus $G_{k,\ell}$ has $n = 2k + 3\ell + 4$ vertices. The edge set of $G_{k,\ell}$ is $E_{k,\ell} = \{\{w, v\}\} \cup \{\{v, x_i\} : 0 \leq i \leq \ell\} \cup \{\{v, y_i\} : 1 \leq i \leq \ell\} \cup \{\{v, s_i\} : 1 \leq i \leq k + \ell - 1\} \cup \bigcup_{i=1}^{k} \{\{u_i, s_j\} : i \leq j \leq i + \ell - 1\} \cup \{\{a, x_i\} : 0 \leq i \leq \ell\} \cup \{\{b, y_i\} : 1 \leq i \leq \ell\}$. Finally all pairs of vertices in the set $S = \{s_i : 1 \leq i \leq k + \ell - 1\} \cup \{x_i : 0 \leq i \leq \ell\} \cup \{y_i : 1 \leq i \leq \ell\}$ are adjacent in $G_{k,\ell}$ with the exception of the following pairs $\{\{s_i, x_0\} : i \bmod \ell \neq 0\} \cup \{\{x_i, y_j\} : 1 \leq i, j \leq \ell\}$. See also the graph $G_{k,5}$ in Fig. 1.

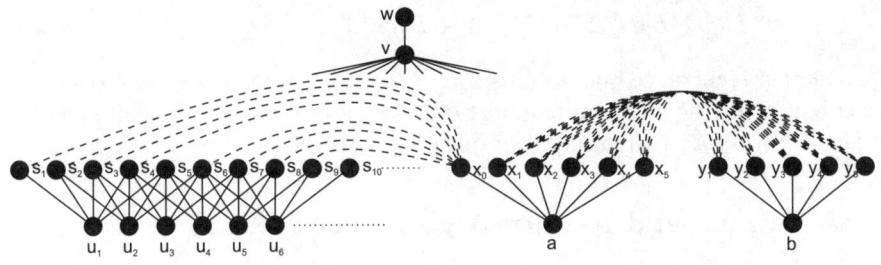

Fig. 1. Graph $G_{k,5}$ (Dashed lines represent non edges of $G_{k,5}$.)

Let C be a dominating clique of $G_{k,\ell}$. Clearly, C contains the vertex b or one of its neighbors. However, $b \in C$ is impossible since a could not have a neighbor in C in this case. By the construction of $G_{k,\ell}$, all vertices y_i have the same neighborhood, thus w.l.o.g. we may assume $y_1 \in C$. This implies $x_0 \in C$ since x_0 is the only common neighbor of a and y_1. Consequently, and that is important for the analysis, the vertex x_0 belongs to each dominating clique, and therefore no vertex s_i with $i \bmod \ell \neq 0$ belongs to a dominating clique of $G_{k,\ell}$.

Consider the algorithm DomClq of [2]. When applied to a graph $G_{k,\ell}$, first a vertex of minimum degree is determined and this is w and then the algorithm branches on v which is the unique neighbor of w in $G_{k,\ell}$. Thus DomClq($G_{k,\ell}, \{v\}, S, \{u_i : 1 \leq i \leq k\} \cup \{a, b\}$) is executed. Hence DomClq chooses a vertex in $F = \{u_i : 1 \leq i \leq k\} \cup \{a, b\}$ with smallest number of neighbors in S. Suppose that each time the algorithm has to do such a choice it chooses the free vertex u_i with the smallest possible index i. Then the algorithm branches on the neighbors of u_i. Note that, according to the above remark, no vertex s_i with $i \bmod \ell \neq 0$ belongs to a dominating clique, and thus DomClq really branches on each of the five neighbors of u_i.

Consider now the search tree obtained after branching on the neighbors of $u_1, u_2, \ldots, u_{t-1}$, $2 \leq t \leq k$. When branching on the neighbors of u_t the set of vertices already discarded from the original graph $G_{k,\ell}$ is $\{w, v\} \cup \{u_j : 1 \leq j < t\} \cup \{s_j : 1 \leq j < t\}$. How many subproblems are obtained by branching on the neighbors of u_t?

- when branching on s_t (the first neighbor of u_t), the algorithm discards 2 vertices (since $N[s_t] = \{s_t, u_t\}$),
- when branching on s_{t+1} (the second neighbor of u_t), the algorithm discards 4 vertices: $s_t, s_{t+1}, u_t, u_{t+1}$,
- ...
- when branching on $s_{t+\ell-1}$ (the last neighbor of u_t), the algorithm removes 2ℓ vertices: $s_t, s_{t+1}, s_{t+2}, \ldots, s_{t+\ell-1}, u_t, u_{t+1}, u_{t+2}, \ldots, u_{t+\ell-1}$.

We denote by $L[q]$ the maximum number of leaves of the search tree when applying DomClq to an induced subgraph of $G_{k,\ell}$ on $q \leq n$ vertices. According to our previous analysis, we obtain the following recurrence

$$L[q] \geq L[q-2] + L[q-4] + L[q-6] + \cdots + L[q-2\ell].$$

Solving this recurrence one obtains $L[q] \geq 1.4142^q$. Consequently, 1.4142^n is a lower bound for the maximum number of leaves of the search tree of an execution of algorithm DomClq on an input graph with n vertices.   □

3   A Branch and Reduce Algorithm for MinDC

In this section we present an algorithm to solve the MinDC problem for an input graph $G = (V, E)$. We may assume G to be connected, otherwise G cannot have a dominating clique.

Our algorithm uses four different types of vertices, and thus maintains a partition of V into four pairwise disjoint subsets:

- S is the set of *selected* vertices, i.e. those vertices that have already been chosen for the potential solution (and thus S is a clique);
- D is the set of *discarded* vertices, i.e. those vertices that have already been removed from the (input) graph;
- $A = \bigcap_{s \in S} N(s) \setminus D$ is the set of *available* vertices, i.e. those vertices that might still be added to the potential solution S;
- $F = V \setminus (N[S] \cup D)$ is the set of *free* vertices, i.e. those vertices are still to be dominated (hence for each free vertex at least one of its available neighbors must be added to S).

Our recursive algorithm mdc finds the size of an optimum solution SOL, i.e. a dominating clique of smallest possible cardinality such that the following properties are fulfilled: (i) $S \subseteq SOL$, (ii) $D \cap SOL = \emptyset$, and (iii) $F \cap SOL = \emptyset$. If no such dominating clique exists then mdc returns "∞". For a discussion of halting, reduction and branching rules of the Branch & Reduce algorithm we refer to the next section.

Algorithm mdc(G,S,D,A,F)
Input: A graph $G = (V, E)$ **and a partition** (S, D, A, F) **of its vertex set.**
Output: The minimum cardinality of a dominating clique of G
       respecting the partition, if one exists.

if $\exists u \in F$ s.t. $d_A(u) = 0$ then
    return ∞ **(H1)**

else if $F = \emptyset$ then
    return $|S|$ **(H2)**

else if $\exists v \in A$ s.t. $d_F(v) = 0$ then
    return $\text{mdc}(G, S, D \cup \{v\}, A \setminus \{v\}, F)$ **(R1)**

else if $\exists u \in F$ s.t. $d_A(u) = 1$ then
    let $v \in A$ be the unique neighbor of u in A
    return $\text{mdc}(G, S \cup \{v\}, D \cup \overline{N_A[v]} \cup N_F(v), A \setminus \overline{N_A(v)}, F \setminus N_F(v))$ **(R2)**

else if $\exists v_1, v_2 \in A$ s.t. $N(v_1) \subseteq N(v_2)$ then
    return $\text{mdc}(G, S, D \cup \{v_1\}, A \setminus \{v_1\}, F)$ **(R3)**

else if $\exists u_1, u_2 \in F$ s.t. $N_A(u_1) \subseteq N_A(u_2)$ then
    return $\text{mdc}(G, S, D \cup \{u_2\}, A, F \setminus \{u_2\})$ **(R4)**

else if $\exists u \in F$ s.t. $\big(\forall v' \in N_A(u), d_F(v') = 1$ and
           $\exists v \in N_A(u)$ s.t. $A \setminus N_A(u) \subseteq N_A(v)\big)$ then
    let $v \in N_A(u)$ be a vertex verifying the condition
    return $\text{mdc}(G, S \cup \{v\}, D \cup (N_A(u) \setminus \{v\}) \cup \{u\}, A \setminus N_A(u), F \setminus \{u\})$
                                                           (R5)

else
    choose $v \in A$ of maximum F-degree
    if $d_F(v) = 1$ then
        let u be a vertex of F
        return $\min_{v \in N_A(u)} \{\text{mdc}(G, S \cup \{v\}, D \cup (N_A(u) \setminus \{v\}) \cup \{u\} \cup$
        $\overline{N_A[v]}, A \setminus (\overline{N_A(v)} \cup N_A(u)), F \setminus \{u\})\}$ **(B1)**
    else
        return
        $\min \big(\text{mdc}(G, S \cup \{v\}, D \cup \overline{N_A[v]} \cup N_F(v), A \setminus \overline{N_A(v)}, F \setminus N_F(v)),$
        $\text{mdc}(G, S, D \cup \{v\}, A \setminus \{v\}, F)\big)$ **(B2)**

To compute a minimum dominating clique of a given graph $G = (V, E)$, $\text{mdc}(G, \{v\}, \emptyset, N(v), V \setminus N[v])$ is called for each vertex $v \in N[w]$, where w is a vertex of G of minimum degree. To observe correctness, assume that C is a minimum dominating clique of G. Since C is a dominating set of G it must contain a vertex of $N[w]$, say v_0. Thus choosing $S = \{v_0\}$ and $D = \emptyset$ implies that only neighbors of v_0 are still available and all non neighbors of v_0 become free vertices, and still have to be dominated (by adding vertices to S). Thus $\text{mdc}(G, \{v_0\}, \emptyset, N(v_0), V \setminus N[v_0])$ returns the size of a minimum dominating clique of G. Clearly if G has no dominating clique then for all $v \in N[w]$, $\text{mdc}(\{v\}, \emptyset, N(v), V \setminus N[v])$ returns "∞".

Note that with a slight modification mdc could also output a minimum dominating clique SOL, if one exists, instead of the minimum cardinality $|SOL|$.

4   Analysis of the Algorithm

Correctness. The algorithm halts (i.e. the subproblem corresponds to a leaf in the search tree) if either there is a free vertex with no available neighbor (H1), and thus there is no dominating clique for the current instance, or there are no free vertices (H2), and thus S is a dominating clique of minimum size for this instance.

Otherwise the algorithm possibly performs some reduction rules on the problem instance, and then it branches using (B1) or (B2) on two or more subproblems, which are solved recursively. In each subproblem the algorithm selects an available vertex and adds it to S and/or discards vertices of G (i.e. adds them to D), and those changes of S and D imply updates on A and F.

The correctness of the reduction rules is not hard to see. Let (S, D, A, F) be the current partition.

(R1) If an available vertex v has no free neighbor then v can be discarded.

(R2) If $v \in A$ is the unique free neighbor of $u \in F$ then v must be selected (and added to S).

(R3) If $v_1, v_2 \in A$ such that $N(v_1) \subseteq N(v_2)$ then for any dominating clique C containing v_1 there is the dominating clique $C' = (C - \{v_1\}) \cup \{v_2\}$ with $|C'| \leq |C|$. Thus we may safely discard v_1.

(R4) If $u_1, u_2 \in F$ such that $N_A(u_1) \subseteq N_A(u_2)$ we may discard $u_2 \in F$. To see this, notice that any dominating clique respecting the partition $(S, D \cup \{u_2\}, A, F \setminus \{u_2\})$ contains a neighbor $v \in A$ of u_1, and thus also a neighbor of u_2.

(R5) Let $u \in F$ be a free vertex such that all its neighbors $v \in A$ have u as unique free neighbor. If one of these available neighbors, say v_0, is adjacent to all vertices in $A \setminus N_A(u)$, then there exists a minimum dominating clique containing v_0 respecting the partition. To see this, notice that a dominating clique must contain a vertex of $N_A(u)$, that only one of those will be chosen (all others will be discarded immediately by rule (R1)), and that v_0 is the best choice since it does not force removal of any remaining A-vertices (i.e. those in $A \setminus N_A(u)$).

Now we consider the correctness of the branching rules. Let (S, D, A, F) be the partition of the vertices of G when applying a branching rule. Note that the minimum F-degree of an available vertex is at least 1 since any available vertex v with $d_F(v) = 0$ would have been discarded according to reduction rule (R1).

(B1) If all available vertices have exactly one free neighbor, then (B1) chooses any free vertex $u \in F$. Clearly any dominating clique respecting (S, D, A, F) must contain precisely one neighbor $v \in A$ of u. Thus for each neighbor $v \in A$ of u, (B1) branches to a subproblem by selecting v. Clearly, the minimum cardinality among all dominating cliques obtained is the minimum cardinality of a dominating clique respecting partition (S, D, A, F).

(B2) For an available vertex v of F-degree at least 2 either v is selected or discarded which is trivially correct.

Analysis of the running time. In order to bound the progress made by our algorithm at each branching step, we apply the Measure & Conquer approach which was introduced in [3] (see also [4]). The following non standard measure on the size of the input of a (sub)problem is used:

$$\mu = \mu(G, S, D, A, F)$$

$$= \sum_{\substack{v \in A, \\ d_F(v)=1}} a_1 + \sum_{\substack{v \in A, \\ d_F(v)=2}} a_2 + \sum_{\substack{v \in A, \\ d_F(v) \geq 3}} a_{\geq 3}$$

$$+ \sum_{\substack{v \in F, \\ d_A(v)=2}} f_2 + \sum_{\substack{v \in F, \\ d_A(v)=3}} f_3 + \sum_{\substack{v \in F, \\ d_A(v)=4}} f_4 + \sum_{\substack{v \in F, \\ d_A(v) \geq 5}} f_{\geq 5}$$

To each vertex v of G, we assign a weight depending on its number of free neighbors if v is an available vertex, or depending on its number of available neighbors if v is a free vertex. Since (S, D, A, F) is a partition of the vertices of $G = (V, E)$, it is easy see that $\mu(G, S, D, A, F) \leq |A \cup F| \leq |V| = n$.

To simplify the analysis we impose that $0 \leq a_1, a_2, a_{\geq 3}, f_2, f_3, f_4, f_{\geq 5} \leq 1$, and $a_1 \leq a_2 \leq a_{\geq 3}$ and $f_2 \leq f_3 \leq f_4 \leq f_{\geq 5}$. Furthermore we introduce the following quantities:

$$\Delta a_i = \begin{cases} 0 & \text{if } i \geq 4 \\ a_i - a_{i-1} & \text{if } 2 \leq i \leq 3 \\ a_1 & \text{if } i = 1 \end{cases} \quad \text{and} \quad \Delta f_i = \begin{cases} 0 & \text{if } i \geq 6 \\ f_i - f_{i-1} & \text{if } 3 \leq i \leq 5 \\ f_2 & \text{if } i = 2 \end{cases}$$

Let $P[\mu]$ denote the maximum number of subproblems recursively solved by mdc to compute a solution on an instance of size μ.

First let us consider the reduction rules. For each of the five reduction rules, when applying it either a vertex is selected or a non empty set of vertices is discarded. Thus the number of consecutive applications of reduction rules to a subproblem (without intermediate branching) is at most n. Since each reduction rule can easily be implemented such that its execution is done in polynomial time, the running time of mdc on a subproblem (which corresponds to a node of the search tree) is polynomial. Furthermore, we want to emphasize that due to the choice of the measure no application of a reduction rule to a problem instance will increase its measure.

Let us now consider the more interesting part: the branching rules (B1) and (B2). In the classical analysis of the running time of Branch & Reduce algorithms with a so-called standard measure, i.e. number of vertices in case of graphs, the two branching rules would be fairly easy to analyse and one would obtain a few linear recurrences. Using Measure & Conquer this analysis is very interesting (and can be quite tedious). On the other hand, Measure & Conquer allows relatively simple algorithms with competitive running times since the tricky case analysis that had been part of Branch & Reduce algorithms (see e.g. [21]) is now 'transferred' to the time analysis of the algorithm.

(B1) The algorithm mdc chooses a free vertex u and for each of its available neighbors v, it calls itself recursively with v being selected (i.e. v added to S).

Recall that (B1) is applied only when all available vertices have F-degree 1. Thus when mdc selects an available neighbor v of u, then all other available neighbors of u would decrease their F-degree to 0 and would be discarded and removed from A by reduction rule (R1). Finally we observe that due to reduction rule (R5), each available neighbor of u must be non-adjacent to at least one vertex of $A \setminus N_A(u)$. Consequently for every $d_A(u) \geq 2$, we obtain a recurrence:

- if $d_A(u) = i$ and $2 \leq i \leq 4$,

$$P[\mu] \leq 1 + 2 \cdot P[\mu - 2a_1 - f_i - a_1] \tag{1}$$

- if $d_A(u) \geq 5$,

$$P[\mu] \leq 1 + d_A(u) \cdot P[\mu - d_A(u)a_1 - f_{\geq 5} - a_1] \tag{2}$$

(B2) Algorithm mdc chooses an available vertex $v \in A$ of maximum F-degree. Since neither (R1) nor (B1) had been applied we may conclude that $d_F(v) \geq 2$. Using (B2) we branch into two subproblems by either selecting v (branch IN) or discarding v (branch OUT).

To obtain subproblem (branch IN), v is removed from A, all free neighbors of v are removed from F and for all $w \in N_A(N_F(v)) \setminus \{v\}$ their F-degree will decrease. Moreover if a vertex $x \in N_A(N_F(v))$ has only one free neighbor, then x will be removed when reduction rule (R1) is applied to subproblem (branch IN). Hence we obtain

$$\Delta_{IN} = a_{d_F(v)} + \sum_{u \in N_F(v)} f_{d_A(u)} + \sum_{w \in N_A(N_F(v)) \setminus \{v\}} \Delta a_{d_F(w)} \tag{3}$$

For the subproblem (branch OUT), v is discarded and thus removed from A. The A-degree of all free neighbors of v decreases. When applying (R2) to subproblem (branch OUT), the free neighbors having only one available neighbor will be removed from F. To dominate those vertices, the algorithm has to select their only remaining neighbors in A. For $y \in A$, we denote by $N_{F2}(y) = \{u \in N_F(y) : d_A(u) = 2\}$ the set of free neighbors of v having precisely two available neighbors and one of those is y. Consequently, we obtain

$$\Delta_{OUT} = a_{d_F(v)} + \sum_{u \in N_F(v)} \Delta f_{d_A(u)} + \sum_{w \in N_A(N_{F2}(v)) \setminus \{v\}} a_{d_F(w)} \tag{4}$$

Finally using equations (3) and (4) we obtain the recurrence

$$P[\mu] \leq 1 + P[\mu - \Delta_{OUT}] + P[\mu - \Delta_{IN}]. \tag{5}$$

Optimizing the values of the parameters. To optimize the parameters one has (as an important step) to solve the system of recurrences obtained when putting into recurrence (5) all possible values of $d_F(v) \geq 2$, $d_A(u) \geq 2$ for every $u \in N_F(v)$ and $d_F(w)$ such that $1 \leq d_F(w) \leq d_F(v)$. The number of recurrences is infinite; fortunately one can restrict to the recurrences with $2 \leq d_F(v) \leq 4$,

$2 \leq d_A(u) \leq 6$ for every $u \in N_F(v)$, and $1 \leq d_F(w) \leq d_F(v)$. Indeed, since $\Delta a_{d_F(v)} = 0$ for $d_F(v) \geq 4$ and $\Delta f_{d_A(v)} = 0$ for each free vertex u with $d_A(u) \geq 6$, all recurrences with $d_F(v) > 4$ or $d_A(u) > 6$ for some vertex $u \in N_F(v)$ will be majorized by some recurrence with $d_F(v) = 4$ or $d_A(u) = 6$. Similarly considering recurrence (2), under the assumption $a_1 \geq 0.75$ and $f_{\geq 5} = 1$, all recurrences with $d_A(u) \geq 7$ will be majorized by the one with $d_A(u) = 6$.

We use a program to generate those linear recurrences that removes superfluous ones. For any fixed and valid choice of the 7 parameters this system of 420 recurrences is easy to solve. Finally using a program based on random local search, values of the 7 parameters are searched as to minimize the bound on the running time. We numerically obtained the following values: $a_1 = 0.7632$, $a_2 = 0.9529$, $a_{\geq 3} = 1$, $f_2 = 0.3869$, $f_3 = 0.7579$, $f_4 = 0.9344$, $f_{\geq 5} = 1$. Using these values an upper bound of $O(1.3390^n)$ is established.

Theorem 2. *Algorithm* mdc *solves problem* MinDC *in time* $O(1.3390^n)$.

It is not unlikely that the worst case running time of the Branch & Bound algorithm mdc is $O(\alpha^n)$ for some $\alpha < 1.3390$. Significant improvements of the upper bound would require a more clever choice of the measure or new techniques to analyse the running time of Branch & Reduce algorithms.

5   An Exponential Lower Bound

Since our upper bound on the running time of mdc might overestimate the worst case running time, it is natural to ask for a lower bound that may give an idea of how far is the established upper bound of $O(1.3390^n)$ from the real worst case running time of mdc.

Theorem 3. *The worst case running time of algorithm* mdc *is* $\Omega(1.2599^n)$.

Proof. To prove the claimed lower bound we consider the graphs G_k for integers $k \geq 1$ (see also Fig. 2).

The vertex set of G_k is $V_k = \{w, v\} \cup \bigcup_{\substack{1 \leq i \leq k \\ 1 \leq j \leq 5}} \{v_{i,j}\} \cup \bigcup_{1 \leq i \leq k} \{u_i\}$. The edge set E_k of the graph G_k consists of the edge $\{w, v\}$ and the union of $\bigcup_{\substack{1 \leq i \leq k \\ 1 \leq j \leq 5}} \{\{v, v_{i,j}\}, \{u_i, v_{i,j}\}\}$ and $\bigcup_{\substack{1 \leq i < i' \leq k \\ 1 \leq j,j' \leq 5}} \{\{v_{i,j}, v_{i',j'}\} : (i', j') \neq (i+1, j)\}$. Notice that the graph G_k has precisely $6k + 2$ vertices.

To establish an exponential lower bound of the worst case running time, we lower bound the number of leaves of a search tree obtained by an execution of mdc on G_k. (Notice that ties will be broken such as to maximise the number of leaves.) A vertex of minimum degree is w, and for our analysis it suffices to consider mdc$(G_k, S_1, D_1, A_1, F_1)$ with $S_1 = \{v\}$, $D_1 = \emptyset$, $A_1 = N_{G_k}(v)$ and $F_1 = V_k \setminus N_{G_k}[v] = \{u_1, u_2, \ldots, u_k\}$.

Notice that no reduction rules can be applied. For each i, every vertex $v_{i,j}$, $1 \leq j \leq 5$, has the unique non neighbor $v_{i+1,j}$ in $\{v_{i+1,j} : 1 \leq j \leq 5\}$. Furthermore since all available vertices have only one free neighbor, branching rule (B1) will be

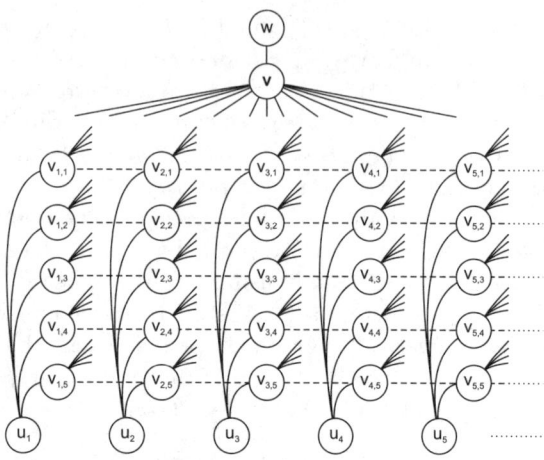

Fig. 2. Graph G_k (Dashed lines represent non edges.)

applied. We assume that whenever `mdc` branches on a subproblem using (B1), then it branches on the neighborhood of the free vertex u_i with the smallest possible value of i.

Let us consider the first branching. By our rule, the algorithm branches on the neighborhood of u_1, i.e. $v_{1,j}$, $1 \le j \le 5$, into 5 subproblems. Consider a subproblem, say v_{1,j_1} had been selected, thus the partition of the subproblem is (S_2, D_2, A_2, F_2) with $S_2 = S_1 \cup \{v_{1,j_1}\}$, $D_2 = D_1 \cup \{v_{2,j_1}\} \cup \bigcup_{1 \le j \le 5}\{v_{1,j} : j \ne j_1\}$, $A_2 = A_1 \setminus (\{v_{2,j_1}\} \cup \bigcup_{1 \le j \le 5}\{v_{1,j}\})$ and $F_2 = F_1 \setminus \{u_1\}$.

All five subproblems have the same sets A_2 and F_2 and thus the graph remaining after removal of discarded vertices is essentially the same. By our construction the previous arguments apply to any partition (S_2, D_2, A_2, F_2), and thus `mdc` branches using (B1) on all *four* available neighbors of u_2. (Note that v_{2,j_1} is not adjacent to v_{1,j_1} and thus not available.)

Inductively one can show that `mdc` branches succesively on the neighborhood of $u_1, u_2, u_3, \ldots u_{k-2}$, and obtains always 4 subproblems (actually 5 for u_1). Thus the number of leaves in the search tree is $\Omega(4^{n/6}) = \Omega(1.2599^n)$, where $n = 6k+2$ is the number of vertices of G_k.   □

6   Concluding Remarks

It is worth noting that our $O(1.3390^n)$ time algorithm for `MinDC` can also be applied to solve `ExDC`.

The problems `MaxDC` and `ExDC` are polynomial time solvable on chordal graphs by enumerating all the at most n maximal cliques and verifying which of them are dominating sets. On the other hand, `MinDC` is NP-hard on split graphs. Using the Minimum Set Cover algorithms given in [3] one easily obtains an $O(1.2303^n)$ algorithm for `MinDC` on split graphs. This approach can be extended to chordal graphs without increasing the running time.

References

1. Cozzens, M. B., and L. L. Kelleher, Dominating cliques in graphs, *Discrete Math.* **86** (1990), pp. 101–116.
2. Culberson, J., Y. Gao, and C. Anton, Phase Transitions of Dominating Clique Problem and Their Implications to Satisfiability Search, *Proc. IJCAI 2005*, pp. 78–83.
3. Fomin, F. V., F. Grandoni, and D. Kratsch, Measure and conquer: Domination - A case study, *Proc. ICALP 2005, LNCS* **3380** (2005), pp. 192–203.
4. Fomin, F. V., F. Grandoni, and D. Kratsch, Some new techniques in design and analysis of exact (exponential) algorithms, *Bull. EATCS* **87** (2005), pp. 47–77.
5. Fomin, F. V., D. Kratsch, and G. J. Woeginger, Exact (exponential) algorithms for the dominating set problem, *Proc. WG 2004, LNCS* **3353** (2004), pp. 245–256.
6. Garey, M. R. and D. S. Johnson, *Computers and intractability. A guide to the theory of NP-completeness.* W.H. Freeman and Co., San Francisco, 1979.
7. Gaspers, S., and M. Liedloff, A branch-and-reduce algorithm for finding a minimum independent dominating set in graphs, *Proc. WG 2006*, to appear.
8. Grandoni, F., A note on the complexity of minimum dominating set, *J. Discrete Algorithms* **4** (2006), pp. 209–214.
9. Haynes T. W., S. T. Hedetniemi, and P. J. Slater, *Fundamentals of domination in graphs.* Marcel Dekker Inc., New York, 1998.
10. Haynes T. W., S. T. Hedetniemi, and P. J. Slater, *Domination in graphs: Advanced Topics.* Marcel Dekker Inc., New York, 1998.
11. Held M. and R.M. Karp, A dynamic programming approach to sequencing problems, *Journal of SIAM* (1962), pp. 196–210.
12. Iwama K., Worst-case upper bounds for ksat, *Bull. EATCS* **82** (2004), pp. 61–71.
13. Johnson, D. S., M. Yannakakis, and C. H. Papadimitriou, On generating all maximal independent sets, *Inf. Process. Lett.*, **27**, (1988), pp. 119–123.
14. Kratsch, D., Algorithms, in *Domination in Graphs: Advanced Topics*, T. Haynes, S. Hedetniemi, P. Slater, (eds.), Marcel Dekker , 1998, pp. 191–231.
15. Lawler E. L., A note on the complexity of the chromatic number problem. *Inform. Process. Lett.* **5(3)** (1976), pp. 66–67.
16. Moon, J. W., and L. Moser, On cliques in graphs, *Israel J. Math.*, **3**, (1965), pp. 23–28.
17. Randerath, B., and I. Schiermeyer, Exact algorithms for Minimum Dominating Set, Technical Report zaik-469, Zentrum fur Angewandte Informatik, Köln, Germany, April 2004.
18. Razgon, I., Exact computation of maximum induced forest, *Proc. SWAT 2006*, to appear.
19. Robson J. M., Algorithms for maximum independent sets, *J. Algorithms* **7** (1986), pp. 425–440.
20. Schöning U., Algorithmics in exponential time, *Proc. STACS 2005, LNCS* **3404** (2005), pp. 36–43.
21. Tarjan R. and A. Trojanowski, Finding a maximum independent set, *SIAM J. Comput.* **6(3)** (1977), pp. 537–546.
22. Villanger Y., Improved exponential-time algorithms for treewidth and minimum fill-in. *Proc. LATIN 2006 LNCS* **3887** (2006), pp. 800–811.
23. Woeginger, G. J., Exact algorithms for NP-hard problems: A survey, *Combinatorial Optimization - Eureka, You Shrink!, LNCS* **2570** (2003), pp. 185–207.
24. Woeginger, G. J., Space and time complexity of exact algorithms: Some open problems, *Proc. IWPEC 2004, LNCS* **3162** (2004), pp. 281–290.

EDGE DOMINATING SET:
Efficient Enumeration-Based Exact Algorithms

Henning Fernau[1,2]

[1] Univ.Trier, FB 4—Abteilung Informatik, 54286 Trier, Germany
[2] Univ.Tübingen, WSI für Informatik, Sand 13, 72076 Tübingen, Germany
`fernau@informatik.uni-trier.de`

Abstract. We analyze EDGE DOMINATING SET from a parameterized perspective. More specifically, we prove that this problem is in $\mathcal{FPT}$ for general (weighted) graphs. The corresponding algorithms rely on enumeration techniques. In particular, we show how the use of compact representations may speed up the decision algorithm.

1 Introduction

Graphs and line graphs. It is a common observation that problems that are hard for general graphs become easier when considered on *line graphs*, i.e., graphs whose adjacency relation can be thought of as originating from the edge-to-edge neighborhood of another graph. More specifically, if $G = (V, E)$ is some graph, then its line graph $L(G)$ has E has the set of "vertices," and there is an "edge" (in $L(G)$) between $e_1, e_2 \in E$ if e_1 and e_2 share a common endpoint in G. For example, while VERTEX COVER is $\mathcal{NP}$-complete on general graphs, it can be solved in polynomial time on line graphs, since this corresponds to the EDGE COVER problem. The same comment applies to the more general problem WEIGHTED VERTEX COVER. However, DOMINATING SET remains $\mathcal{NP}$-complete even when restricted to line graphs, see [23], even when restricted to planar cubic graphs [15]. Does this mean that there is actually "no difference" between general graphs and line graphs with respect to DOMINATING SET? This is in fact the case from a classical perspective, but the picture changes when one considers two prominent approaches of how to deal with computationally hard problems:

- while DOMINATING SET is hard to approximate on general graphs [7] (it cannot be approximated better than $\ln n$ unless $\mathcal{NP} \subseteq \mathcal{DTIME}(n^{\ln \ln n})$), EDGE DOMINATING SET is constant-factor approximable (also in the weighted case), see [3,13,18]; moreover, it is $\mathcal{MAXSNP}$-hard and hence there is no polynomial-time approximation scheme to be expected [3,23];
- while DOMINATING SET is W[2]-hard on general graphs, EDGE DOMINATING SET is in $\mathcal{FPT}$; in this paper, we are going to show that EDGE DOMINATING SET, when parameterized by the number of elements k in the dominating set, can be solved in time $\mathcal{O}^*(2.62^k)$.

H.L. Bodlaender and M.A. Langston (Eds.): IWPEC 2006, LNCS 4169, pp. 142–153, 2006.
© Springer-Verlag Berlin Heidelberg 2006

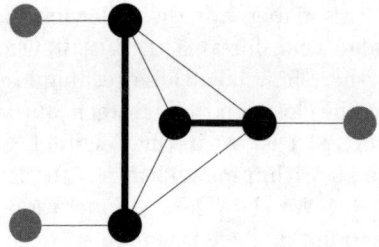

Fig. 1. The two thick-line edges form an edge dominating set

Problem statements. Let us formulate the problem dealt with in this paper without reference to line graphs. An *edge dominating set* of a graph is a subset D of edges such that each edge is either in D or incident to an edge in D. An instance of EDGE DOMINATING SET (EDS) is given by a graph $G = (V, E)$, and the parameter, a positive integer k. The task is: Is there an *edge dominating set* $D \subseteq E$ with $|D| \leq k$? Consider Fig. 1 for an illustration; the two thick edges form an edge dominating set. We will also consider a weighted variant: An instance of WEIGHTED EDGE DOMINATING SET (WEDS) is given by a graph $G = (V, E)$ with edge weights $\omega : E \to \mathbb{R}_{\geq 1}$, and a positive integer k (the parameter). We ask: Is there an edge dominating set $D \subseteq E$ with $\omega(D) \leq k$?

We will analyze these problems in the framework of *parameterized complexity* [6]. A *parameterized problem* P is a subset of $\Sigma^* \times \mathbb{N}$, where Σ is a fixed alphabet and $\mathbb{N}$ is the set of all non-negative integers. Therefore, each instance of the parameterized problem P is a pair (I, k), where the second component k is called the *parameter*. The language $L(P)$ is the set of all YES-instances of P. We say that the parameterized problem P is *fixed-parameter tractable* [6] if there is an algorithm that decides whether an input (I, k) is a member of $L(P)$ in time $f(k)|I|^c$, where c is a fixed constant and $f(k)$ is a function independent of the overall input length $|I|$. The class of all fixed-parameter tractable problems is denoted by $\mathcal{FPT}$. We will make use of the $\mathcal{O}^*$-notation that has now become standard in exact algorithmics: in contrast to the better known $\mathcal{O}$-notation, it not only suppresses constants but also polynomial parts of the run time denoted this way. We are dealing with (sometimes weighted) undirected (hyper-)graphs throughout this paper and use according standard notations.

Results. In previous work, it was claimed that EDGE DOMINATING SET on bipartite graph (this problem is also known as MATRIX DOMINATION SET) belongs to $\mathcal{FPT}$. More precisely, this was posed as an exercise on the kernelization chapter in [6]. Rather recently, Weston [22] exhibited how to obtain a kernel of exponential size. Moreover, he shows a (not quite convincing) enumeration-based (this is our interpretation of the paper) search tree approach. E. Prieto in her PhD thesis [21] obtained a kernel of quadratic size for MINIMUM MAXIMAL MATCHING (i.e., INDEPENDENT DOMINATING SET in line graphs) on general graphs, a problem that is basically the same as EDGE DOMINATING SET, see [3,17,23].

The contributions of this paper are the following ones: (1) We generalize the enumeration-based algorithm for MATRIX DOMINATION SET to EDGE DOMINATING SET. (2) We further show how these techniques may also apply to the weighted case; notice that the close relation between MINIMUM MAXIMAL MATCHING and EDGE DOMINATING SET is lost in the weighted case. (3) Incidentally, we also present a search tree algorithm for WEIGHTED HITTING SET, parameterized by the number of edges. (4) We show how enumeration-based algorithms that are usually deemed to be quite inefficient can be sped up by exploiting that not all minimal vertex covers (in our case) need to be "completely" enumerated. Rather, certain non-determinism can be left to the following computation phase in the leaves of the search tree. (5) We show that auxiliary vertex cover structures can be quite useful for other algorithmic problems. The speed-up technique mentioned in (4) can then be formalized as *compact representations* of vertex covers. This can be seen as extensions of ideas presented by Damaschke [4]. (6) We also use the mentioned vertex cover structures to obtain quadratic kernels for the considered problems.

2   Relating to VERTEX COVER and to WEIGHTED HITTING SET

First, we observe that to every EDGE DOMINATING SET instance $G = (V, E)$ with a solution $D \subseteq E$ of size at most k, there corresponds a vertex cover of size at most $2k$: take all vertices incident to the at most k edges: $C_D := \bigcup_{e \in D} e$.[1] This observation is also true for WEIGHTED EDGE DOMINATING SET: by our problem definitions, a solution of weight at most k can contain at most k edges. Conversely, any vertex cover C can be extended to some edge set D with $C \subseteq C_D$ with $|C_D| \leq 2|C|$; since C is a vertex cover, D will be an edge dominating set.

This gives the following idea: (1) we cycle through all minimal vertex covers C of size up to $2k$ of the given graph $G = (V, E)$; (2) we use this additional structure to solve the problem of finding an edge dominating set D of size at most k (or weight at most k) that contains all vertices of C, i.e., $C \subseteq C_D$.

As to step (1), it is known how to list all minimal vertex covers up to size $2k$ in time $\mathcal{O}^*(4^k)$, see [4]. To see that the second step works, consider the following auxiliary hypergraph $G' = (V', E')$: G' contains the edges E of G as its vertices, i.e., $V' = E$, and for every x contained in the cover C, we introduce a hyperedge h_x that contains all edges of G that are incident with x. Hence, $|E'| \leq |C| \leq 2k$.

Fomin, Kratsch and Woeginger [12] recently came up with an efficient parameterized algorithm for the following problem: An instance of MINIMUM HITTING SET, PARAMETERIZED BY # EDGES (HSE) is given by a hypergraph $G = (V, E)$, and the parameter, $|E|$. The task is: Find a minimum hitting set $C \subseteq V$!

We will generalize that algorithm in the following so that it can also cope with WEIGHTED HITTING SET. The algorithm uses a technique known as *dynamic programming on subsets*. To this end, given a hypergraph $G = (V, E)$ with $V = \{v_1, \ldots, v_n\}$, the algorithm maintains a 2-dimensional array F that contains, for

---

[1] This observation was also the basis of a first, simple factor-4 approximation algorithm for MINIMUM EDGE DOMINATING SET presented in [3].

Algorithm 1. A dynamic programming algorithm for MINIMUM HITTING SET, PARAMETERIZED BY # EDGES, called HSE

Input(s): a hypergraph $G = (V, E)$ with vertex weights ω, $V = \{v_1, \ldots, v_n\}$
Output(s): a hitting set $C \subset V$ of minimum weight

  for all $E' \subseteq E$ **do**
    $F[E, 0] = \infty$
  end for
  $F[\emptyset, 0] = 0$
  for $j = 1, \ldots, n$ **do**
    for all $E' \subseteq E$ **do**
      Let $E'' := \{e \in E' \mid v_j \in e\}$.
      $F[E', j] := \min\{F[E', j - 1], F[E' \setminus E'', j - 1] + w(v_j)\}$
      {Two cases arise: either v_j is not belonging to a minimum hitting set for E', then, $F[E', j] = F[E', j - 1]$; or v_j belongs to a minimum hitting set C for E', but then, $C \setminus \{v_j\}$ is a minimum hitting set for $E' \setminus E''$ relative to V_{j-1}, so that $F[E', j] = F[E' \setminus E'', j - 1] + w(v_j)$.}
    end for
  end for

$E' \subseteq E$ and for $j = 0, \ldots, n$, in $F[E', j]$ the minimum weight of a subset C of $V_j := \{v_1, \ldots, v_j\}$ that covers E' (C is also called a *hitting set for E' (relative to V_j)*); if no such cover exists, set $F[E', j] = \infty$. More details on how to construct the entries for F are contained in Alg. 1. There, also the basic reasoning for the inductive step in the correctness proof of that algorithm is given.

Theorem 1. MINIMUM HITTING SET, PARAMETERIZED BY # EDGES *can be solved in time* $\mathcal{O}^*(2^{|E|})$ *on a hypergraph* $G = (V, E)$ *with vertex weights* ω.

Corollary 1. WEIGHTED EDGE DOMINATING SET *can be solved in time* $\mathcal{O}^*(16^k)$.

3 Replacing the Hitting Set Phase

In this section, we are going to explain that the second HITTING SET phase can be replaced by a polynomial-time computation, which already considerably improves on the run time stated in Cor. 1.

If we have a *minimal* vertex cover C of size at most $2k$, we first compute a maximum matching M in the induced graph $G[C]$. Clearly, $|M| \leq k$. There might be a set $C' \subseteq C$ of vertices not matched by M. Since M is maximal, no two vertices of C' are neighbors. Hence, for all $x \in C'$, we can take any edge incident with x into the edge dominating set to be constructed.

Theorem 2. *Alg. 2 runs in time* $\mathcal{O}^*(4^k)$ *and solves* EDGE DOMINATING SET.

Proof. The run time is dominated by the enumeration of at most 4^k minimal vertex covers of size $2k$, see [4]. The correctness is based on the fact that the following problem can be solved in polynomial time by matching techniques: An

---

Algorithm 2. EDS-enum: An enumeration-based search tree algorithm for EDS

Input(s): a graph $G = (V, E)$, a positive integer k
Output(s): if possible: a subset $D \subset E$, $|D| \leq k$, that dominates all edges or
NO if no such set exists.

Create a list L of minimal vertex covers C of G with $|C| \leq 2k$, see [4].
{This hides the search tree part.}
for all $C \in L$ **do**
  Set $D :=$GECmatch(G, C)
  if $|D| \leq k$ **then**
    return D
  end if
end for
return NO

---

instance of GENERALIZED EDGE COVER (GEC) is given by a graph $G = (V, E)$ together with a set of red vertices $R \subseteq V$. The task is to construct a minimum set $C \subseteq E$ of edges that cover all vertices from R. Namely, a maximum matching of $G[R]$ plus adding edges to cover the hitherto unmatched vertices will do. This describes the procedure GECmatch that gets as arguments the graph and the set of red vertices. Moreover, if R is a vertex cover, then a GEC solution will be an edge dominating set.   ∎

Remark 1. Observe the crucial difference between the weighted and the unweighted case here. For example, notice that the thick edges in Fig. 1 (that form a minimum edge dominating set) can be obtained by computing a maximum matching from a vertex cover formed by the black vertices (or from a minimal vertex cover formed by the "outermost" three black vertices). However, if the edges that connect the outermost black vertices with the black center vertex have edge weight one and all other edges weight three, then those three edges with weight one will form a minimum edge dominating set, in there is only one minimum edge dominating set in this weighted graph. However, starting with a minimum matching in the graph induced by the outermost three black vertices will never find the mentioned minimum edge dominating set, since some edge that connects the outermost black vertices will be contained in such a solution.

It is known (see [19]) that MINIMUM WEIGHTED EDGE COVER can be optimally solved in polynomial (cubic) time. Based on this result, Plesník gave a cubic time algorithm for a generalization of the problem we are interested in, see [20]; below, we give a shorter construction for a cubic time algorithm for GENERALIZED MINIMUM WEIGHTED EDGE COVER, directly based on [19].

So, let $G = (V, E)$ be a graph with edge weight function ω and with a set R of distinguished vertices. The task is to find a minimum weight edge set that covers R. As explained above, we cannot restrict our attention to $G[R] = (R, E_R)$. In addition, we consider the following edges. For each $x \in R$, let e_x be the edge incident with x that has lowest weight amongst all edges incident with x and not contained in E_R. If all edges incident with x are contained in E_R, then e_x is undefined. So, we add all such edges e_x to $G[R]$ to obtain a graph $G' = (V', E')$ with $R \subseteq V'$ and $E_R \subseteq E'$. Now, we identify all vertices in $V' \setminus R$ to get one new vertex r, so that we obtain a graph $G'' = (V'', E'')$ with $V'' = R \cup \{r\}$. The edge weight function is accordingly adapted and also called ω. Let f be the function that maps edges from E' onto edges from E'' (by the described identification).

Proposition 1. *(G, ω, R) has a minimum generalized weighted edge cover of weight ω_{opt} if and only if either $(G[R], \omega)$ or (G'', ω) has a minimum weighted edge cover of weight ω_{opt}.*

Proof. Let C be a minimum generalized weighted edge cover covering R. Without loss of generality, edges from C that are not contained in $G[R]$ could be the edges of the form e_x chosen as described above. If C contains no edges of the form e_x, then C is also a minimum weighted edge cover for $G[R]$. If C contains edges of the form e_x, then $f(C)$ is a minimum weighted edge cover for G''.

Conversely, a set C that is a feasible edge cover for $G[R]$ is also a feasible generalized edge cover for (G, R). Similarly, an edge set C (containing no edges outside of E') such that $f(C)$ is a feasible edge cover of G'' is also a feasible generalized edge cover for (G, R). Taking the cover of lowest weight ensures to produce a minimum generalized weighted edge cover for (G, ω, R).   ∎

Hence, Theorem 2 is also valid in the weighted case.

4   A Closer Look at the Search Tree

Can we further improve on the running time of Theorem 2 ? To this end, reconsider the main idea behind the enumeration of all minimal vertex covers C of size up to $2k$: starting from such a cover, we are going to construct a corresponding edge dominating set D of minimum cardinality among all edge dominating sets D' with $C \subseteq C_{D'}$.

It is usually a good strategy to look at small-degree vertices to find nice branching scenarios (or kernelizations). Assume that we do not branch at vertices of degree one. If ℓ is a bound on the vertex cover size, this gives the recurrence

$$T(\ell) \leq T(\ell - 1) + T(\ell - 2)$$

for the running time, i.e., $T(\ell) \leq 1.6181^{\ell}$, i.e., with $\ell = 2k$, 2.6181^k is an upperbound for the run time in this case.[2]

So, after branching we are left with a set of vertices C that covers all vertices of the originally given graph $G = (V, E)$ but an edge set E' that is the set of edges of $G[V \setminus C]$ that has maximum degree one. Now, we can produce a corresponding edge dominating set for G from this structure by finding an edge set D with the following properties:

---

[2] The related so-called enumerate-and-expand speed-up technique was independently developed by Mölle, Richter and Rossmanith (see Proc. CSR and COCOON, appearing in 2006), worked out with the example of CONNECTED VERTEX COVER.

- every $v \in C$ is covered by some $e \in D$, i.e., $C \subseteq C_D$;
- every $e' \in E'$ is dominated by some $e \in D$, i.e., $\forall e' \in E' \exists e \in D : e' \cap e \neq \emptyset$.

We will call a D satisfying both properties (C, E')-satisfying. Coming back to our HITTING SET model from the beginning, one can see that these properties can be modeled by aiming at constructing a minimum hitting set for a hypergraph whose vertex set is E and which has the following edges:

- for every $v \in C$, introduce the hyperedge $\{\{u, v\} \mid u \in N(v)\}$;
- for every $e' \in E'$, introduce the hyperedge $\{e \mid e \cap e' \neq \emptyset\}$.

Notice that there cannot be more than $2k$ hyperedges in the constructed HITTING SET instance (unless we face a NO-instance), since in the previously described complete vertex enumeration scenario, each of the edges from E' would have been "resolved" by a further branching step.

We can go one step further; namely, notice that we can ignore the possibility to include any $e' \in E'$ into the corresponding edge dominating set we construct, as long as there is any edge $e \in E$ with $|e \cap e'| = 1$. To exclude that special case, we employ in the very beginning of the algorithm (before even starting the vertex cover enumeration phase) the following reduction rule as long as possible:

Reduction rule 1. *(isolated edges) Let $(G = (V, E), k)$ be an instance of* EDGE DOMINATING SET. *If $e \in E$ such that $\forall \hat{e} \in E : \hat{e} \cap e \neq \emptyset \Rightarrow \hat{e} = e$, then delete e from the graph instance and decrease the parameter k by one.*

After having performed the vertex cover enumeration phase (without branching at vertices of degree one) on such a reduced instance G (without isolated edges), we have arrived (at each leave of the search tree) at a partial cover set C plus the above-mentioned edge set E'. Now, we form a new graph G' from G by contracting all edges from E'. Moreover, let M be the set of $|E'|$ vertices obtained by merging endpoints of edges from E'. Let $C' = C \cup M$. We claim that D is a minimum edge dominating set for G that is (C, E')-satisfying if and only if there is a minimum general edge cover D' for G' (with red vertex set C') with $|D'| = |D|$. Namely, if D' is a general edge cover D' for G' covering all vertices from C', then after "unmerging" we recover the graph $G = (V, E)$ in which we can view the edges from D' as elements from E. Then, D' is a (C, E')-satisfying edge set that is an edge dominating set according to our previous reasoning. Conversely, if D is a (C, E')-satisfying edge dominating set, then first we can transform D into a (C, E')-satisfying edge dominating set D' with $D' \cap E' = \emptyset$; namely, since G contains no isolated edges, we can replace any $e \in D \cap E'$ by some (arbitrarily chosen) incident edge e'. The edge set D' constructed this way can be interpreted as an edge set of G', and now D' (with $|D'| = |D|$ if we assume minimality of D) is a general edge cover for G'.

Theorem 3. EDGE DOMINATING SET *can be solved in time $\mathcal{O}^*((2.6181)^k)$.*

For MINIMUM MAXIMAL MATCHING we can conclude (based on [3,17,23]):

Corollary 2. *The problem* MMM *can be solved in time $\mathcal{O}^*((2.6181)^k)$.*

Algorithm 3. Reducing MATRIX DOMINATION SET to EDGE DOMINATING SET.

Input(s): a matrix instance (M, k) of MATRIX DOMINATION SET.

Output(s): a graph instance (G, k) of EDGE DOMINATING SET such that (M, k) is a YES-instance iff (G, k) is a YES-instance.

Let C be the set of columns of M.
Let R be the set of rows of M.
Form the vertex set $V = C \cup R$ of $G = (V, E)$.
for all $i \in R$, $j \in C$ **do**
    Put $\{i, j\} \in E$ iff entry (i, j) of M is one.
end for

The sketched unweighted case can be transferred to the weighted case. More precisely, the yet uncovered edges (left over from the vertex cover enumeration phase) can be modelled by first introducing a fresh vertex u and connecting u to all vertices $[v, w]$ obtained by merging v and w as described above. The new edge $\{u, [v, w]\}$ will get the same weight as the former edge $\{v, w\}$ had, and all other edge weights will stay the same.

Corollary 3. *The problem* WEDS *can be solved in time* $\mathcal{O}^*((2.6181)^k)$.

Let us finally consider a related problem, also mentioned in [23]: An instance of MATRIX DOMINATION SET (MDS) is given by an $n \times n$ matrix with entries from $\{0, 1\}$, and a positive integer k (the parameter). We ask: Is there a set D of one-entries in the matrix, where $|D| \leq k$, such that every other one-entry has at least one row or one column in common with some one-entry from D? Observe that this problem can be also seen as a chess piece domination problem: interpret the matrix as a chessboard showing places where it is allowed to place a rook or where not (by having a one- or a zero-entry in the corresponding position).

Lemma 1 (Yannakakis/Gavril). MATRIX DOMINATION SET *can be reduced (via* $\mathcal{FPT}$ *reduction) to* EDGE DOMINATING SET.

The corresponding reduction is formulated in Alg. 3, hence making explicit the remark in [8, p. 249] that MDS can be solved via EDS. In [6, Exercise 3.2.9], solving MATRIX DOMINATION SET by means of a kernelization and search tree based algorithm is proposed as an exercise. [3]

Hence, MDS is in one-to-one correspondence to EDS, restricted to bipartite graphs. Since our solution of EDGE DOMINATING SET is based on VERTEX COVER, and the latter (in its decision version!) is known to be easier on bipartite graph, the following corollary might see some improvements; however, we did not manage to get improvements in a straightforward manner, since we are rather relying on the enumeration than on the decision version of VC.

Corollary 4. MATRIX DOMINATION SET *can be solved in time* $\mathcal{O}^*((2.6181)^k)$.

---

[3] In [22], a $\mathcal{O}^*(c^k)$ algorithm is proposed for MATRIX DOMINATION SET with $c < 2$; however, this is based on a wrong interpretation of our results on CONSTRAINT BIPARTITE VERTEX COVER as detailed in [11].

5   Compact Representations

We have shown that auxiliary vertex cover structures can be quite useful to solve EDGE DOMINATING SET and related problems. The speed-up described in the previous section can be also interpreted as enumerating compact representations of vertex covers that can be formalized similar to regular expressions.

1. $\emptyset$ is an expression denoting a compact representation that denotes no sets at all, i.e., $C(\emptyset) = \{\emptyset\}$.
2. If a is a vertex, then a is an (atomic) compact representation of the cover collection $C(a)$ only containing the cover $\{a\}$, i.e., $C(a) = \{\{a\}\}$.
3. If $e = \{a, b\}$ is an edge, then $\hat{e}$ is an (atomic) compact representation of the cover collection $C(\hat{e})$ only containing the covers $\{a\}$ and $\{b\}$,
4. If A and B are compact representations that represent cover collections $C(A)$ and $C(B)$, resp., then $A + B$ represents the cover collections

$$C(A + B) = \{X \cup Y \mid X \in C(A), Y \in C(B)\}.$$

5. If A and B are compact representations that represent cover collections $C(A)$ and $C(B)$, then $A \cup B$ represents the cover collection $C(A \cup B) = C(A) \cup C(B)$.
6. Nothing else are compact representations.

Example 1. For example, the minimal vertex covers of the graph

$$(\{1, \ldots, k\} \times \{1, 2\}, \{\{(i, 1), (i, 2)\} \mid 1 \leq i \leq k\})$$

can be written as

$$\{(1, \widehat{1}), (1, 2)\} + \{(2, \widehat{1}), (2, 2)\} + \cdots + \{(k, \widehat{1}), (k, 2)\}.$$

For instance, if $k = 3$,

$$\begin{aligned}
&\{(1, \widehat{1}), (1, 2)\} + \{(2, \widehat{1}), (2, 2)\} + \{(3, \widehat{1}), (3, 2)\} \\
&= \{\{(1, 1)\}, \{(1, 2)\}\} + \{\{(2, 1)\}, \{(2, 2)\}\} + \{\{(3, 1)\}, \{(3, 2)\}\} \\
&= \{\{(1, i), (2, j), (3, \ell)\} \mid 1 \leq i, j, \ell \leq 2\}
\end{aligned}$$

Theorem 4. *Representations of all minimal vertex covers of size up to k (and possibly some more non-minimal cover representations) can be listed in time $\mathcal{O}^*(1.6181^k)$.*

Proof. (Sketch) The usual enumeration algorithm for listing all minimal vertex covers up to size k can be modified by avoiding branches at vertices of degree one. After this branching, the remaining graph has maximum degree one, and the cover of an edge $e = \{x, y\}$ can be described by $\hat{e}$. ∎

The difference to the results of Damaschke [4] is that he insists on enumerating only minimal vertex covers up to size k; hence, his running times are worse. Since vertex cover structure already found many applications, we believe that such representations (or possibly similar ones) may yield interesting improvements in the development of exact graph algorithms. For example, in [10], the problem of finding a *total vertex cover* of size up to k was discussed, where a vertex cover C is called total if every $x \in C$ satisfies $N(x) \cap C \neq \emptyset$. By a vertex cover enumeration phase yielding covers C in the leaves, followed by a hitting set phase (with hyperedge set $\{N(x) \mid x \in C\}$), this problem can be solved in time $\mathcal{O}^*(4^k)$. With compact representations, this can be readily improved to $\mathcal{O}^*(3.2361^k)$: for $\hat{e}$ in the compact representation of some covers with $e = \{x, y\}$, we can introduce a hyperedge $N(x) \cup N(y)$. Choosing $z \in N(x) \cup N(y)$ also determines whether x or y is put into the cover. However, different techniques allowed to further lower the constants for that problem to $\mathcal{O}^*(2.3655^k)$ in [10].

6 Kernels

We have solved the problems dealt with in this paper by a search-tree technique based on enumerating minimal vertex covers. This approach can be also used to show quadratic kernels for all these problems. Namely, as explained in [9], Buss' kernelization rules are also valid for the enumeration task. Hence, we can assume that the reduced graph has no more than $2(2k)^2 = 8k^2$ vertices in the enumeration phase, where k is the parameter of the say EDS instance. We can turn this into a kernel for EDS by the following observations: (a) We keep each vertex in the vertex cover enumeration kernel in the EDS kernel. (b) For each vertex v that was put into each vertex cover by Buss' rule, we have to put v plus an arbitrary neighbor u of v that is not yet in the EDS kernel, provided that v does not have already neighbors in the EDS kernel. This possibly complicated-looking rule allows to cover v by some edge also in the reduced instance. Since the number of vertices added by this special treatment of vertices put into the vertex cover by Buss' rule is smaller than the quantity in the general case, we get an upper bound of $8k^2$ vertices for the number of vertices in the EDS kernel.

Lemma 2. *Given an instance (G, k) of* EDGE DOMINATING SET, *it is possible to produce a kernel (G', k') of* EDS *with $|V(G')| \leq 8k^2$ and $k' \leq k$. Similar results are true for* WEDS, MMM, *and* MDS.

Notice that a kernel of size $4k(k + 2)$ was obtained by Prieto for MMM by adapting crown reduction techniques, see [21]. We can improve on the kernel size for EDS at the cost of introducing annotations (marking vertices that should go into the vertex cover):[4] (a) isolated edges are to be put into the edge dominating set anyways; (b) vertices of degree two with two neighbors of degree one comprise a component that can be solved by arbitrarily taking one of either edges of the component into the dominating set; (c) vertices of degree one that are not covered by rules (a) and (b) need not be put into the vertex covers considered in the

---

[4] The notion of *annotated kernel* is further discussed in joined work of Abu-Khzwam and Fernau, also contained in these proceedings.

enumeration phase but rather their unique neighbors; this is reflected by deleting the degree-one vertex and by marking its unique neighbor. By (c), also marked vertices have minimum degree of two. Hence, the kernel (possibly containing some marked vertices doomed to go into the vertex cover) contains at most $4k^2$ vertices: Buss' rule yields that there are at most $(2k)^2$ many edges in the graph instance, and knowing that both marked and unmarked vertices have minimum degree of two means: there are also at most $(2k)^2$ many vertices in the graph.

7   Conclusions and Open Problems

We have shown how ideas stemming from the area of parameterized enumeration can be useful to obtain efficient parameterized algorithms for decision problems. More precisely, we derived an $\mathcal{O}^*(2.62^k)$ algorithms for EDGE DOMINATING SET and many variants. It would be interesting to see if these constants could be further improved. In particular, it might be possible to avoid branching at degree-two vertices in the enumeration phase, as argued before Theorem 3 for degree-one vertices. Notice that in terms of approximation factors, there seems to be no difference between VERTEX COVER, TOTAL VERTEX COVER, EDGE DOMINATING SET, and FEEDBACK VERTEX SET; however, in terms of search tree based parameterized algorithms, VERTEX COVER appears to be the easiest of the three, while FEEDBACK VERTEX SET seems to be the hardest one, see [1,5,14,18].

It would be interesting to see if and how the ideas presented in this paper can be applied to solve WEIGHTED MINIMUM MAXIMAL MATCHING. Notice that the main problem is to allow Alg. 1 to cope with the additional independence condition. In the literature, several other variants of EDGE DOMINATING SET have been considered that might deserve further studies from the viewpoint of parameterized complexity; recent papers are [2,16].

Acknowledgments. We thank M. R. Fellows, R. Niedermeier, E. Prieto Rodríguez, B. Randerath, U. Stege, and M. Weston for some discussions.

References

1. V. Bafna, P. Berman, and T. Fujito. A 2-approximation algorithm for the undirected feedback vertex set problem. *SIAM Journal of Discrete Mathematics*, 12:289–297, 1999.
2. A. Berger and O. Parekh. Linear time algorithms for generalized edge dominating set problems. Technical Report TR-2005-002-A, Emory University Math/CS department, 2005.
3. R. Carr, T. Fujito, G. Konjevod, and O. Parekh. A 2 1/10 approximation algorithm for a generalization of the weighted edge-dominating set problem. *Journal of Combinatorial Optimization*, 5:317–326, 2001.
4. P. Damaschke. Parameterized enumeration, transversals, and imperfect phylogeny reconstruction. In R. Downey, M. Fellows, and F. Dehne, editors, *International Workshop on Parameterized and Exact Computation IWPEC 2004*, volume 3162 of *LNCS*, pages 1–12. Springer, 2004.

5. F. Dehne, M. R. Fellows, M. Langston, F. Rosamond, and K. Stevens. An $o(2^{O(k)}n^3)$ fpt algorithm for the undirected feedback vertex set problem. In *Proc. 11th International Computing and Combinatorics Conference COCOON*, volume 3595 of *LNCS*, pages 859–869. Springer, 2005.

6. R. G. Downey and M. R. Fellows. *Parameterized Complexity*. Springer, 1999.

7. U. Feige. A threshold of $\ln n$ for approximating set cover. *Journal of the ACM*, 45:634–652, 1998.

8. M. R. Fellows, C. McCartin, F. A. Rosamond, and U. Stege. Coordinatized kernels and catalytic reductions: an improved FPT algorithm for Max Leaf Spanning Tree and other problems. In S. Kapoor and S. Prasad, editors, *FST TCS 2000*, volume 1974 of *LNCS*, pages 240–251. Springer, 2000.

9. H. Fernau. On parameterized enumeration. In O. H. Ibarra and L. Zhang, editors, *Computing and Combinatorics, Proceedings COCOON 2002*, volume 2383 of *LNCS*, pages 564–573. Springer, 2002.

10. H. Fernau and D. F. Manlove. Vertex and edge covers with clustering properties: Complexity and algorithms. Technical Report TR-2006-210, Department of Computing Science DCS Glasgow, UK, April 2006. Available through: http://www.dcs.gla.ac.uk/publications/paperdetails.cfm?id=8137.

11. H. Fernau and R. Niedermeier. An efficient exact algorithm for constraint bipartite vertex cover. *Journal of Algorithms*, 38(2):374–410, 2001.

12. F. Fomin, D. Kratsch, and G. Woeginger. Exact (exponential) algorithms for the dominating set problem. In J. Hromkovic et al., editors, *30th International Workshop on Graph-Theoretic Concepts in Computer Science WG 2004*, volume 3353 of *LNCS*, pages 245–256. Springer, 2004.

13. T. Fujito and H. Nagamochi. A 2-approximation algorithm for the minimum weight edge dominating set problem. *Discrete Applied Mathematics*, 118:199–207, 2002.

14. J. Guo, J. Gramm, F. Hüffner, R. Niedermeier, and S. Wernicke. Improved fixed-parameter algorithms for two feedback set problems. In *Proceedings of the 9th Workshop on Algorithms and Data Structures WADS*, volume 3608 of *LNCS*, pages 158–168. Springer, 2005.

15. J. D. Horton and K. Kilakos. Minimum edge dominating sets. *SIAM Journal of Discrete Mathematics*, 6:375–387, 1993.

16. C. L. Lu, M.-T. Ko, and C. Y. Tang. Perfect edge domination and efficient edge domination in graphs. *Discrete Applied Mathematics*, 119(3):227–250, 2002.

17. D. F. Manlove. *Minimaximal and maximinimal optimisation problems: a partial order-based approach*. PhD thesis, University of Glasgow, Computing Science, 1998.

18. O. Parekh. Edge domination and hypomatchable sets. In *Symposium on Discrete Algorithms SODA 2002*, pages 287–291. ACM Press, 2002.

19. J. Plesník. Equivalence between the minimum covering problem and the maximum matching problem. *Discrete Mathematics*, 49:315–317, 1984.

20. J. Plesník. Constrained weighted matchings and edge coverings in graphs. *Discrete Applied Mathematics*, 92:229–241, 1999.

21. E. Prieto. *Systematic Kernelization in FPT Algorithm Design*. PhD thesis, The University of Newcastle, Australia, 2005.

22. M. Weston. A fixed-parameter tractable algorithm for matrix domination. *Information Processing Letters*, 90:267–272, 2004.

23. M. Yannakakis and F. Gavril. Edge dominating sets in graphs. *SIAM Journal of Applied Mathematics*, 38(3):364–372, June 1980.

Parameterized Complexity of Independence and Domination on Geometric Graphs

Dániel Marx

Institut für Informatik,
Humboldt-Universität zu Berlin,
Unter den Linden 6, 10099
Berlin, Germany
dmarx@informatik.hu-berlin.de

Abstract. We investigate the parameterized complexity of MAXIMUM INDEPENDENT SET and DOMINATING SET restricted to certain geometric graphs. We show that DOMINATING SET is W[1]-hard for the intersection graphs of unit squares, unit disks, and line segments. For MAXIMUM INDEPENDENT SET, we show that the problem is W[1]-complete for unit segments, but fixed-parameter tractable if the segments are axis-parallel.

1 Introduction

For a set V of geometric objects, the *intersection graph* of V is a graph with vertex set V where two vertices are connected if and only if the corresponding two objects have non-empty intersection. Intersection graphs of disks, rectangles, line segments, and other objects arise in applications such as facility location [5], frequency assignment [3], and map labeling [1].

In this paper we investigate the parameterized complexity of MAXIMUM INDEPENDENT SET and DOMINATING SET restricted to certain geometric graphs. Both of these problems are W[1]-hard on general graphs, but fixed-parameter tractable when restricted to planar graphs. Geometric intersection graphs are in some sense intermediate between these two classes: they still have lot of geometric structure that might be used in algorithms, but we lose some of the simplicity of planar graphs. Therefore, it is an interesting question to investigate the complexity of these problems on different types of geometric graphs.

This line of research was pursued in [4], where MAXIMUM INDEPENDENT SET was proved to be W[1]-complete for unit disk and unit square graphs. Here we extend the results by considering the intersection graphs of line segments and the DOMINATING SET problem.

In Section 2, we introduce a general framework that can be used to prove W[1]-hardness for geometric problems. We give a semi-formal definition of what properties the gadgets of the reduction have to satisfy; in later sections the only thing we have to do for each W[1]-hardness proof is to define the problem-specific gadgets and verify the required properties.

In Section 3, we show that DOMINATING SET is W[1]-hard for unit disk graphs and unit square graphs. In general, DOMINATING SET is W[2]-complete, but it

H.L. Bodlaender and M.A. Langston (Eds.): IWPEC 2006, LNCS 4169, pp. 154–165, 2006.

turns out that DOMINATING SET is in W[1] (hence W[1]-complete) for unit square graphs. As far as we know, this is the first example when DOMINATING SET restricted to some class of graphs is not W[2]-complete, but not fixed-parameter tractable either. Section 4 shows that DOMINATING SET is W[1]-complete also for the intersection graphs of axis-parallel line segments.

Section 5 considers the MAXIMUM INDEPENDENT SET problem for the intersection graphs of line segments. If the segments are axis-parallel (or more generally, if they belong to at most d different directions), then the problem is fixed-parameter tractable. However, if there is no restriction on the number of different directions, then the problem becomes W[1]-complete, even if every segment has the same length.

2   General Framework

All the W[1]-hardness proofs in the paper follow the same general framework. In this section we present a general reduction technique that can be used to prove hardness of a geometric or planar problem. The reduction creates an instance that consists of some number of gadgets, and connections between gadgets. The exact details of the gadgets and the connections are problem specific, and will be given in later sections separately for each problem. However, we show here that if for a particular problem there is a gadget satisfying certain properties, then the problem is W[1]-hard.

The W[1]-hardness proof is by parameterized reduction from MAXIMUM CLIQUE. Given a graph G and an integer k, it has to be decided if G has a clique of size k. For convenience, we assume that G has n vertices and n edges. The set of vertices and the set of edges are identified with the set $\{1, 2, \ldots, n\}$.

The constructed instance contains k^2 copies of the gadget, arranged in k rows and k columns. The gadget in row i and column j will be denoted by $G_{i,j}$. Adjacent gadgets in the same row are connected by a *horizontal connection* and adjacent gadgets in the same column are connected by a *vertical connection*.

Let $\iota : \{1, \ldots, n^2\} \to \{1, \ldots, n\} \times \{1, \ldots, n\}$ be an arbitrary one-to-one mapping, and let $\iota(s) = (\iota_1(s), \iota_2(s))$ for every s. For technical reasons, in this paper we always use the mapping defined by $s = (\iota_1(s) - 1)n + \iota_2(s)$. The crucial property of the gadget is that in every optimum solution it represents an integer number between $1 \le s \le n^2$, which can be also interpreted as the pair $\iota(s)$. The role of the horizontal connections is to ensure that if the values of the two gadgets are s and s', then $\iota_1(s) = \iota_1(s')$, i.e., they agree in the first component. Therefore, in an optimum solution the same value v_i will be represented by the first component of every gadget in row i. Similarly, the vertical connections ensure that if s and s' are the values of two adjacent gadgets in a column, then $\iota_2(s) = \iota_2(s')$. Thus the second component has the same value v'_j in column j.

Now we encode the graph G into the instance by restricting certain gadgets. Restricting a gadget to the subset $S \subseteq \{1, 2, \ldots, n^2\}$ means that the gadget is modified such that it can represent values only from S. For every $1 \le i \le k$, we restrict the gadget $G_{i,i}$ to the set $\{s : \iota_1(s) = \iota_2(s)\}$. This ensures that the first component in row i is the same as the second component in column i, i.e., $v_i = v_i'$ for every $1 \leq i \leq k$. To encode the structure of the graph, we restrict $G_{i,j}$ (for every $i \neq j$) to the set $\{s : \iota_1(s) \text{ and } \iota_2(s) \text{ are adjacent vertices}\}$. It is clear that if every gadget has a value that respects these restrictions, then $v_1, v_2, \ldots, v_k$ are all distinct and they form a clique of size k: if v_i and v_j are not adjacent, then the value $(v_i, v_j) = (v_i, v_j')$ does not respect the restriction on gadget $G_{i,j}$. On the other hand, if $v_1, v_2, \ldots, v_k$ is a clique of size k, then we can assign value $\iota^{-1}((v_i, v_j))$ to gadget $G_{i,j}$. This assignment respects the restrictions on the gadgets and the connections.

In summary, the gadgets have to satisfy the following requirements:

Definition 1 (Matrix Gadget). *A gadget satisfies the following properties:*

1. **(The gadget)** *In every solution of the constructed instance, each gadget represents a number between 1 and n^2.*
2. **(Restriction)** *The gadget can be restricted to a set $\emptyset \neq S \subseteq \{1, \ldots, n^2\}$ such that in every solution the gadget represents a number in S.*
3. **(Horizontal connection)** *If two gadgets are connected by a horizontal connection, then the values they represent agree in the first component.*
4. **(Vertical connection)** *If two gadgets are connected by a vertical connection, then the values they represent agree in the second component.*
5. **(Constructing a solution)** *If it is possible to assign values to the gadgets such that this assignment respects the restrictions and respects the connections, then the constructed instance has a solution.*

The first four requirements ensure that if the instance described above has a solution, then G has a clique of size k. The other direction of the reduction follows from the last requirement: if $v_1, \ldots, v_k$ is a clique of size k, then giving the value (v_i, v_j) to gadget $G_{i,j}$ respects the restrictions and the connections, thus there is a solution.

3   Dominating Set for Squares and Disks

The first problem we consider is DOMINATING SET: given a graph G, the task is to find a set S of k vertices such that each vertex of the graph is either in S or is a neighbor of a member of S. In this section we prove hardness results for the problem in the case of unit disk graphs and unit square graphs.

Theorem 1. DOMINATING SET *is W[1]-hard for axis-parallel unit squares.*

Proof. The proof uses the framework of Section 2. Let $\epsilon < 1/3n^2$. In this proof it does not matter if the squares are open or closed. In the constructed instance of DOMINATING SET the lower left corner of each square is of the form $(i + \alpha\epsilon, j + \beta\epsilon)$, where i and j are integers, and $-n \leq \alpha, \beta \leq n$. If two squares have the same i, j values, then they belong to the same *block;* the blocks form a partition of the squares. If the lower left corner of a square S is $(i + \alpha\epsilon, j + \beta\epsilon)$ then α (resp., β) is the *horizontal* (resp., *vertical*) offset of S, and we set offset$(S) = (\alpha, \beta)$.

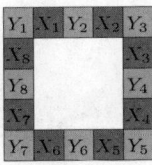

Fig. 1. The gadget used in the proof of Theorem 1

The gadget. The gadget used in the reduction is shown in Figure 1. It consists of 16 blocks $X_1, \ldots, X_8, Y_1, \ldots, Y_8$. Each block X_i contains n^2 squares $X_{i,1}, \ldots, X_{i,n^2}$, while each block Y_i contains $n^2 + 1$ squares $Y_{i,0}, \ldots, Y_{i,n^2}$. The offsets of the squares are defined as follows:

$$\text{offset}(X_{1,j}) = (j, -\iota_2(j)) \qquad \text{offset}(Y_{1,j}) = (j + 0.5, j + 0.5)$$

$$\text{offset}(X_{2,j}) = (j, \iota_2(j)) \qquad \text{offset}(Y_{2,j}) = (j + 0.5, -n)$$

$$\text{offset}(X_{3,j}) = (-\iota_1(j), -j) \qquad \text{offset}(Y_{3,j}) = (j + 0.5, -j - 0.5)$$

$$\text{offset}(X_{4,j}) = (\iota_1(j), -j) \qquad \text{offset}(Y_{4,j}) = (-n, -j - 0.5)$$

$$\text{offset}(X_{5,j}) = (-j, \iota_2(j)) \qquad \text{offset}(Y_{5,j}) = (-j - 0.5, -j - 0.5)$$

$$\text{offset}(X_{6,j}) = (-j, -\iota_2(j)) \qquad \text{offset}(Y_{6,j}) = (-j - 0.5, n)$$

$$\text{offset}(X_{7,j}) = (\iota_1(j), j) \qquad \text{offset}(Y_{7,j}) = (-j - 0.5, j + 0.5)$$

$$\text{offset}(X_{8,j}) = (-\iota_1(j), j) \qquad \text{offset}(Y_{8,j}) = (n, j + 0.5)$$

Observe that two squares can intersect only if they belong to the same or adjacent blocks. For example, the squares in block X_2 have positive vertical offsets and the squares in X_3 have negative vertical offset, hence they do not intersect. The crucial property of the construction is that two squares X_{i,j_1}, X_{i+1,j_2} dominate every square of Y_{i+1} if and only if $j_1 \geq j_2$. This follows from the fact that X_{i,j_1} dominates exactly $Y_{i+1,0}, \ldots, Y_{i+1,j_1-1}$ from block Y_{i+1} and X_{i+1,j_2} dominates exactly $Y_{i+1,j_2}, \ldots, Y_{i+1,n^2}$ from block Y_{i+1}.

Lemma 1. *Assume that a gadget is part of an instance such that none of the blocks Y_i are intersected by squares outside the gadget. If there is a dominating set D of the instance that contains exactly 8 squares from the gadget, then there is a dominating set D' with $|D'| \leq |D|$, and there is an integer $1 \leq j \leq n^2$ such that D' contains exactly the squares $X_{1,j}, \ldots, X_{8,j}$ from the gadget.*

Proof. If D contains no square from any X_i, then it has to contain at least one square from each Y_i. Remove these squares, and add the squares $X_{1,1}, \ldots, X_{8,1}$ to D instead. This does not increase the size of D, and every square of the gadget will be dominated. Furthermore, as a square from Y_i cannot dominate anything outside the gadget, the modified set is also a dominating set, and we are done.

We show that D' can be chosen such that it contains exactly one square from each X_i, and consequently, it contains no squares from the blocks Y_i. Observe that the squares in Y_i cannot be all dominated by squares only from X_{i-1} or by squares only from X_i (the indices of the blocks are modulo 8). This implies that if D contains no square from Y_i, then D contains at least one square from X_{i+1}

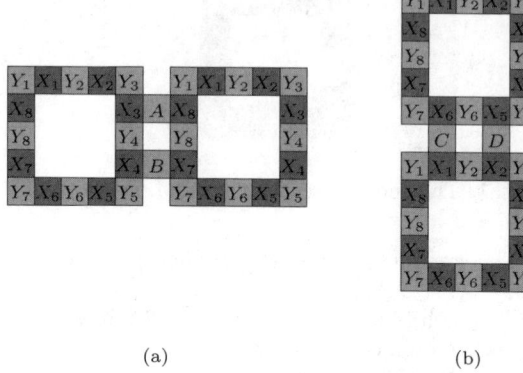

(a)   (b)

Fig. 2. The horizontal (a) and vertical (b) connections used in the proof of Theorem 1

and at least one square from X_{i-1}. Assume that $D \cap X_i = \emptyset$ for some i, but $D \cap (X_1 \cup \cdots \cup X_8)$ is maximal. Since D contains a square from some X_i, there are integers a, b such that $D \cap X_a \neq \emptyset$, $D \cap X_b \neq \emptyset$, and $D \cap X_i = \emptyset$ for every $a < i < b$. Therefore, $D \cap Y_i \neq \emptyset$ for every $a < i \leq b$. Let $X_{a,j}$ be a member of $X_a \cap D$. Set $D' := (D \setminus (Y_{a+1} \cup \cdots \cup Y_b)) \cup X_{a+1,j} \cup \cdots \cup X_{b,j}$. Clearly, $|D'| \leq |D|$, and D is also a dominating set: the squares in Y_i are dominated by $X_{i-1,j}$ and $X_{i,j}$ for every $a < i \leq b$. This contradicts the maximality of $D \cap (X_1 \cup \cdots \cup X_8)$.

Assume that D contains squares $X_{1,j_1}, \ldots, X_{8,j_8}$, this means that D contains no other square from the gadget. As we have observed above, the squares in Y_i are dominated only if $j_{i-1} \geq j_i$. This gives the chain of inequalities $j_1 \geq j_2 \geq \cdots \geq j_8 \geq j_1$, thus all these values are the same integer j. Thus D contains exactly the squares $X_{1,j}, \ldots, X_{8,j}$ from the gadget.   □

The constructed instance contains k^2 copies of the gadget, and it will be true that gadgets are connected to the rest of the instance only via the X_i blocks. The new parameter (the size of the dominating set to be found) is $k' = 8k^2$. At least 8 squares are required to dominate the Y_i blocks of a gadget, thus every solution has to contain exactly 8 squares from each gadget. In this case, Lemma 1 defines a number j for each gadget, which will be called the *value* of the gadget. Therefore, Property 1 of Definition 1 is satisfied.

Restriction. Let $S \subseteq \{1, 2, \ldots, n^2\}$ be an arbitrary set. We restrict the gadget by removing every square $X_{i,j}$ for $1 \leq i \leq 8$ and $j \notin S$. It can be checked that Lemma 1 remains true for gadgets modified this way. Obviously, if $X_{1,j}$ is removed, then the gadget cannot represent value j, thus the value represented by the gadget will be a member of S.

Horizontal connections. The horizontal connections required by Property 3 are shown in Figure 2a. We add a block A that is adjacent to block X_3 of the first gadget and block X_8 of the second, and we add a block B adjacent to X_4 of the first gadget and X_7 of the second. Blocks A and B contain $n+1$ squares each:

square A_j has offset $(-j - 0.5, -n^2 - 1)$ and square B_j has offset $(j + 0.5, n^2 + 1)$ $(0 \leq j \leq n)$. These blocks do not intersect the Y_i blocks.

Assume that a dominating set does not contain any of the squares from A and B, it contains exactly the squares $X_{1,j}, \ldots, X_{8,j}$ from the first gadget, and it contains exactly the squares $X_{1,j'}, \ldots, X_{8,j'}$ from the second gadget. We claim that $\iota_1(j) = \iota_1(j')$. If $\iota_1(j) > \iota_1(j')$, then $X_{3,j}$ of the first gadget dominates the squares $A_{\iota_1(j)}, \ldots, A_{n^2}$ and $X_{8,j}$ of the second gadget dominates squares $A_0, \ldots, A_{\iota_1(j')-1}$, thus $A_{\iota_1(j')}$ is not dominated. If $\iota_1(j) < \iota_1(j')$, then no square dominates $B_{\iota_1(j)}$ of block B. Thus $\iota_1(j) = \iota_1(j')$, and the values of the two gadgets agree in the first component.

Vertical connections. Vertical connections are defined analogously (see Figure 2b). Square C_j of block C has offsets $(n^2 + 1, -\iota_2(j))$ and square D_j has offsets $(-n^2 - 1, \iota_2(j))$ $(0 \leq j \leq n)$.

Constructing a solution. It is straightforward to see that if every gadget has a correct value, then a dominating set of size $8k^2$ can be found: if the value of a gadget is j, then select the 8 squares $X_{1,j}, \ldots, X_{8,j}$ from the gadget.   □

The same reduction shows hardness for unit disks: it can be shown that if each square in the constructed instance is replaced by a disk and ϵ is sufficiently small, then the intersection structure does not change. Details omitted.

Theorem 2. *Maximum independent set is* W[1]*-hard for the intersection graphs of unit disks in the plane.*   □

For general graphs DOMINATING SET set is W[2]-complete, therefore Theorem 1 leaves open the question whether the problem is W[1]-complete or W[2]-complete when restricted to these graph classes. For unit squares (and more generally, for axis-parallel rectangles) we show that dominating set is in W[1]. This is the first example when a restriction of dominating set is easier than the general problem, but it is not fixed-parameter tractable.

Theorem 3. DOMINATING SET *is in* W[1] *for the intersection graphs of axis-parallel rectangles.*

Proof. We prove membership in W[1] by reducing DOMINATING SET to SHORT TURING MACHINE COMPUTATION. We construct a Turing machine (with unbounded nondeterminism) that accepts the empty string in k' steps if and only if there is a dominating set of size k. Henceforth $L(S)$ (resp., $R(S)$) denotes the x-coordinate of the left (resp., right) edge of open rectangle S, and $T(S)$ (resp., $B(S)$) denotes the y-coordinate of the top (resp., bottom) edge.

The tape alphabet of the Turing machine consists of one symbol for each rectangle in the instance plus two special symbols 0 and 1. In the fist k steps the machine nondeterministically writes k symbols $x_1, \ldots, x_k$ on the tape, which is a guess at a size k dominating set. Next $4k^2$ symbols $h_{1,1}, \ldots, h_{k,k}, h'_{1,1}, \ldots, h'_{k,k}$, $v_{1,1}, \ldots, v_{k,k}, v'_{1,1}, \ldots, v'_{k,k}$ are written, each of these symbols is either 0 or 1. The intended meaning of $h_{i,j}$ is the following: it is 1 if and only if $R(x_i) \leq L(x_j)$. Similarly, we will interpret $h'_{i,j} = 1$ as $R(x_i) \leq R(x_j)$. The symbols $v_{i,j}$ and $v'_{i,j}$ have similar meaning, but with B and T instead of L and R.

The rest of the computation is deterministic. First we check the consistency of the symbols $h_{i,j}$ with the symbols x_i, x_j. For each $1 \leq i, j \leq k$, we make a full scan of the tape, and store in the internal state of the machine the symbols $h_{i,j}$, x_i, x_j. If these symbols are not consistent (e.g., $R(x_i) > L(x_j)$ but $h_{i,j} = 1$) then the machine rejects. The length of the tape is $k + 4k^2$, and we repeat the check for k^2 pairs i, j, thus the checks take a constant number of steps.

For technical reasons we add four dummy rectangles D_L, D_R, D_T, D_B. The rectangle D_L is to the left of the other rectangles, i.e., $R(D_L) \leq L(S)$ for every other rectangle S. Similarly, the rectangles D_R, D_T, D_B are to the right, top, bottom of the other rectangles, respectively. Instead of testing whether the k rectangles $x_1, \ldots, x_k$ form a dominating set, we will test whether $x_1, \ldots, x_k$, D_L, D_R, D_T, D_B are dominating. Clearly, the answer is the same.

We say that the $k + 4$ selected rectangles contain an *invalid window* if there are four selected rectangles S_L, S_R, S_T, S_B with the following properties.

- $R(S_L) \leq L(S_R)$ and $T(S_B) \leq B(S_T)$. Let A be the rectangle with left edge $R(S_L)$, right edge $L(S_R)$, bottom edge $T(S_B)$, top edge $B(S_T)$.
- There is no selected rectangle that intersects A.
- There is a rectangle S that is completely contained in A.

If the selected rectangles contain an invalid window, then they are not dominating since rectangle S is not dominated. On the other hand, if there is a rectangle S which is not dominated, then the selected squares contain an invalid window: by extending S into the four directions until we reach the edge of some selected rectangles, we obtain the window A. The four rectangles that stopped us from further extending A can be used as S_L, S_R, S_T, S_B.

In the rest of the computation, the Turing machine checks whether the selected rectangles contain an invalid window. For each quadruple i_L, i_R, i_T, i_B it has to be checked whether the rectangles x_{i_L}, x_{i_R}, x_{i_T}, x_{i_B} form an invalid window. Using the symbols $h_{i,j}$ etc. on the tape, it can be tested in a constant number of steps whether a selected rectangle intersects the window determined by these four rectangles. If not, then the machine reads into its internal state the four values x_{i_L}, x_{i_R}, x_{i_T}, x_{i_B}, and rejects if there is a rectangle in the window determined by these squares. There are $(k + 4)^4$ possible quadruples and each check can be done in a constant number of steps; therefore, the whole computation takes a constant number k' of steps. □

4   Dominating Set for Line Segments

In this section we use the framework of Section 2 to prove that DOMINATING SET is W[1]-complete also for axis-parallel line segments.

Theorem 4. DOMINATING SET *is* W[1]-*complete for axis-parallel segments.*

Proof. Membership in W[1] follows from Theorem 3. Therefore, only W[1]-hardness has to be proven here. In the constructed instance of DOMINATING

SET, there are k^2 gadgets, and the new parameter is $k' = 12k^2$. Every dominating set has to contain at least 12 segments from each gadget, hence every solution contains exactly 12 segments from each gadget.

The gadget. The gadget satisfying the requirements of Definition 1 is shown in Figure 3. Unless stated otherwise, the segments are open in this proof. The line segments in the gadget can be dominated only by at least 12 segments, since a segment can dominate at most one of a', b', ..., ℓ'. Furthermore, we claim that there are exactly n^2 dominating sets of size 12: they are of the form $a_s, b_s, \ldots, \ell_s$ for $1 \le s \le n^2$. First, it is easy to see that a size 12 dominating set has to contain exactly one segment from $a_1, \ldots, a_{n^2}$, exactly one segment from $b_1, \ldots, b_{n^2}$, etc. For example, if none of $a_1, \ldots, a_{n^2}$ is selected, then we have to select both a' and a'', which makes the size of the dominating set greater than 12. Assume that $a_{s_a}, b_{s_b}, \ldots, \ell_{s_\ell}$ is a dominating set. Segment a_{s_a} dominates $b_1, b_2, \ldots,$ b_{s_a-1} (see Fig. 3) and c_{s_c} dominates $b_{s_c+1}, \ldots, b_{n^2}$. Therefore, if $s_c > s_a$ then neither b_{s_a} nor b_{s_c} are dominated. At most one of these two segments can be selected, thus there would be a segment that is neither selected nor dominated. We can conclude that $s_c \le s_a$. Moreover, it is also true that if $s_c = s_a$, then neither a_{s_a} nor c_{s_c} dominates $b_{s_a} = b_{s_c}$, hence $s_b = s_a = s_c$ follows. By a similar argument, $s_e \le s_c$ with equality only if $s_c = s_d = s_e$. Continuing further we obtain $s_a \ge s_c \ge s_e \ge s_g \ge s_i \ge s_k \ge s_a$, thus there are equalities throughout, implying $s_a = s_b = s_c = \cdots = s_\ell$, as required. This means that the gadget represents a value between 1 and n^2 in every solution.

Restriction. Restricting a gadget to set S is implemented by removing the segments $a_s, b_s, \ldots, \ell_s$ from the gadget for every $s \notin S$.

Horizontal connections. Figure 4a shows how to connect two adjacent gadgets by a horizontal connection. We add $2n$ new segments $x_1, \ldots, x_n, y_1, \ldots,$ y_n. The right end point of h_i (resp., j_i) in the first gadget is modified to be its intersection with $x_{\iota_1(i)}$ (resp., $y_{\iota_1(i)}$), and this end point is set to be a closed end point. The left end point of d_i and b_i are similarly modified, but these end points are set to be open. Assume that there is a dominating set that contains 12 segments from each of the gadgets and contains none of the segments $x_1, \ldots,$ $x_n, y_1, \ldots, y_n$. Furthermore, assume that the pair $i = (\iota_1(i), \iota_2(i))$ is the value of the first gadget and the $i' = (\iota_1(i'), \iota_2(i'))$ is the value of the second gadget. In particular, this means that h_i, j_i are selected in the first gadget, and $b_{i'}$, $d_{i'}$ are selected in the second. Now if $\iota_1(i) < \iota_1(i')$, then $x_{\iota_1(i')}$ is not dominated, and if $\iota_1(i) > \iota_1(i')$, then $y_{\iota_1(i')}$ is not dominated, thus $\iota_1(i) = \iota_1(i')$ follows.

Vertical connections. Done analogously, see Figure 4b.   □

5 Maximum Independent Set for Line Segments

In this section we turn our attention to the MAXIMUM INDEPENDENT SET problem. The problem is fixed-parameter tractable for axis-parallel line segments, or more generally, if the lines have only a fixed number of different directions:

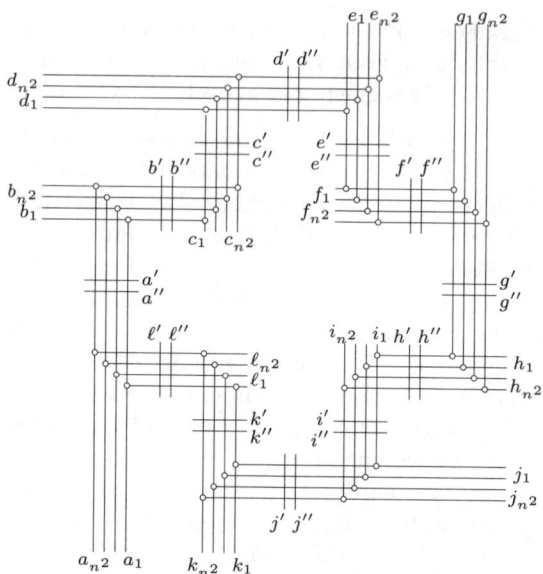

Fig. 3. The gadget used in the proof of Theorem 4

Theorem 5. MAXIMUM INDEPENDENT SET *for the intersection graphs of line segments in the plane can be solved in* $2^{O(k^2 d^2 \log d)} n \log n$ *time if the lines are allowed to have at most d different directions.*

Proof. Let $L_1, L_2, \ldots, L_d$ be the partition of the line segments according to their directions. The segments in L_i lie on n_i parallel lines $\ell_{i,1}, \ldots, \ell_{i,n_i}$. If $n_i \geq k$, then we can select k parallel segments from L_i that are on different lines, hence we have an independent set of size k. Thus it can be assumed that $n_i < k$ for every i. Therefore, the $n_1 + n_2 + \cdots + n_d$ lines have at most $\binom{d}{2}(k-1)^2$ intersection points, which will be called the *special points*. Apart from the special points, every point in the plane is covered by segments of at most one direction only. In a solution a special point is either not covered, or covered by a segment in one of $L_1, L_2, \ldots, L_d$. We try all $d^{\binom{d}{2}(k-1)^2}$ possibilities: each special point is assigned to one of the d directions. After deleting the segments that cross a special point from the wrong direction, we get d independent problems: segments with different directions do not cross each other. Furthermore, problem L_i consists of n_i independent problems: the parallel lines do not intersect. Therefore, the solution for this case can be obtained by selecting from each line as many independent segments as possible. It is well-known that this can be done in $O(n \log n)$ time.   □

A similar result was independently obtained by Kára and Kratochvíl (see [2] elsewhere in this volume). Their algorithm is somewhat faster and works even if only the intersection graph is given (not the segments themselves).

However, the problem is W[1]-hard with arbitrary directions:

Theorem 6. MAXIMUM INDEPENDENT SET *is* W[1]-*complete for intersection graphs of unit line segments.*

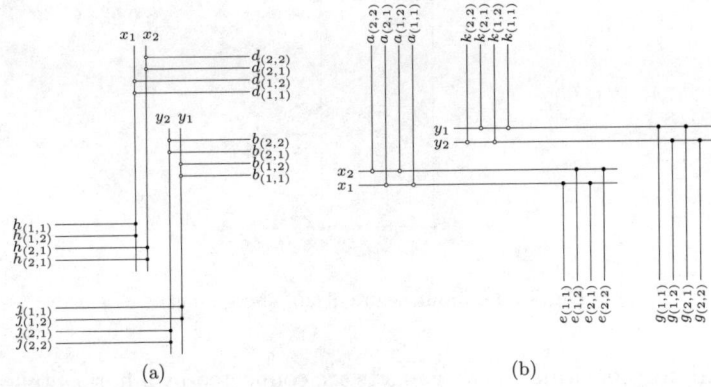

Fig. 4. Connecting two gadgets in the same row (a) or column (b)

Proof. The proof uses the framework of Section 2. The new parameter $k' :=$ $4k^2 + 2k(k-1)$ is 4 times the number of gadgets plus 2 times the number of connections. It is not possible to select more than 4 (resp., 2) independent segments from a gadget (resp., connection), hence every solution has to contain exactly that many segments from every gadget and connection.

The gadget. Henceforth we assume that the line segments are open. Each gadget consists of $4n^2$ line segments. For the gadget *centered at point* (x, y) the segments a_1, b_1, c_1, d_1 are arranged as shown in Figure 5. Set $\theta = 1/2n^6$. For $2 \leq i \leq n^2$, the lines a_i, b_i, c_i, d_i are obtained by rotating counterclockwise the four lines in Figure 5 around (x, y) by $(i-1)\theta$ radians. As discussed above, the parameter of the MAXIMUM INDEPENDENT SET problem is set in such a way that every solution contains 4 independent segments of the gadget. We say that the gadget represents value i in a solution if these four segments are a_i, b_i, c_i, d_i. The following lemma shows that every gadget represents a value in a solution:

Lemma 2. *At most 4 segments can be selected from each gadget. If S is an independent set of size 4 in a gadget, then $S = \{a_i, b_i, c_i, d_i\}$ for some $1 \leq i \leq n^2$.*

Proof. Since a_i and $a_{i'}$ intersect each other, at most one segment can be selected from $\{a_i : 1 \leq i \leq n^2\}$. Similarly, we can select at most one segment from the b_i's, c_i's, and d_i's, hence an independent set cannot have size more than 4.

Assume now that $a_{i_a}, b_{i_b}, c_{i_c}, d_{i_d}$ is an independent set in the gadget. First we show that $i_a \leq i_b$. It is sufficient to show that every a_j with $j > 1$ intersects b_1, since a_{i_a} and b_{i_b} has the same relation as $a_{i_a - i_b + 1}$ and b_1. The upper end point of a_j has y-coordinate greater than $y + 0.5$, while the y-coordinate of the other end point is smaller than y, thus it is easy to see that it intersects b_1. Similar arguments show that $i_a \leq i_b \leq i_c \leq i_d \leq i_a$, hence $i_a = i_b = i_c = i_d$ ☐

Restriction. To restrict the gadget to a set S, we remove a_i, b_i, c_i, d_i from the gadget for every $i \notin S$.

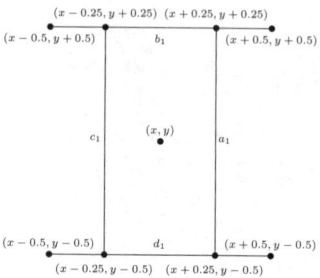

Fig. 5. The four segments of the gadget

Horizontal connections. If two gadgets are connected by a horizontal connection, then their distance is $1 + \delta$ (where the constant $\delta > 0$ is to be determined later), i.e., they are centered at (x_0, y_0) and $(x'_0, y'_0) = (x_0 + 1 + \delta, y_0)$. Let A_i be the intersection of the line $y = y_0 + 0.1$ and segment a_i of the first gadget. Let C_i be the intersection of the same line and segment c_i of the second gadget. We want to add n segments in such a way that segment e_j $(1 \leq j \leq n)$ intersects only segments $a_1, \ldots, a_{(j-1)n}$ of the first gadget and segments $c_{jn+1}, \ldots, c_{n^2}$ of the second gadget. This can be achieved if (open) segment e_j $(1 \leq j \leq n)$ connects $A_{(j-1)n+1}$ and C_{jn}. The segments e_j have different lengths, but it is possible to modify the x-coordinates of the end points and set δ such that every e_j has unit length (details omitted).

The e_j's ensure that if a_i, b_i, c_i, d_i are selected from the first gadget, $a_{i'}$, $b_{i'}$, $c_{i'}$, $d_{i'}$ are selected from the second gadget, and a segment e_j is also selected, then $\iota_1(i) \geq \iota_1(i')$. Recall that $i = (\iota_1(i) - 1)n + \iota_2(i)$ and $i' = (\iota_1(i') - 1)n + \iota_2(i')$. As e_j intersects $a_1, \ldots, a_{(j-1)n}$, it follows that $\iota_1(i) \geq j$, otherwise e_j would intersect a_i. Segment e_j intersects segments $c_{jn+1}, \ldots, c_{n^2}$ of the second gadget, hence $i' \leq (j-1)n + n$ and $\iota_1(i') \leq j \leq \iota_1(i)$ follows.

In a similar way, we add segments $f_1, \ldots, f_n$, whose job is to ensure that $\iota_1(i') \geq \iota_1(i)$. We want to define the segments in such a way that f_j intersects $a_{jn+1}, \ldots, a_{n^2}$ of the first gadget and $c_1, \ldots, c_{(j-1)n}$ of the second gadget. This can be done analogously to the definition of the segments e_j, but this time we intersect the a_i's and c_i's with the line $y = y_0 - 0.1$. It can be shown, that if a_i of the first gadget, $c_{i'}$ of the second gadget, and segment f_j are independent, then $\iota_1(i') \geq \iota_1(i)$. Therefore, the horizontal connection effectively forces that $\iota_1(i) = \iota_2(i')$ if i and i' are the values represented by the two adjacent gadgets.

Vertical connections. The vertical connection consists of two sets of segments $g_1, \ldots, g_n$ and $h_1, \ldots, h_n$, where every g_i intersects every $g_{i'}$, and every h_j intersects every $h_{j'}$. These segments are defined in such a way that

- g_{j_1} intersects b_i of the lower gadget if and only if $\iota_2(i) > j_1$,
- g_{j_1} intersects $d_{i'}$ of the upper gadget if and only if $\iota_2(i') < j_1$,
- h_{j_2} intersects b_i of the lower gadget if and only if $\iota_2(i) < j_2$,
- h_{j_2} intersects $d_{i'}$ of the upper gadget if and only if $\iota_2(i') > j_2$.

It is easy to see that these segments do what is required from a vertical connection: if b_i of the first gadget, $d_{i'}$ of the second gadget, and g_{j_1}, h_{j_2} are independent segments, then $\iota_2(i) = \iota_2(i') = j_1 = j_2$. The only question is how to construct the segments such that they have the intersection structure defined above.

We modify the gadget centered at (x_0, y_0) as follows. Set $\gamma = 1/n^3$. For each segment b_i, consider the line ℓ_i containing this segment, and shift b_i along ℓ such that the x-coordinate of the right end point of b_i becomes $x_0 + 0.5 + \iota_2(i)\gamma - \gamma/2$. The b_i's are "almost horizontal," thus it can be verified (details omitted) that

- the y-coordinate of the right end point of b_i is between $y_0 + 0.5$ and $y_0 + 0.5 + \gamma$,
- the x-coordinate of the left end point of b_i is between $x_0 - 0.5 + \iota_2(i)\gamma - 0.6\gamma$ and $x_0 - 0.5 + \iota_2(i)\gamma - 0.4\gamma$,
- the y-coordinate of the left end point of b_i is between $y_0 + 0.5 - \gamma$ and $y_0 + 0.5$.

In a symmetrical way, we can ensure that

- the x-coordinate of the left end point of d_i is $x_0 - 0.5 - \iota_2(i)\gamma + \gamma/2$.
- the y-coordinate of the left end point of d_i is between $y_0 - 0.5 - \gamma$ and $y_0 - 0.5$,
- the x-coordinate of the right end point of d_i is between $x_0 + 0.5 - \iota_2(i)\gamma + 0.4\gamma$ and $x_0 + 0.5 - \iota_2(i)\gamma + 0.6\gamma$,
- the y-coord. of the right end point of d_i is between $y_0 - 0.5 + \gamma$ and $y_0 - 0.5$.

In the vertical connection between the two gadgets centered at (x_0, y_0) and $(x_0, y_0 + 1.5)$, the segment g_j is a unit length segment that goes through the points $(x_0 + 0.5 + j\gamma, y_0 + 0.5)$, $(x_0 + 0.5 - (j-1)\gamma, y_0 + 1 + \gamma)$, and the center point of g_j has $y_0 + 0.75$ as y-coordinate. As $\gamma < 1/n^2$, segment g_j is almost vertical; in particular, it reaches the line $y = y_0 + 0.5 + \gamma$ with an x-coordinate greater than $x_0 + 0.5 + j\gamma - \gamma/2$, and it reaches the line $y = y_0 + 1$ with x-coordinate less than $x_0 + 0.5 + (j-1)\gamma + 0.4\gamma$. This means that g_j does not intersect a segment b_i if its right end point has x-coordinate at most $x_0 + j\gamma - \gamma/2$ (i.e., $\iota_2(i) > j$) and it does intersect a g_j if the x-coordinate of its right end point is greater than $x_0 + j\gamma$ (i.e., $\iota_2(i) \le j$). Similarly, in the upper gadget, g_j intersects every d_i with $\iota_2(i) < j$, and does not intersect d_i if $\iota_2(i) > j$.   □

References

1. P. K. Agarwal, M. van Kreveld, and S. Suri. Label placement by maximum independent set in rectangles. *Comput. Geom.*, 11(3-4):209–218, 1998.
2. J. Kára and J. Kratochvíl. Fixed parameter tractability of Independent Set in segment intersection graphs. Accepted to *IWPEC 2006*.
3. E. Malesińska. *Graph-Theoretical Models for Frequency Assignment Problems*. PhD thesis, Technical University of Berlin, 1997.
4. D. Marx. Efficient approximation schemes for geometric problems? In *Proceedings of ESA 2005*, pages 448–459, 2005.
5. D. W. Wang and Y.-S. Kuo. A study on two geometric location problems. *Inform. Process. Lett.*, 28(6):281–286, 1988.

Fixed Parameter Tractability of Independent Set in Segment Intersection Graphs*

Jan Kára and Jan Kratochvíl

Department of Applied Mathematics and Institute for Theoretical Computer Science,
Faculty of Mathematics and Physics, Charles University, Prague, Czech Republic
{kara, honza}@kam.mff.cuni.cz

Abstract. We present a fixed parameter tractable algorithm for the Independent Set problem in 2-DIR graphs and also its generalization to d-DIR graphs. A graph belongs to the class of d-DIR graphs if it is an intersection graph of segments in at most d directions in the plane. Moreover our algorithms are robust in the sense that they do not need the actual representation of the input graph and they answer correctly even if they are given a graph from outside the promised class.

1 Introduction

Let $G = (V, E)$ be a simple undirected graph. In the following we use n for the number of vertices $|V|$ and m for the number of edges $|E|$. Let $\deg v$ denote the number of edges incident with v, $N(v)$ the open neighborhood of v (i. e. $N(v) = \{u \in V : \{u, v\} \in E\}$) and $N[v]$ the closed neighborhood of v (i. e. $N[v] = N(v) \cup \{v\}$). If $U \subseteq V$, then we write $G[U]$ for the subgraph induced on the set U (i. e., the graph $G' = (U, E')$ where $E' = \{\{u, v\} \in E : u \in U, v \in U\}$). We write $G - v$ instead of $G[V \setminus \{v\}]$ for the sake of brevity. A set of vertices U is called *independent* if $G[U]$ contains no edges, and it is called a *clique* if $G[U]$ is a complete graph. The symbols $\alpha(G)$ and $\omega(G)$ denote the largest size of an independent set and of a clique, respectively.

A graph $G = (V, E)$ is a *segment intersection graph* (and SEG denotes this class of graphs) if each vertex $v \in V$ can be assigned a straight line segment s_v in the plane in such a way that s_u and s_v intersect if and only if $\{u, v\} \in E$. We call a family of segments $R_G = \{s_v : v \in V\}$ a SEG *representation of G* or simply a *representation* if the class is clear from the context. If there is no danger of confusion, we sometimes identify the segments with the vertices they are representing, i.e., we assume that $V = R_G$. If G has a representation such that the segments use only d different directions (the segments in the same direction can intersect), we call such a graph a *segment intersection graph in d directions* (and we denote by d-DIR the class of such graphs). Note that the 1-DIR graphs are exactly the *interval graphs*. An interval graph is a *proper interval graph* if it has a representation such that no segment properly contains another

---

* Supported by project 1M0021620808 of the Ministry of Education of the Czech Republic.

H.L. Bodlaender and M.A. Langston (Eds.): IWPEC 2006, LNCS 4169, pp. 166–174, 2006.

one (note that this is also equivalent to the requirement that all the segments have the same length [10]). It is a well known fact that interval graphs can be recognized in $O(m+n)$ time [2]. On the other hand, recognition of 2-DIR graphs or SEG graphs is NP-hard [5,6]. In this connection we note that the algorithms we develop in this paper are robust in the sense that they do not need the actual representations of the input graphs.

The problem of finding an independent set of maximum size is a well known NP-hard problem, which remains NP-hard for many restricted graph classes. For instance for planar cubic graphs [4]. Though it is solvable in polynomial time in interval graphs (1-DIR graphs), it is NP-hard already for 2-DIR graphs [7]. Also from the fixed parameter complexity point of view, the problem is hard. It is W[1]-complete when parameterized by the size of the independent set k [3]. Therefore it is reasonable to ask for its parametrized complexity for restricted graph classes. As observed many times (and noted e.g., in the recent monograph of Niedermeier [9]), the problem is trivial for planar graphs (when parameterized by k) since every planar graph is guaranteed to have $\alpha(G) \geq \frac{n}{4}$. Fellows asked [private communication, 2005] what is the fixed parameter complexity of this problem when restricted to 2-DIR graphs. Note that the answer is not as obvious as for planar graphs, since 2-DIR do not guarantee independent sets of any nontrivial size — all cliques are 2-DIR graphs (in fact even 1-DIR graphs).

In our paper we answer this question in affirmative, even in a more general setting for d-DIR graphs. This complements independent result of Marx [8] showing that finding an independent set in segment intersection graphs (i.e. with unlimited number of directions) is W[1]-complete. The author also presents a fixed parameter tractable algorithm for the case when the number of directions is bounded but his algorithm requires a segment representation of an input graph.

In Section 3 we present a fixed parameter tractable algorithm (i. e., an algorithm running in time $O(f(k)n^{O(1)})$) for the Independent Set problem in 2-DIR graphs. In Section 4 we generalize this algorithm to an FPT algorithm for d-DIR graphs (where both d and k are parameters). Our algorithms are designed to be "robust" — i. e., they either output an independent set of size k or answer that there is no independent set of size k or detect that the given graph is not a 2-DIR (d-DIR) graph. Hence they answer correctly even if the input graph is not from the required class.

2   Reduction Step

Lemma 1. *Let $G = (V, E)$ be an arbitrary graph and $u, v \in V$ two of its vertices such that $N[u] \subseteq N[v]$. Then $\alpha(G) = \alpha(G - v)$.*

Proof. Clearly $\alpha(G) \geq \alpha(G - v)$ and so we concentrate on the second inequality. Note that $N[u] \subseteq N[v]$ implies $\{u, v\} \in E$. Hence at most one of the vertices u, v can be in any independent set. If an independent set S contains v, the set $S \setminus \{v\} \cup \{u\}$ is also independent and of the same size. That proves the statement of the observation.

We call a graph *reduced* if no closed neighborhoods are in inclusion. By consecutive application of the reduction step described in Lemma 1 we reduce the input graph G to a reduced graph G' such that $\alpha(G) = \alpha(G')$. Such reduction can be performed in time $O(mn)$ (and independent of k). In pseudocode of algorithms presented below we use procedure Reduce to perform this graph reduction.

3   Algorithm for 2-DIR Graphs

In this section we present an FPT algorithm for Independent Set in 2-DIR graphs. We begin with an easy observation:

Lemma 2. *Let $G = (V, E)$ be a 2-DIR graph. Then for each $v \in V$, the graph $G[N(v)]$ is an interval graph.*

Proof. Consider a 2-DIR representation R_G of the graph G. The vertex v is represented as a segment s_v in one of the two directions. Consider the set of segments $I_v = \{s \cap s_v : s \in R_G, s \cap s_v \neq \emptyset\}$. This set clearly contains one segment for each neighbor of v and all the segments have the same direction (some of them can be a single point). Hence I_v induces an interval graph. Moreover, it can be easily verified that I_v is an interval representation of $G[N(v)]$ (for all $x, y \in I_v$, $s_x \cap s_y \subseteq s_v$).

Next we derive an important general lemma about d-DIR graphs.

Lemma 3. *Let R_G be a d-DIR-representation of a reduced graph $G = (V, E)$. If $|V| > ((k-1)(d-1)+1)d(k-1)^2$, then either there are segments $s_1, \ldots, s_k \subseteq R_G$ such that all of them are parallel with the same direction and no two of them lie on the same line, or there is a line l which contains k pairwise non-intersecting segments.*

Proof. Suppose that R_G does not contain k parallel segments lying on k different lines. Let l be a line containing at least one segment. Denote by V_l the set of vertices corresponding to the segments lying on the line l and let G_l be the (interval) graph $G[V_l]$. Since G is reduced, G_l is a proper interval graph (no interval can be contained in another one).

Let $c := (k - 1)(d - 1) + 1$. First we prove that $\omega(G_l) \leq c$. Suppose for contradiction that G_l contains a clique of size $c + 1$. Order the vertices from left to right by the left endpoints of their intervals (or by the the order of their right endpoints — since this is a proper interval representation, these two orderings are the same). Let $C_2, \ldots, C_{c+1}$ be the lexicographically minimal cliques of sizes $2, \ldots, c+1$, respectively, and denote their vertices $C_i = \{v_1^i, v_2^i, \ldots, v_i^i\}$ (in the ordering). Consider the last two vertices v_c^{c+1}, v_{c+1}^{c+1} of C_{c+1}. The vertex v_c^{c+1} must have a neighbor u_{c+1} (in the graph G) that is not a neighbor of v_{c+1}^{c+1} (otherwise v_c^{c+1} would be thrown out in the reduction step). The vertex u_{c+1} cannot lie on the line l since v_c^{c+1} lies before v_{c+1}^{c+1} and so the vertex u_{c+1} would have to lie before v_{c+1}^{c+1} and we would get a lexicographically smaller clique $C_{c+1} \setminus \{v_{c+1}^{c+1}\} \cup \{u_{c+1}\}$ of the size $c + 1$. The same argument for cliques $C_2, \ldots, C_c$ yields vertices $u_2, \ldots, u_c$ in the directions different from the direction of l. Let $r_i = s_{v_{i-1}^i} \setminus s_{v_i^i}$ be the part of the segment representing v_{i-1}^i that is not part of the segment representing v_i^i, for $2 \leq i \leq c + 1$. Since the segments r_i do not overlap and each segment representing u_i intersects v_{i-1}^i in r_i, each of the vertices $u_2, \ldots, u_{c+1}$ lies on a different line. But by the choice of $c = (k - 1)(d - 1) + 1$ (applying the pigeon-hole principle), there must be a direction with at least k segments from $u_2, \ldots, u_{c+1}$. This is a contradiction with the assumption that there are no such k segments. See Figure 1 for an example of a situation in one clique and Figure 2 for an example of the whole situation on line l.

Since G_l is an interval graph, it is in particular a perfect graph and hence $\alpha(G_l) \cdot \omega(G_l) \geq |V_l|$. As we assumed that $|V| > cd(k - 1)^2$ and because for every direction, there are at most $k - 1$ lines containing at least one segment in this direction (otherwise we easily find k parallel segments), there must be a line l with at least $c(k - 1) + 1$ segments. Hence, for this line l, we conclude that $\alpha(G_l) \geq \lceil \frac{c(k-1)+1}{c} \rceil = k$ as claimed by the statement of the lemma.

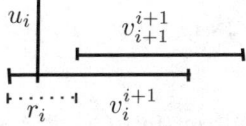

Fig. 1. A situation of the last two vertices of the i-th clique

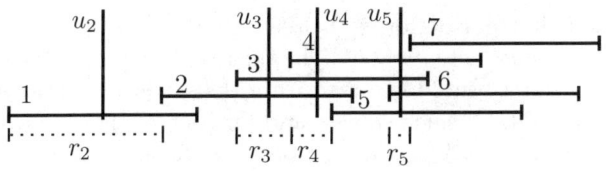

Fig. 2. A situation in a graph G_l for $c = 4$. The lexicographically minimal cliques are: $C_2 = \{1, 2\}, C_3 = \{2, 3, 4\}, C_4 = \{2, 3, 4, 5\}, C_5 = \{3, 4, 5, 6, 7\}$.

Corollary 1. Let $G = (V, E)$ be a reduced 2-DIR graph. Then either $|V| \leq 2k(k - 1)^2$ or $\alpha(G) \geq k$.

Proof. The corollary trivially follows from Lemma 3.

Algorithm 1
   Let $G = (V, E)$ be the input graph.
   Reduce (G)
   if $|V| \le 2k^2(k^2 - 1)^2$ then begin
      Compute the adjacency matrix of G.
      for each $I \subseteq V$, $|I| = k$ do
         if I is independent then
            Output I and exit.
      Output that there is no independent set of size k in G.
   end
   else begin
      Find greedily an inclusion-wise maximal independent set I.
      if $|I| \ge k$ then
         Output I and exit.
      for each $v \in I$ do begin
         Compute $G' = G[N(v)]$.
         if G' is not an interval graph then
            Output that G is not a 2-DIR graph and exit.
         Find a maximum independent set I of G' (using the fact that
            G' is an interval graph).
         if $|I| \ge k$ then
            Output I and exit.
      end
      Output that G is not a 2-DIR graph.
   end

Theorem 1. *Algorithm 1 finds in time $O(k^{6k+2} + mn + n^2 + k(m + n))$ for a given graph $G = (V, E)$ either an independent set of size at least k or answers that there is no independent set of size (at least) k or detects that G is not a 2-DIR graph.*

Proof. The algorithm first performs the reduction step as described in Section 2. This takes $O(mn)$ time. Let $G' = G[V']$ be the resulting graph. If $|V'| \le 2k^2(k^2 - 1)^2 = O(k^6)$, we run a brute-force algorithm that tries all subsets of vertices of size k — there are $O(k^{6k})$ such sets — and for each of them, we check whether it is an independent set or not (this can be easily done in time $O(k^2)$ if we have precomputed adjacency matrix of the graph). Hence in this case we are done in the time $O(k^{6k+2} + mn + n^2)$.

   If $|V'| > 2k^2(k^2 - 1)^2$, then Corollary 1 asserts that either G' is not a 2-DIR graph or it contains an independent set of size k^2. We find an (inclusion-wise) maximal independent set I in G' — this can be done in $O(m + n)$ time by a greedy algorithm. If it has size at least k we are done. Otherwise we claim that there must be a vertex $v \in I$ such that there is an independent set of size k in $N(v)$ (if G was a 2-DIR graph). This follows from the fact that every vertex not in I is adjacent to at least one vertex in I. Hence if J is an independent set of size k^2, then some vertex in I must be adjacent to at least k vertices from $J \setminus I$.

So it is enough to find a maximum independent set in $G'[N(v)]$ for each $v \in I$. From Lemma 2 we know that each of the graphs $G'[N(v)]$ is an interval graph, and thus its independence number can be computed in polynomial time. Hence, for every vertex v of I, we verify that $G'[N(v)]$ is an interval graph (and reject G as not being 2-DIR if it is not) in $O(m+n)$ time and we compute a maximum independent set in time $O(m+n)$. If none of these independent sets has size at least k, we again reject G as not being a 2-DIR graph.

Note that if we get a 2-DIR representation of the graph G as part of the input, the algorithm from Theorem 1 would be much simpler. In the kernelization step we would suffice to have enough vertices guaranteeing an independent set of size k (and not k^2) and we could find it by simply checking the types of independent sets described in Lemma 3.

4   Algorithm for d-DIR Graphs

In this section we present an algorithm for Independent Set in d-DIR graphs. The algorithm is in fact simpler than the one for 2-DIR graphs but its correctness is less obvious and its running time is worse. Due to Lemma 3, a large enough reduced d-DIR graph must contain a big independent set. Hence we use a similar trick as for 2-DIR graphs and for d-DIR graphs whose number of vertices does not guarantee an independent set of size k^{3d}, we run a brute force algorithm to decide the existence of an independent set of size k. For larger graphs we indirectly show that there is a sufficiently limited number of cliques in whose neighborhoods it suffices to search for independent sets of size k, and that those neighborhoods have a simple structure. Now we formalize the above statements:

Corollary 2. *Let $G = (V, E)$ be a reduced d-DIR graph. Then either $|V| \leq ((k-1)(d-1)+1)d(k-1)^2$ or $\alpha(G) \geq k$.*

Proof. The statement trivially follows from Lemma 3.

Lemma 4. *Let R_G be a d-DIR representation of a reduced graph $G = (V, E)$. Let d_v be the direction of the segment representing a vertex $v \in V$. Denote by G' the graph obtained from $G[N(v)]$ by the reduction procedure of Section 2. Then G' contains at most two vertices that were represented by segments in the direction d_v in R_G.*

Proof. Let l be the line containing s_v. Note that G' contains only neighbors of v. Let V_l be the set of neighbors of v that are represented by segments parallel with direction d_v, and let S_l be the set of segments representing them. Then all segments of S_l lie on the line l. Since G was reduced, no segment in S_l can contain another one (including s_v itself) and hence each segment in S_l contains exactly one endpoint of the segment representing v. Let $v_1, \ldots, v_t$ be the vertices of S_l whose segments contain the left endpoint of s_v, ordered from left to right. Since apart from S_l, $G[N(v)]$ contains only vertices corresponding to the segments crossing s_v, we see that $N[v_1] \subseteq N[v_2] \subseteq \ldots \subseteq N[v_t]$, and hence only the vertex

v_1 survives the reduction step and is included in G'. Similarly, at most one of the vertices represented by segments containing the right endpoint of s_v remains in G'. See Figure 3 for an example of a reduction of a neighborhood.

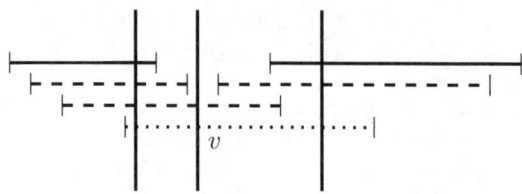

Fig. 3. A vertex v (dotted interval) and its neighbors. Dashed intervals are the vertices that are removed in the reduction step on $G[N(v)]$. Vertical thick lines represent points of intersection with segments in other directions.

Theorem 2. *There is an algorithm running in time $O(d^{2k}k^{9dk+2}+n^2+k^{3d}mn)$ that given a graph $G = (V, E)$ either finds an independent set of size k or answers that there is no independent set of size k or detects that G is not a d-DIR graph.*

Proof. The algorithm first performs the reduction step as described in Section 2. This takes $O(mn)$ time. Let $G' = G[V']$ be the resulting graph. If $|V'| \leq ((k^{3d} - 1)(d-1) + 1)d(k^{3d} - 1)^2 = O(d^2 k^{9d})$ we run a brute-force algorithm that tries all subsets of vertices of size k — there are $O(d^{2k}k^{9kd})$ such sets — and for each set checks whether it is an independent set or not (that can be easily done in time $O(k^2)$ provided we have precomputed the adjacency matrix of G'). Hence in this case we are done in time $O(d^{2k}k^{9dk+2} + n^2 + mn)$.

If $|V'| > ((k^{3d} - 1)(d-1) + 1)d(k^{3d} - 1)^2$ we know by Corollary 2 that either G' is not a d-DIR graph or G' contains an independent set of size k^{3d}. Now we start the following recursive procedure:

Algorithm 2
```
  procedure FindIndependentSet(G, depth)
  begin
     Find a maximal independent set I in the given graph G.
     if |I| ≥ k then
        Output I and abort (at all levels of recursion).
     if depth ≥ 3d then
        Output that G is not a d-DIR graph and abort.
     for each v ∈ I do begin
        G' := G[N(v)]
        Reduce(G')
        if |V(G')| ≥ k then
           FindIndependentSet(G', depth + 1)
     end
  end
```

If the recursive procedure did not find any independent set, we answer that G is not a d-DIR graph. Hence the overall algorithm for d-DIR graphs looks as follows:

Algorithm 3
    Let $G = (V, E)$ be the input graph.
    `Reduce`(G)
    `if` $|V| \leq ((k^{3d} - 1)(d - 1) + 1)d(k^{3d} - 1)^2$ `then begin`
        Compute the adjacency matrix of G.
        `for each` $I \subseteq V$, $|I| = k$ `do`
            `if` I is independent `then`
                Output I and `exit`.
        Output that there is no independent set of size k in G.
    `end`
    `else begin`
    `FindIndependentSet`$(G,0)$
    Output that G is not a d-DIR graph.
    `end`

The time complexity of the recursive call of `FindIndependentSet` is obviously $O(k^{3d}mn)$. What remains to be proved is that if G is a d-DIR graph, then the recursion always finds an independent set of size k (recall that if G is a d-DIR graph it must contain such an independent set). Correctness of the other answers easily follows from this fact. If G is a d-DIR graph, it contains an independent set of size k^{3d}. Hence some vertex from the maximal independent set must have an independent set of size k^{3d-1} in its neighborhood (by a similar argument as in Theorem 1). By induction and because the reduction does not change the size of a maximum independent set we argue that there is a branch of recursion whose graph considered in the depth i of the recursion contains an independent set of size k^{3d-i}. By Lemma 4 we know that during the recursion we can choose at most three vertices from each direction. Thus in the depth of recursion $3d - 1$ every obtained graph can contain at most one vertex. Hence we see that if G is a d-DIR graph, then the recursion must stop before the depth $3d - 1$ as otherwise we get a contradiction. There are only two ways of stopping the recursion before the depth $3d$. Either we get a graph with less than k vertices or we find an

Fig. 4. A recursion tree with a depth of the recursion written on the left and guaranteed independent set size written on the right

independent set of size k. As there must be a branch of recursion whose graph at the depth i contains an independent set of size k^{3d-i}, we know that this branch cannot stop because of the lack of vertices and hence it must stop because it found an independent set of size k.

5   Conclusion

We have presented efficient FPT algorithms for the Independent Set problem restricted to intersection graphs of segments with segments lying in a bounded number of direction (where both the size of the sought independent set k and the number of directions d are considered as parameters). Given the simplicity of the situation for two directions, it might be interesting to determine whether there is a fixed parameter tractable algorithm running in time $2^{O(k)} \cdot p(m, n)$ in this case.

It is worth mentioning the parallel to the (classical) complexity of the Clique problem restricted to intersection graphs. The Clique problem is known to be NP-hard for string graphs and also for intersection graphs of convex sets, polynomial time solvable (but not FPT with regard to d as parameter) for d-DIR graphs and its complexity still remains open for SEG graphs. This question was asked in [7], for a recent survey on its development and related questions cf. [1].

References

1. J. Bang-Jensen, B. Reed, M. Schacht, R. Šámal, B. Toft, U. Wagner: On six problems posed by Jarik Nešetřil, in: Topics in Discrete Mathematics (Dedicated to Jarik Nešetřil on the occasion of his 60th birthday), M. Klazar, J. Kratochvíl, M. Loebl, J. Matoušek, P. Valtr, R. Thomas, eds., Sringer Verlag, 2006 (to appear)
2. K.S. Booth, G.S. Lueker: Testing for the consecutive ones property, interval graphs, and graph planarity using PQ-tree algorithms, *J. Comput. Syst. Sci.* 13, pp. 335–379 (1976).
3. R.G. Downey, M.R. Fellows: Parametrized Complexity, Springer Verlag, 1999.
4. M. Garey, D. Johnson: Computers and Intractability. A Guide to the Theory of NP-Completeness, Harry Freeman, 1979.
5. J. Kratochvíl: A special planar satisfiability problem and a consequence of its NP-completeness, *Discrete Appl. Math.* 52 (1994), pp. 233–252.
6. J. Kratochvíl, J. Matoušek: Intersection graphs of segments, *Journal of Combinatorial Theory Series B* 62 (1994), pp. 289–315.
7. J. Kratochvíl, J. Nešetřil: INDEPENDENT SET and CLIQUE problems in intersection defined classes of graphs, Comment. Math. Univ. Carolin. 31 (1990), 85-93
8. D. Marx: Parametrized Complexity of Independence and Domination on Geometric Graphs, IWPEC 2006
9. R. Niedermeier: Invitation to Fixed-Parameter Algorithms, Oxford University Press, 2006.
10. F.S. Roberts: Indifference graphs, *Proof Techniques in Graph Theory*, F.Harary, ed.,Academic Press, New York 1969, pp. 139–146.

On the Parameterized Complexity of
d-Dimensional Point Set Pattern Matching

Sergio Cabello[1,*], Panos Giannopoulos[2,**], and Christian Knauer[3]

[1] IMFM, Department of Mathematics, Jadranska 19, SI-1000 Ljubljana, Slovenia
sergio.cabello@imfm.uni-lj.si
[2] Humboldt-Universität zu Berlin, Institut für Informatik, Unter den Linden 6,
D-10099 Berlin, Germany
panos@informatik.hu-berlin.de
[3] Institut für Informatik, Freie Universität Berlin, Takustraße 9,
D-14195 Berlin, Germany
Christian.Knauer@inf.fu-berlin.de

Abstract. Deciding whether two n-point sets $A, B \in \mathbb{R}^d$ are congruent is a fundamental problem in geometric pattern matching. When the dimension d is unbounded, the problem is equivalent to graph isomorphism and is conjectured to be in FPT.

When $|A| = m < |B| = n$, the problem becomes that of deciding whether A is congruent to a subset of B and is known to be NP-complete. We show that point subset congruence, with d as a parameter, is W[1]-hard, and that it cannot be solved in $O(mn^{o(d)})$-time, unless SNP $\subset$ DTIME($2^{o(n)}$). This shows that, unless FPT = W[1], the problem of finding an isometry of A that minimizes its directed Hausdorff distance, or its Earth Mover's Distance, to B, is not in FPT.

Keywords: Fixed Parameter Tractability, Geometric Point Set Matching, Congruence, Unbounded Dimension.

1 Introduction

Geometric pattern matching has been a topic of considerable research in computational geometry with applications in computer vision, and is usually modeled as the following optimization problem: given two sets A and B of geometric primitives, an appropriate distance measure, and a transformation group, find a transformation of A that minimizes its distance to B; see the surveys by Alt and Guibas [2] and Hagedoorn and Veltkamp [15]. Typical geometric primitives include points, segments, disks, while typical transformations include isometries, and scaling, that is, combinations of translations, rotations, reflection, and scaling.

* Partially supported by the European Community Sixth Framework Programme under a Marie Curie Intra-European Fellowship and by the Slovenian Research Agency, project J1-7218-0101.
** Partially supported by the DFG project "Parametrische Komplexitätstheorie", no. GR 1492/7-2.

H.L. Bodlaender and M.A. Langston (Eds.): IWPEC 2006, LNCS 4169, pp. 175–183, 2006.

For sets of points, several distance measures have been extensively studied in this framework, such as, the bottleneck distance [3,12,13], the (directed) Hausdorff distance [9,10,16] and the Earth Mover's Distance [7,11]. The case of the bottleneck distance with respect to isometries leads to the fundamental decision problems of whether A is congruent to B or to a subset of B; a formal definition will be given shortly.

The complexity of these two problems for point sets in unbounded dimensions has been already studied within the classical complexity theory: the former is graph-isomorphism-hard and the latter is NP-complete. In this paper we study subset congruence and related problems from the parameterized complexity point of view, with the dimension as the parameter.

In general, geometric optimization problems have been always studied with respect to parameters related to properties of the input objects, and the literature is scattered with algorithms that, from the parameterized complexity theory point of view, can be seen as being in FPT. One such famous example is Megiddo's algorithm [18] for linear programming in d dimensions with n constraints, which runs in $O(2^{2^d} n)$ time. However, there is only a handful of results concerning the fixed-parameter intractability of certain hard geometric problems, and these concern only standard parameterizations of optimization problems, i.e., where the parameter measures the size of the solution; see, for example, Marx [17] for such results concerning geometric graph problems. The dimension of the input objects is a "structural" parameter important to many (apparently) intractable geometric problems, and, thus, a good candidate for parameterized complexity.

Preliminaries. For a point $a \in \mathbb{R}^d$, let $a(r)$ denote its rth component. The origin is denoted by o. For two points $a, b \in \mathbb{R}^d$, let $||a - b|| = (\sum_{r=1}^d (a(r) - b(r))^2)^{1/2}$ be their Euclidean distance. A map $\mu : \mathbb{R}^d \to \mathbb{R}^d$ is an *isometry* if it preserves distances, that is, if $||\mu(a) - \mu(b)|| = ||a - b||$ for all $a, b \in \mathbb{R}^d$.

Let $A = \{a_1, \ldots, a_m\}$ and $B = \{b_1, \ldots, b_n\}$ be point sets in $\mathbb{R}^d$ with $m \leq n$; for simplicity, we will sometimes assume the obvious ordering on the elements of sets defined in this way, e.g., a_1 is the first point in A and so on. We use the notation $\mu(A) = \{\mu(a_1), \ldots, \mu(a_m)\}$. Sets A and B are said to be *congruent* if there is an isometry μ for which $\mu(A) = B$.

We denote by CONGRUENCE the problem of finding whether two sets A and B are congruent. When $|A| < |B|$, the problem becomes the one of finding whether A is congruent to a subset of B, and is denoted by SUBSET-CONGRUENCE. We are interested in the parameterized version of these problems, referred to as p-CONGRUENCE and p-SUBSET-CONGRUENCE, with the dimension d being the input parameter.

Related work. There is a variety of algorithms that solve CONGRUENCE in $O(n \log n)$ time for $d = 2, 3$; see Alt et al. [3] and references therein. In higher dimensions, the currently best time bound is $O(n^{\lceil d/3 \rceil} \log n)$ [5]; Akutsu [1] claims that a bound of $O(n^{d/4+O(1)})$ (randomized) is possible, but he gives no direct proof. It is conjectured [5] that p-CONGRUENCE is in FPT. Moreover, for unbounded dimension, CONGRUENCE is polynomially equivalent to graph isomorphism [1,20].

SUBSET-CONGRUENCE can be solved in time of $O(mn^{4/3}\log n)$ and $O(mn^{7/4}\log n\,\beta(n))$ (randomized, where $\beta(n)$ is an extremely slow growing function), for $d = 2$ and $d = 3$ respectively [4]. In higher dimensions, the best known algorithm runs in $O(mn^d)$ time [12], and the problem is NP-complete when d is unbounded [1]. It is an open question whether the high-dimensional bound can be improved [4]; see also Brass and Pach [6] for a survey on computational and combinatorial problems related to geometric patterns.

Model of computation. We assume the standard TM model of computation. The coordinates of the points used in our reduction are rational, with denominators and numerators bounded by a polynomial in n.

Results. We show that p-SUBSET-CONGRUENCE is W[1]-hard. Our reduction from p-CLIQUE, with d being linear in the size of the clique, shows in addition that an $O(mn^{o(d)})$-time algorithm for former problem exists only if SNP $\subset$ DTIME$(2^{o(n)})$. Moreover, for $|A| < |B|$ and any point set distance D for which $D(A, B) = 0 \Leftrightarrow A \subset B$, our hardness result implies that minimizing D under isometries is not in FPT (unless FPT = W[1]).

2   Parameterized Subset Congruence

First, it is easy to see that p-SUBSET-CONGRUENCE is in W[P]; an accepting certificate of size $O(d\log n)$ for a NDTM can be given by guessing a bijection between a d-point subset of A and a d-point subset of B, with $\log n$ bits needed to represent each point. Then, one can check in polynomial time whether the bijection is an isometry and, if yes, whether this isometry maps the rest of the points in A to points in B.

Next, we reduce p-CLIQUE, which is known to be W[1]-complete [14], to p-SUBSET-CONGRUENCE.

Theorem 1. p-SUBSET-CONGRUENCE *is* W[1]-*hard.*

Proof: Let k be the size of the clique being looked for in a simple and connected graph $G([n], E)$ with $|E| = m$. We first construct a set $\mathcal{L} = \{L_i | i = 1,\ldots,k\}$ of auxiliary point sets L_i - referred to as *level* sets. Each level set lies on a two-dimensional plane in $\mathbb{R}^{2k}$, with all k planes being pairwise orthogonal, and contains n points that lie on a unit circle centered at the origin o. A detailed description of a level set follows.

Let $L_i = \{l_{ij} \in \mathbb{R}^{2k} | j = 1,\ldots,n\}$. First, we have $l_{ij}(r) = 0$ for $r \neq 2i - 1, 2i$ and $\sum_{r=1}^{2k} l_{ij}^2(r) = 1$; this implies that $||l_{ij} - l_{i'j'}|| = \sqrt{2}$ for every j, j' and every i, i' with $i \neq i'$. It will be convenient to choose the points on the unit circle from a short arc such that the distance between any two points is at most $\sqrt{2}/6$, that

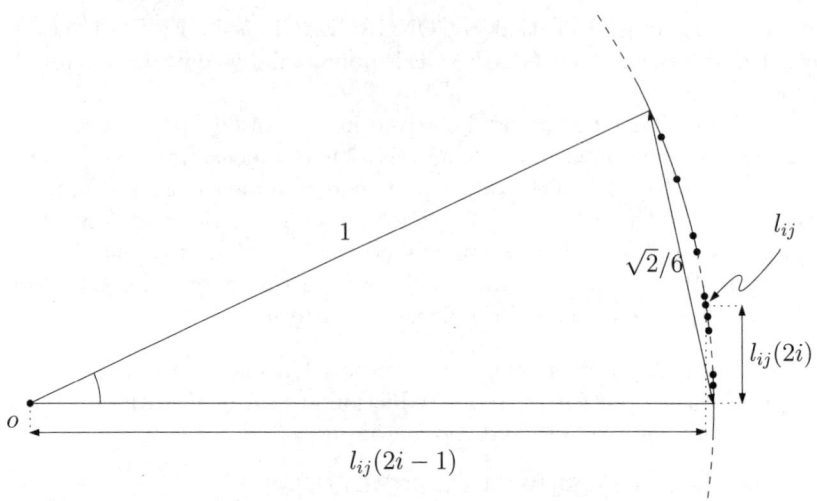

Fig. 1. The points of each level set lie on a circular arc of chord length $\sqrt{2}/6$

is, $\|l_{ij} - l_{ij'}\| < \sqrt{2}/6$ for every i, j and j'; see Fig. 1. This is done according to the following lemma.

Lemma 1. *For any $n > 0$, there exist n distinct points on the unit circle such that they all have rational coordinates with the numerators and denominators bounded by a polynomial in n, and the distance between any two points is at most $\sqrt{2}/6$.*

Proof: Each point $p_i, i = 1, \ldots, n$, has rational non-zero coordinates generated by Pythagorean triples α_i, β_i and γ_i:

$$p_i(1) = (\beta_i/\gamma_i) \quad \text{and} \quad p_i(2) = (\alpha_i/\gamma_i),$$

with

$$\alpha_i = 2(8+i), \quad \beta_i = (8+i)^2 - 1 \quad \text{and} \quad \gamma_i = (8+i)^2 + 1.$$

Defined this way, all points have distinct coordinates and $p_i^2(1) + p_i^2(2) = 1$. Moreover, since $\sin^{-1}(p_i(2)) < 2\sin^{-1}(\sqrt{2}/12)$, all points lie on an arc with chord length $\sqrt{2}/6$, and so, any two of them can be at most $\sqrt{2}/6$ apart.   ▱

Next, using the level sets, we define a point set $A \subset \mathbb{R}^{2k}$, with $|A| = \binom{k}{2}$, one point for each (unordered) pair of level sets. Each point is the *middle*-point of a straight line segment whose end-points belong to two distinct level sets. The choice of the end-points is such that any two middle-points satisfy one of the following *conditions*:

    (i) one common end-point and second end-point in different levels,
    (ii) all four end-points in different levels.

In addition, for every $j \leq k$ only the end-point l_{jj} is used. We have

$$A = \{\frac{l_{11} + l_{22}}{2}, \frac{l_{11} + l_{33}}{2}, \ldots, \frac{l_{11} + l_{kk}}{2}\}$$
$$\cup \{\frac{l_{22} + l_{33}}{2}, \ldots, \frac{l_{22} + l_{kk}}{2}\}$$
$$\ddots$$
$$\cup \{\frac{l_{k-1k-1} + l_{kk}}{2}\}.$$

Observe that there are only two distinct inter-point distances in A: $\sqrt{2}/2$ and 1. For each point $(l_{jj} + l_{qq})/2$ there are $(k-2)$ points $(l_{jj} + l_{q'q'})/2$ with $q \neq q'$, and $(k-2)$ points $(l_{j'j'} + l_{qq})/2$ with $j \neq j'$, such that

$$\|\frac{l_{jj} + l_{qq}}{2} - \frac{l_{jj} + l_{q'q'}}{2}\| = \frac{1}{2}\|l_{qq} - l_{q'q'}\| = \|\frac{l_{jj} + l_{qq}}{2} - \frac{l_{j'j'} + l_{qq}}{2}\|$$
$$= \frac{1}{2}\|l_{jj} - l_{j'j'}\| = \frac{\sqrt{2}}{2};$$

such pairs correspond to condition (i) above. Also, for each point $(l_{jj} + l_{qq})/2$ there are $\binom{k-2}{2}$ points $(l_{j'j'} + l_{q'q'})/2$ with j, j', q, q' pairwise disjoint, such that

$$\|\frac{l_{jj} + l_{qq}}{2} - \frac{l_{j'j'} + l_{q'q'}}{2}\| = \frac{1}{2}\|l_{jj} + l_{qq} - l_{j'j'} - l_{q'q'}\| = \frac{1}{2}\sqrt{1 + 1 + 1 + 1} = 1;$$

such pairs correspond to condition (ii) above.

Next, using the level sets and graph G, we define a set $B \subset \mathbb{R}^{2k}$, with $|B| = m\binom{k}{2}$. For any edge $\{j, j'\} \in E$ we assume without loss of generality that $j < j'$; this is always possible since G has no loops. Each edge $\{j, j'\}$ generates a set $B_{jj'}$ of $\binom{k}{2}$ points with

$$B_{jj'} = \{\frac{l_{1j} + l_{2j'}}{2}, \frac{l_{1j} + l_{3j'}}{2}, \ldots, \frac{l_{1j} + l_{kj'}}{2}\}$$
$$\cup \{\frac{l_{2j} + l_{3j'}}{2}, \ldots, \frac{l_{2j} + l_{kj'}}{2}\}$$
$$\ddots$$
$$\cup \{\frac{l_{(k-1)j} + l_{kj'}}{2}\}.$$

We set $B = \cup_{\{j,j'\} \in E} B_{jj'}$.

We now prove that if G has a k-clique then there is an isometry μ such that $\mu(A) \subset B$. Let $\{j_1, j_2, j_3 \ldots, j_k\}$ be such a clique with $j_1 < j_2 < \cdots < j_k$. Consider the set $B_c \subset B$ defined by

$$B_c = \{\frac{l_{1j_1} + l_{2j_2}}{2}, \frac{l_{1j_1} + l_{3j_3}}{2}, \ldots, \frac{l_{1j_1} + l_{kj_k}}{2}\}$$
$$\cup \{\frac{l_{2j_2} + l_{3j_3}}{2}, \ldots, \frac{l_{2j_2} + l_{kj_k}}{2}\}$$

$$\ddots$$

$$\cup \{\frac{l_{(k-1)j_{k-1}} + l_{kj_k}}{2}\}.$$

B_c has $\binom{k}{2}$ points, the first one being the first point of $B_{j_1 j_2}$, the second one being the second point of $B_{j_1 j_3}$ and so on, with the last one being the last point of $B_{j_{k-1} j_k}$. Observe that the points in B_c are middle-points that satisfy the conditions (i), (ii), regarding their defining endpoints, as the points in A do. Hence, it is straightforward to check that the bijection $\mu : B_c \to A$ that maps the first point of B_c to the first point of A and so on, is isometric.

Conversely, we can prove that an isometry μ such that $\mu(A) = B_c$ for some set $B_c \subset B$ implies that G has a k-clique. Since μ is an isometry, $||b - b'|| \in \{\sqrt{2}/2, 1\}$, for any two points $b, b' \in B_c$. Let $b = (l_{ij} + l_{pq})/2$ and $b' = (l_{i'j'} + l_{p'q'})/2$. Since $B_c \subset B$, we have that $i \neq p$ and $i' \neq p'$. First, using the facts that any two points in the same level set are at most $\sqrt{2}/6$ apart, and that the distance between any two points in different level sets is $\sqrt{2}$, we exclude the following cases from the valid combinations of level sets, which the endpoints of the points in B_c can belong to:

(1) $i = i', p = p'$; then, we have

$$||b - b'|| = ||\frac{(l_{ij} + l_{pq})}{2} - \frac{(l_{ij'} + l_{pq'})}{2}|| \leq \frac{1}{2}||l_{ij} - l_{ij'}|| + \frac{1}{2}||l_{pq} - l_{pq'}|| < \sqrt{2}/2.$$

(2) $i = i', p \neq p'$ with $j \neq j'$; then, we have

$$||b - b'|| = ||\frac{(l_{ij} + l_{pq})}{2} - \frac{(l_{ij'} + l_{p'q'})}{2}|| \leq \frac{1}{2}||l_{ij} - l_{ij'}|| + \frac{1}{2}||l_{pq} - l_{p'q'}||$$
$$< \sqrt{2}/12 + \sqrt{2}/2 < 1;$$

also, in this case, since from Lemma 1, $l_{ij}(2i - 1) \neq l_{ij'}(2i - 1)$ and $l_{ij}(2i) \neq l_{ij'}(2i)$,

$$||b - b'|| = \frac{1}{2}\sqrt{((l_{ij}(2i - 1) - l_{ij'}(2i - 1))^2 + (l_{ij}(2i) - l_{ij'}(2i))^2 + l_{pq}^2(2p - 1) + l_{pq}^2(2p)}$$
$$+ l_{p'q'}^2(2p' - 1) + l_{p'q'}^2(2p'))$$
$$> \sqrt{1 + 1}/2 = \sqrt{2}/2.$$

Only two combinations of level sets to which the end-points can belong are left, both of which are valid, the same as the ones for the points in A, given by conditions (i) and (ii) above. Namely, $||b - b'|| = \sqrt{2}/2$ implies that $i = i', j = j'$ and $p \neq p'$, and $||b - b'|| = 1$ implies that i, i', p, p' are pairwise disjoint. Since $i \neq p$ and $i' \neq p'$ as well, we conclude that only one point from each level set can be present as an end-point in B_c.

Since B_c and A are congruent, for any point $b \in B_c$ there must be $\binom{k-2}{2}$ points in B_c whose distance to b is 1. Therefore, all possible $\binom{k-2}{2}$ remaining (unordered) pairs of level sets must appear in B_c. Without loss of generality, we fix such a point $b = (l_{1j_1} + l_{2j_2})/2$. Then,

$$\{\frac{l_{3j_3} + l_{4j_4}}{2}, \frac{l_{3j_3} + l_{5j_5}}{2}, \ldots, \frac{l_{3j_3} + l_{kj_k}}{2}\}$$
$$\cup \{\frac{l_{4j_4} + l_{5j_5}}{2}, \ldots, \frac{l_{4j_4} + l_{kj_k}}{2}\}$$
$$\ddots$$
$$\cup \{\frac{l_{(k-1)j_{k-1}} + l_{kj_k}}{2}\} \subset B_c,$$

with all $j_1, \ldots, j_k$ being distinct. The $\binom{k}{2} - \binom{k-2}{2} - 1 = 2(k-2)$ remaining points in B_c must be such that their distance to b equals $\sqrt{2}/2$. The only valid combinations of level sets left, give the following two sets

$$\{\frac{l_{1j_1} + l_{3j_3}}{2}, \frac{l_{1j_1} + l_{4j_4}}{2}, \ldots, \frac{l_{1j_1} + l_{kj_k}}{2}\},$$

and

$$\{\frac{l_{2j_2} + l_{3j_3}}{2}, \frac{l_{2j_2} + l_{4j_4}}{2}, \ldots, \frac{l_{2j_2} + l_{kj_k}}{2}\},$$

of $k - 2$ points each, all in B_c as well.

Since each point $(l_{ij} + l_{i'j'})/2 \in B_c$ with $j, j' \in \{j_1, \ldots, j_k\}$, is generated by a distinct edge $\{i, j\} \in E$, it is easy to check that $\{j_1, \ldots, j_k\}$ is indeed a k-clique. $\quad\square$

Since in the above fpt-reduction $d = 2k$, an $O(mn^{o(d)})$-time algorithm for p-SUBSET-CONGRUENCE implies an $O(n^{o(k)})$-time algorithm for p-CLIQUE, which in turn implies that SNP $\subset$ DTIME$(2^{o(n)})$ [8].

Corollary 1. p-SUBSET-CONGRUENCE *can be solved in* $O(mn^{o(d)})$ *time, only if* SNP $\subset$ DTIME$(2^{o(n)})$.

Consider a point set distance D for which $D(A, B) = 0 \Leftrightarrow A \subset B$ for every $A, B \in \mathbb{R}^d$ with $|A| < |B|$; this is a desired property for any distance that is used to find small patterns into larger ones, e.g., directed Hausdorff distance, Earth Mover's Distance. Then, $\min_\mu D(\mu(A), B) = 0 \Leftrightarrow \mu'(A) \subset B$ for some isometry μ'. Hence, we have the following.

Corollary 2. *Given two point sets* $A, B \in \mathbb{R}^d$, *with* $|A| < |B|$, *and a distance* D *for which* $D(A, B) = 0$ *if and only if* $A \subset B$, *the problem of minimizing* D *under isometries, when* d *is part of the input, is not in FPT, unless* FPT$=$W$[1]$.

3   Concluding Remarks

We have studied the parameterized complexity of some fundamental point set pattern matching problems with respect to the dimension. We proved that subset congruence is $W[1]$-hard, which also implies that minimizing under isometries the directed Hausdorff distance, or the Earth Mover's Distance between two point sets in unbounded dimension is not in FPT (unless FPT=$W[1]$).

There are quite a few other geometric optimization problems whose complexity due to unbounded dimension has been studied; see, for example, Megiddo [19]. We believe that for such problems the dimension is an interesting parameter to be studied within the framework of parameterized complexity theory.

References

1. T. Akutsu. On determining the congruence of point sets in d-dimensions. *Comput. Geom. Theory Appl.*, 9(4):247–256, 1998.
2. H. Alt and L.J. Guibas. Discrete geometric shapes: Matching, interpolation, and approximation. In J.R. Sack and J. Urrutia, editors, *Handbook of Computational Geometry*, pages 121–153. Elsevier Science Publishers B.V. North-Holland, Amsterdam, 1999.
3. H. Alt, K. Mehlhorn, H. Wagener, and E. Welzl. Congruence, similarity, and symmetries of geometric objects. *Discrete Comput. Geom.*, 3:237–256, 1988.
4. P. Braß. Exact point pattern matching and the number of congruent triangles in a three-dimensional point set. In *Proc. of the 8th Annu. European Sympos. Algorithms (ESA)*, volume 1879 of *LNCS*, pages 112–119, 2000.
5. P. Braß and C. Knauer. Testing the congruence of d-dimensional point sets. *Int. J. of Comp. Geom. Appl.*, 12(1):115–124, 2002.
6. P. Braß and J. Pach. Problems and results on geometric patterns. In D. Avis, A. Hertz, and O. Marcotte, editors, *Graph Theory and Combinatorial Optimization*, Gerad 25th Anniversary, pages 17–36. Springer, 2005.
7. S. Cabello, P. Giannopoulos, C. Knauer, and G. Rote. Matching point sets with respect to the Earth Mover's Distance. In *Proc. of the 13th Annu. European Sympos. Algorithms (ESA)*, volume 3669 of *LNCS*, pages 520–531, 2005.
8. Jianer Chen, Benny Chor, Mike Fellows, Xiuzhen Huang, David Juedes, Iyad A. Kanj, and Ge Xia. Tight lower bounds for certain parameterized NP-hard problems. *Information and Computation*, 201(2):216–231, 2005.
9. L.P. Chew, D. Dor, A. Efrat, and K. Kedem. Geometric pattern matching in d-dimensional space. *Discrete Comput. Geom.*, 21, 1999.
10. L.P. Chew, M. Goodrich, D. P. Huttenlocher, K. Kedem, J. M. Kleinberg, and D. Kravets. Geometric pattern matching under Euclidean motion. *Comput. Geom. Theory Appl.*, 7:113–124, 1997.
11. S.D. Cohen and L.J. Guibas. The Earth Mover's Distance under transformation sets. In *Proceedings of the 7th IEEE International Conference on Computer Vision*, pages 173–187, September 1999.
12. P.J. de Rezende and D.T. Lee. Point set matching in d-dimensions. *Algorithmica*, 13:387–404, 1995.
13. A. Efrat, A. Itai, and M.J. Katz. Geometry helps in bottleneck matching and related problems. *Algorithmica*, 31:1–28, 2001.
14. J. Flum and M. Grohe. *Parameterized Complexity Theory*, volume XIV of *Series: Texts in Theoretical Computer Science. An EATCS Series*. Springer-Verlag, 2006.
15. M. Hagedoorn and R.C. Veltkamp. State-of-the-art in shape matching. Technical Report UU-CS-1999-027, Institute of Information and Computing Sciences, Utrecht University, The Netherlands, 1999.

16. D.P. Huttenlocher, K. Kedem, and M. Sharir. The upper envelope of Voronoi surfaces and its applications. *Discrete Comput. Geom.*, 9:267–291, 1993.

17. D. Marx. Efficient approximation schemes for geometric problems? In *Proc. of the 13th Annu. European Sympos. Algorithms (ESA)*, volume 3669 of *LNCS*, pages 448–459, 2005.

18. N. Megiddo. Linear programming in linear time when the dimension is fixed. *J. ACM*, 31:114–127, 1984.

19. N. Megiddo. On the complexity of some geometric problems in unbounded dimension. *J. Symb. Comput.*, 10(3-4):327–334, 1990.

20. C. Papadimitriou and S. Safra. The complexity of low-distortion embeddings between point sets. In *Proc. of the 16th Annu. ACM-SIAM Sympos. Discrete Algorithms (SODA)*, pages 112–118. SIAM, 2005.

Finding a Minimum Feedback Vertex Set in Time $\mathcal{O}(1.7548^n)^\star$

Fedor V. Fomin, Serge Gaspers, and Artem V. Pyatkin**

Department of Informatics, University of Bergen,
N-5020 Bergen, Norway
{Fedor.Fomin, Serge.Gaspers, Artem.Pyatkin}@ii.uib.no

Abstract. We present an $\mathcal{O}(1.7548^n)$ algorithm finding a minimum feedback vertex set in a graph on n vertices.

Keywords: minimum feedback vertex set, maximum induced forest, exact exponential algorithm.

1 Introduction

The problem of finding a minimum feedback vertex set has many applications and its history can be traced back to the early '60s (see the survey of Festa at al. [2]). It is also one of the classical NP-complete problems from Karp's list [5]. There is quite a dramatic story of obtaining faster and faster parameterized algorithms with a chain of improvements (see e.g. [7]) concluding with $2^{\mathcal{O}(k)}n^{\mathcal{O}(1)}$-time algorithms obtained independently by different research groups [1,4].

A feedback vertex set of a graph on n vertices can be trivially found in time $\mathcal{O}(2^n n)$ by trying all possible vertex subsets. For a long time, despite attacks of many researchers, no faster exponential time algorithm was known. Very recently Razgon [8] broke the 2^n barrier with an $\mathcal{O}(1.8899^n)$ time algorithm. The algorithm of Razgon is based on the Branch & Reduce paradigm and its analysis is nice and clever.

In this paper we show how to find a minimum feedback vertex set in time $\mathcal{O}(1.7548^n)$. Our improvement is based on Razgon's idea of measuring the progress of the branching algorithm. The most significant improvement in the running time of our algorithm is due to a new branching rule which is based on Proposition 2. This rule works nicely except one case, which, luckily, can be reduced to finding an independent set of maximum size.

2 Preliminaries

Let $G = (V, E)$ be an undirected graph on n vertices. For $V' \subseteq V$ we denote by $G[V']$ the graph induced by V' and by $G \setminus V'$ the graph induced by $V \setminus V'$. For

---

* Additional support by the Research Council of Norway.
** The work was partially supported by grants of the Russian Foundation for Basic Research (project code 05-01-00395), INTAS (project code 04–77–7173).

H.L. Bodlaender and M.A. Langston (Eds.): IWPEC 2006, LNCS 4169, pp. 184–191, 2006.

a vertex $v \in V$ let $N(v)$ be the sets of its neighbors. We denote by $\Delta(G)$ the maximum vertex degree of G.

The set $X \subseteq V$ is called a *feedback vertex set* or an FVS if $G \backslash X$ is a forest. Thus the problem of finding a minimum FVS is equivalent to the problem of finding a maximum induced forest or an MIF. For the description of the algorithm it is more convenient to work with MIF than with FVS.

We call a subset $F \subseteq V$ *acyclic* if $G[F]$ is a forest and *independent* if every component of $G[F]$ is an isolated vertex; F is a *maximum independent set* of G if has a maximum cardinality among all independent sets. If F is acyclic but not independent then every connected component on at least two vertices is called *non-trivial*. If T is a non-trivial component then we denote by $\mathrm{Id}(T, t)$ the operation of contracting all edges of T into one vertex t and removing appeared loops. Note that this operation may create multiedges in G. We denote by $\mathrm{Id}^*(T, t)$ the operation $\mathrm{Id}(T, t)$ followed by the removal of all vertices connected with t by multiedges.

For an acyclic subset $F \subseteq V$, denote by $\mathcal{M}_G(F)$ the set of all maximum acyclic supersets of F in G (we omit the subindex G when it is clear from the context which graph is meant). Let $\mathcal{M} = \mathcal{M}(\emptyset)$. Then the problem of finding a MIF can be stated as finding an element of $\mathcal{M}$. We solve a more general problem, namely finding an element of $\mathcal{M}(F)$ for an arbitrary acyclic subset F.

To simplify the description of the algorithm, we suppose that F is always an independent set. The next proposition justifies this supposition.

Proposition 1. *Let $G = (V, E)$ be a graph, $F \subseteq V$ be an acyclic subset of vertices and T be a non-trivial component of F. Denote by G' the graph obtained from G by the operation $\mathrm{Id}^*(T, t)$ and let $F' = F \cup \{t\} \setminus T$. Then $X \in \mathcal{M}_G(F)$ if and only if $X' \in \mathcal{M}_{G'}(F')$ where $X' = X \cup \{t\} \setminus T$.*

Proof. If, after the operation $\mathrm{Id}(T, t)$, a vertex v is connected with t by a multi-edge then the set $T \cup \{v\}$ is not acyclic in G. Hence, no element of $\mathcal{M}_G(F)$ may contain v. Therefore, the function $X \mapsto X \cup \{t\} \setminus T$ is a bijection from $\mathcal{M}_G(F)$ to $\mathcal{M}_{G'}(F')$. □

By using the operation Id^* on every non-trivial component of F, we obtain an independent set F'.

The following proposition is used to justify the main branching rule of the algorithm.

Proposition 2. *Let $G = (V, E)$ be a graph, $F \subseteq V$ be an independent subset of vertices and $v \notin F$ be a vertex adjacent to exactly one vertex $t \in F$. Then, there exists $X \in \mathcal{M}(F)$ such that either v or at least two vertices of $N(v) \setminus \{t\}$ are in X.*

Proof. Suppose, for the sake of contradiction, that there is $X \in \mathcal{M}(F)$ such that $v \notin X$ and only one vertex of $N(v) \setminus \{t\}$ is in X, say z. It follows from the maximality of X that $X \cup \{v\}$ is not acyclic. But since v has degree at most 2 in X all the cycles in $X \cup \{v\}$ must contain z. Then the set $X \cup \{v\} \setminus \{z\}$ is in

$\mathcal{M}(F)$ and satisfies the conditions. The case where no vertex of $N(v) \setminus \{t\}$ is in X is even simpler.   □

Consequently, if $N(v) = \{t, v_1, v_2, \ldots, v_k\}$, then there exists $X \in \mathcal{M}(F)$ satisfying one of the following properties:

1. $v \in X$;
2. $v \notin X$, $v_i \in X$ for some $i \in \{1, 2, \ldots, k-2\}$ while $v_j \notin X$ for all $j < i$;
3. $v, v_1, v_2, \ldots, v_{k-2} \notin X$ but $v_{k-1}, v_k \in X$.

   In particular, if $k \leq 1$, then $v \in X$ for some $X \in \mathcal{M}(F)$.
   We also need the following

Proposition 3. *Let* $G = (V, E)$ *be a graph and* F *be an independent set in* G *such that* $G \setminus F = N(t)$ *for some* $t \in F$. *Consider the graph* $G' = G[N(t)]$ *and for every pair of vertices* $u, v \in N(t)$ *having a common neighbor in* $F \setminus \{t\}$ *add an edge* uv *to* G'. *Denote the obtained graph by* H *and let* I *be a maximum independent set in* H. *Then* $F \cup I \in \mathcal{M}_G(F)$.

Proof. Let $X \in \mathcal{M}_G(F)$ and $u, v \in G \setminus F$. If $uv \in E$ then u, v, t form a triangle. If there is a vertex $w \in F \setminus \{t\}$ adjacent to both u and v then $tuwv$ is a 4-cycle. In both cases, X cannot contain u and v at the same time. Therefore, $X \in \mathcal{M}_G(F)$ if and only if $X \setminus F$ is a maximum independent set in H.   □

There are several fast exponential algorithms computing a maximum independent set in a graph. We use the fastest known polynomial space algorithm.

Proposition 4 ([3]). *Let* G *be a graph on* n *vertices. Then a maximum independent set in* G *can be found in time* $\mathcal{O}(1.2210^n)$.

3   The Algorithm

In this section the algorithm finding the maximum size of an induced forest containing a given acyclic set F is presented. This algorithm can easily be turned into an algorithm computing at least one element of $\mathcal{M}_G(F)$. During the work of the algorithm one vertex $t \in F$ is called an *active vertex*. The algorithm branches on a chosen neighbor of t. Let $v \in N(t)$. Denote by K the set of all vertices of F other than t that are adjacent to v. Let G' be the graph obtained after the operation $\mathrm{Id}(K \cup \{v\}, u)$. We say that a vertex $w \in V \setminus \{t\}$ is a *generalized neighbor* of v in G if w is the neighbor of u in G'. Denote by $\mathrm{gd}(v)$ the *generalized degree* of v which is the number of its generalized neighbors.

   The description of the algorithm consists of a sequence of cases and subcases. To avoid a confusing nesting of if-then-else statements let us use the following convention: The first case which applies is used in the algorithm. Thus, inside a given case, the hypotheses of all previous cases are assumed to be false.

   Algorithm $\mathtt{mif}(G, F)$ computing for a given graph G and an acyclic set F the maximum size of an induced forest containing F is described by the following preprocessing and main procedures. (Let us note that $\mathtt{mif}(G, \emptyset)$ computes the maximum size of an induced forest in G.)

Preprocessing

1. If G consists of $k \geq 2$ connected components $G_1, G_2, \ldots, G_k$, then the algorithm is called on each of the components and

$$\mathtt{mif}(G, F) = \sum_{i=1}^{k} \mathtt{mif}(G_i, F_i),$$

where $F_i = G_i \cap F$ for all $i \in \{1, 2, \ldots, k\}$.

2. If F is not independent, then apply operation $\mathrm{Id}^*(T, v_T)$ on every non-trivial component T of F. Moreover, if T contains the active vertex then v_T becomes active. Let G' be the resulting graph and let F' be the independent set in G' obtained from F. Then

$$\mathtt{mif}(G, F) = \mathtt{mif}(G', F') + |F \setminus F'|.$$

Main procedures

1. If $F = V$ then $\mathcal{M}_G(F) = \{V\}$. Thus,

$$\mathtt{mif}(G, F) = |V|.$$

2. If $F = \emptyset$ and $\Delta(G) \leq 1$ then $\mathcal{M}_G(F) = \{V\}$ and

$$\mathtt{mif}(G, F) = |V|.$$

3. If $F = \emptyset$ and $\Delta(G) \geq 2$ then the algorithm chooses a vertex $t \in V(G)$ of degree at least 2. Then t is either contained in a maximum induced forest or not. Thus the algorithm branches on two subproblems and returns the maximum:

$$\mathtt{mif}(G, F) = \max \{ \mathtt{mif}(G, F \cup \{t\}),$$
$$\mathtt{mif}(G \setminus \{t\}, F)\}.$$

4. If F contains no active vertex then choose an arbitrary vertex $t \in F$ as an active vertex. Denote the active vertex by t from now on.

5. If $V \setminus F = N(t)$ then the algorithm constructs the graph H from Proposition 3 and computes a maximum independent set I in H. Then

$$\mathtt{mif}(G, F) = |F| + |I|.$$

6. If there is $v \in N(t)$ with $\mathrm{gd}(v) \leq 1$ then add v to F.

$$\mathtt{mif}(G, F) = \mathtt{mif}(G, F \cup \{v\})$$

7. If there is $v \in N(t)$ with $\mathrm{gd}(v) \geq 4$ then either add v to F or remove v from G.

$$\mathtt{mif}(G, F) = \max \{ \mathtt{mif}(G, F \cup \{v\}),$$
$$\mathtt{mif}(G \setminus \{v\}, F)\}$$

8. If there is $v \in N(t)$ with $gd(v) = 2$ then denote its generalized neighbors by w_1 and w_2. Either add v to F or remove v from G but add w_1 and w_2 to F.

$$\mathtt{mif}(G, F) = \max \{ \mathtt{mif}(G, F \cup \{v\}),$$
$$\mathtt{mif}(G \setminus \{v\}, F \cup \{w_1, w_2\})\}$$

9. If all vertices in $N(t)$ have exactly three generalized neighbors then at least one of these vertices must have a generalized neighbor outside $N(t)$, since the graph is connected and the condition of the case Main 5 does not hold. Denote such a vertex by v and its generalized neighbors by w_1, w_2 and w_3 in such a way that $w_1 \notin N(t)$. Then we either add v to F; or remove v from G but add w_1 to F; or remove v and w_1 from G and add w_2 and w_3 to F.

$$\mathtt{mif}(G, F) = \max \{ \mathtt{mif}(G, F \cup \{v\}),$$
$$\mathtt{mif}(G \setminus \{v\}, F \cup \{w_1\}),$$
$$\mathtt{mif}(G \setminus \{v, w_1\}, F \cup \{w_2, w_3\})\}$$

The behavior of the algorithm is analyzed in the following

Theorem 1. *Let G be a graph on n vertices. Then a maximum induced forest of G can be found in time $\mathcal{O}(1.7548^n)$.*

Proof. Let us consider the algorithm $\mathtt{mif}(G, F)$ described above. The correctness of Preprocessing 1 and Main 1,2,3,4,7 is clear. The correctness of Main 5 follows from Proposition 3, while the correctness of Preprocessing 2 and Main 6,8,9 follows from Proposition 1 and 2 (indeed, applying Proposition 2 to the vertex u of the graph G' shows that for some $X \in \mathcal{M}_G(F)$ either v or at least two of its generalized neighbors are in X).

In order to evaluate the time complexity of the algorithm we use the following measure:

$$\mu = |V \setminus F| + \alpha |V \setminus (F \cup N(t))|$$

where $\alpha = 0.955$. In other words, each vertex in F has weight 0, each vertex in $N(t)$ has weight 1, each other vertex has weight $1 + \alpha$, and the size of the problem is equal to the sum of the vertex weights. We will prove that a problem of size μ can be solved in time $\mathcal{O}(x^\mu)$ where

$$x < 1.333277.$$

Denote by $f(\mu)$ the maximum number of times the algorithm is called recursively on a problem of size μ (i. e. the number of leaves in the search tree). Then the running time $T(\mu)$ of the algorithm is bounded by $\mathcal{O}(f(\mu) \cdot n^{\mathcal{O}(1)})$. We use induction on μ to prove that $f(\mu) \leq x^\mu$. Then $T(\mu) = \mathcal{O}(f(\mu) \cdot n^{\mathcal{O}(1)})$, and since the polynomial is suppressed by rounding the exponential base, we have $T(\mu) = \mathcal{O}(1.333277^\mu)$. Clearly, $f(0) = 1$. Suppose that $f(k) \leq x^k$ for every $k < \mu$ and consider a problem of size μ.

It is clear that the following steps do not contribute to the exponential factor of the running time of the algorithm: Preprocessing 1,2 and Main 1,2,4,6.

If the condition of the case Main 5 holds then the graph H has exactly μ vertices since each vertex that is not in F has weight 1. By Theorem 4, a maximum independent set in H can be found in time $\mathcal{O}(1.2210^\mu)$. Also the algorithm computing a maximum independent set in [3] is a branching algorithm with a number of recursive calls bounded by $1.2210^\mu < 1.333277^\mu$.

In all remaining cases the algorithm is called recursively on smaller problems. We consider these cases separately.

In the case Main 3 every vertex has weight $1 + \alpha$. So, removing v leads to a problem of size $\mu - 1 - \alpha$. Otherwise, v becomes active after the next Main 4 step. Then all its neighbors become of weight 1, and we obtain a problem of size at most $\mu - 1 - 3\alpha$ since v has degree at least 2. Thus

$$f(\mu) \leq f(\mu - 1 - \alpha) + f(\mu - 1 - 3\alpha) \leq (x^{\mu-1-\alpha} + x^{\mu-1-3\alpha}) \leq x^\mu$$

by the induction assumption and the choice of x and α.

In the case Main 7 removing the vertex v decreases the size of the problem by 1. If v is added to F then we obtain a non-trivial component in F, which is contracted into a new active vertex t' at the next Preprocessing 2 step. Those of the generalized neighbors of v that had weight 1 will be connected with t' by multiedges and thus removed during the next Preprocessing 2 step. If a generalized neighbor of v had weight $1 + \alpha$ then it will become a neighbor of t', i. e. of weight 1. Thus, in any case the size of the problem is decreased by at least $1 + 4\alpha$. So, we have that

$$f(\mu) \leq f(\mu - 1) + f(\mu - 1 - 4\alpha) \leq (x^{\mu-1} + x^{\mu-1-4\alpha}) \leq x^\mu.$$

In the case Main 8 we distinguish three subcases depending on the weights of the generalized neighbors of v. Let i be the number of generalized neighbors of v having weight $1 + \alpha$. Adding v to F reduces the weight of a generalized neighbor either from 1 to 0 or from $1 + \alpha$ to 1. Removing v from the graph reduces the weight of both generalized neighbors of v to 0 (since we add them to F). According to this, we obtain three recurrences: for $i \in \{0, 1, 2\}$,

$$f(\mu) \leq f(\mu - (3 - i) - i\alpha) + f(\mu - 3 - i\alpha) \leq (x^{\mu-3+i-i\alpha} + x^{\mu-3-i\alpha}) \leq x^\mu.$$

The case Main 9 is considered analogously to the case Main 8, except that at least one of the generalized neighbors of v has weight $1 + \alpha$, that is $i \geq 1$. In this case, we have for $i \in \{1, 2, 3\}$,

$$f(\mu) \leq f(\mu - (4 - i) - i\alpha) + f(\mu - 2 - \alpha) + f(\mu - 4 - i\alpha)$$
$$\leq (x^{\mu-4+i-i\alpha} + x^{\mu-2-\alpha} + x^{\mu-4-i\alpha}) \leq x^\mu.$$

Thus

$$f(\mu) \leq x^\mu.$$

Since every vertex of G is of weight at most $1 + \alpha$, we have that the running time of the algorithm is

$$T(\mu) = \mathcal{O}(x^{\mu}) = \mathcal{O}(x^{(1+\alpha)n}) = \mathcal{O}(1.333277^{1.955n}) = \mathcal{O}(1.7548^n). \qquad \square$$

Remark 1. The only tight recurrence is the one of case Main 7 when v has degree 4. Thus, an improvement of this case would improve the overall (upper bound of the) running time of the algorithm.

4   Conclusion

We have shown that a few simple changes in the branch-and-reduce algorithm of Razgon [8] together with a flexible measure of the size of a (sub)problem leads to a significant improvement in the proved upper bound of the worst case running time of the algorithm.

Note added in camera-ready: Recently, we also proved that the number of maximal induced forests (and thus the number of minimal feedback vertex sets) in a graph on n vertices is at most 1.8638^n. Schwikowski and Speckenmeyer presented in [6] an algorithm which enumerates all minimal feedback vertex sets of a graph with polynomial time delay. Thus our upper bound implies that all minimal feedback vertex sets (and maximal induced forests) can be enumerated in time $\mathcal{O}(1.8638^n)$.

Acknowledgment. We thank Igor Razgon for sending us a preliminary version of [8].

References

1. F. K. H. A. Dehne, M. R. Fellows, M. A. Langston, F. A. Rosamond, and K. Stevens, *An $O(2^{O(k)}n^3)$ FPT algorithm for the undirected feedback vertex set problem.*, in Proceedings of the 11th Annual International Conference on Computing and Combinatorics (COCOON 2005), vol. 3595 of LNCS, Berlin, 2005, Springer, pp. 859–869.
2. P. Festa, P. M. Pardalos, and M. G. C. Resende, *Feedback set problems*, in Handbook of combinatorial optimization, Supplement Vol. A, Kluwer Acad. Publ., Dordrecht, 1999, pp. 209–258.
3. F. V. Fomin, F. Grandoni, and D. Kratsch, *Measure and conquer: A simple $O(2^{0.288\,n})$ independent set algorithm*, in 17th Annual ACM-SIAM Symposium on Discrete Algorithms (SODA 2006), New York, 2006, ACM and SIAM, pp. 18–25.
4. J. Guo, R. Niedermeier, and S. Wernicke, *Parameterized complexity of generalized vertex cover problems.*, in Proceedings of the 9th International Workshop on Algorithms and Data Structures (WADS 2005), vol. 3608 of LNCS, Springer, Berlin, 2005, pp. 36–48.
5. R. M. Karp, *Reducibility among combinatorial problems*, in Complexity of computer computations, Plenum Press, New York, 1972, pp. 85–103.

6. B. Schwikowski, and E. Speckenmeyer, *On Computing All Minimal Solutions for Feedback Problems*, in Discrete Applied Mathematics 117(1-3), 2002, pp. 253–265.
7. V. Raman, S. Saurabh, and C. R. Subramanian, *Faster fixed parameter tractable algorithms for undirected feedback vertex set*, in Proceedings of the 13th International Symposium on Algorithms and Computation (ISAAC 2002), vol. 2518 of LNCS, Springer-Verlag, Berlin, 2002, pp. 241–248.
8. I. Razgon, *Exact computation of maximum induced forest*, in Proceedings of the 10th Scandinavian Workshop on Algorithm Theory (SWAT 2006), LNCS, Berlin, 2006, Springer, to appear.

The Undirected Feedback Vertex Set Problem Has a Poly(k) Kernel*

Kevin Burrage[1], Vladimir Estivill-Castro[2], Michael Fellows[3],
Michael Langston[4], Shev Mac[1], and Frances Rosamond[5]

[1] Department of Mathematics, University of Queensland, Brisbane, QLD 4072
[2] Griffith University, Brisbane QLD 4111, Australia
[3] School of EE & CS, University of Newcastle, Callaghan NSW 2308, Australia
[4] Department of Computer Science, University of Tennessee, Knoxville TN
37996-3450 and Computer Science and Mathematics Division, Oak Ridge National
Laboratory, Oak Ridge, TN 37831-6164 U.S.A.
[5] The Retreat for the Arts and Sciences, Newcastle, Australia

Abstract. Resolving a noted open problem, we show that the UNDI-
RECTED FEEDBACK VERTEX SET problem, parameterized by the size
of the solution set of vertices, is in the parameterized complexity class
$Poly(k)$, that is, polynomial-time pre-processing is sufficient to reduce
an initial problem instance (G, k) to a decision-equivalent simplified instance (G', k') where $k' \leq k$, and the number of vertices of G' is bounded
by a polynomial function of k. Our main result shows an $O(k^{11})$ kernelization bound.

1 Introduction

One of the most important concrete problems yet to be analyzed from the parameterized perspective is the FEEDBACK VERTEX SET problem, in both its
undirected and directed forms. For a survey of many applications of feedback
sets, see [FPR99]. The problem asks whether there is a set S of at most k vertices of the input graph (digraph) G, such that every cycle (directed cycle) in G
contains at least one vertex in S. The directed form of the problem is notoriously
open as to whether it is fixed-parameter tractable (FPT); the problem remains
open even for the restriction to planar digraphs.

Previous results on FPT algorithms for the UNDIRECTED FEEDBACK VERTEX
SET problem have followed a trajectory of steady improvements in the run times
[DF92, Bod94, DF99, BBG00, RSS02, KPS04, RSS05, GGHNW05, DFLRS05].

---

* This research has been supported in part by the U.S. National Science Foundation under grant CCR–0075792, by the U.S. Office of Naval Research under grant
N00014–01–1–0608, by the U.S. Department of Energy under contract DE–AC05–
00OR22725, and by the Australian Research Council under the auspices of the Australian Centre for Bioinformatics, through Federation Fellowship support of the first
author, and through Discovery Project support of the second and third authors.

H.L. Bodlaender and M.A. Langston (Eds.): IWPEC 2006, LNCS 4169, pp. 192–202, 2006.

The current best results are:

- A practical randomized FPT algorithm due to Becker, et al., runs in time $\mathcal{O}(4^k kn)$ and finds a feedback vertex set of size k (assuming one exists) with probability at least $1 - (1 - 4^{-k})^{c4^k}$ for an arbitrary constant c [BBG00] .
- A deterministic FPT algorithm due independently to Guo, et al., and to Dehne, et al., solves the problem in time $\mathcal{O}^*(10.567^k)$ [GGHNW05, DFLRS05] (the run-time analysis is in the latter reference).

Here we address a noted open problem:

Is there a polynomial-time algorithm that kernelizes FVS on undirected graphs to a kernel of size polynomial in k?

Kernelization bounds for FPT parameterized problems are an area of increasing interest, because of the strong connection between effective kernelization and practical algorithmics [Wei98, Wei00, Nie02]. The issue of efficient kernelization is "completely general" for parameterized complexity because a parameterized problem Π is in FPT if and only if there is a transformation from Π to itself, and a function g, that reduces an instance (x, k) to (x', k') such that:

(1) the transformation runs in time polynomial in $|(x, k)|$,
(2) (x, k) is a yes-instance of Π if and only if (x', k') is a yes-instance of Π,
(3) $k' \leq k$, and
(4) $|x'| \leq g(k)$.

In the situation described above, we say that we have a *kernelization bound* of $g(k)$. The proof of the above "point of view" on FPT that focuses on P-time kernelization is completely trivial, giving a kernelization bound of $g(k) = f(k)$ for an FPT problem solvable in time $f(k)n^c$. But for many important FPT problems, we can do *much* better, and the "pre-processing" routines that produce small kernels seem to have great practical value [ACFLSS04, Nie02, Nie06]. For example, the VERTEX COVER problem can be kernelized in polynomial time to a graph on at most $2k$ vertices [NT75, ACFLSS04, CFJ04]. PLANAR DOMINATING SET also has a problem kernel of linear size [AFN04].

For typical problems that have been classified as fixed-parameter tractable, we see steady improvements both in the $f(k)$ in the best known FPT algorithms solving the problem in time $f(k)n^c$, and also (as an independent issue) in the best known kernelization bounds $g(k)$. In fact, polynomial-time kernelization/pre-processing seems to be both a deeper subject than one might have thought and also one of the most universally relevant ways of dealing with hard computational problems for realistic input distributions [Nie02, Nie06]. The practical importance of efficient kernelization has focused attention on subclasses of FPT, such as the class $Lin(k)$ consisting of the parameterized problems that admit linear problem kernels, and the class $Poly(k)$ of FPT problems with polynomial-sized kernels:

$$Lin(k) \subseteq Poly(k) \subseteq FPT$$

Little is yet known about these natural subclasses of FPT. In fact, there are many famous FPT problems for which membership in $Poly(k)$ is unknown. It is easy to point to examples of problems in FPT that are unlikely to be in $Poly(k)$, but mathematical methods to substantiate such intuitive judgements are currently lacking. (For a concrete example, we think it is unlikely that MIN CUT LINEAR ARRANGEMENT is in $Poly(k)$.) It also seems that devising polynomial-time data reduction algorithms to show membership in $Lin(k)$ and $Poly(k)$ may be fruitful combinatorial ground for novel algorithmic strategies. All of this area seems well-worth investigating because of the close connections to practical computing. The recent Ph.D. dissertations of Prieto [Pr05] and Guo [Guo06] present some of the first systematic investigations of these challenges.

In the next section, we prove our main result, an $O(k^{11})$ kernelization for the UNDIRECTED FEEDBACK VERTEX SET problem. In the concluding section we point to some open problems.

2   UFVS Is in $Poly(k)$

The problem we address is formally defined as follows.

UNDIRECTED FEEDBACK VERTEX SET
Instance: An undirected multigraph $G = (V, E)$ (that is, a graph with loops and multiple edges allowed), and a positive integer k.
Parameter: k
Question: Is there a subset $V' \subseteq V$, $|V'| \leq k$, such that every cycle in G contains at least one vertex of V'?

The basic approach of our algorithm is to compute in polynomial time a specific *structural map* of the problem instance. Some of our polynomial-time data reduction rules are defined *relative to this structural map*. We study the situation by advancing "structural claims" that hold concerning a reduced instance. The reduction rules that are defined relative to the structural map either decide the instance, or reduce the size of the instance, or result in an "improvement" in the quality of the structural map (where the quality is defined so that only polynomially many improvements are possible). (This approach to kernelization is exposited with a number of examples in [Pr05].)

Additionally, we employ the following simple reduction rules that also play a role in the best known FPT algorithms for the problem, and are adapted from that context [GGHNW05, DFLRS05]:

Rule 1: The Degree One Rule. If v is a vertex of degree 1 in G, then delete v. The parameter k is unchanged.

Rule 2: The Degree Two Rule. If v is a vertex of degree 2 in G, with neighbors a and b (allowing possibly $a = b$), then modify G by replacing v and its two incident edges with a single edge between a and b (or a loop on $a = b$). The parameter k is unchanged.

Rule 3: The Loop Rule. If there is a loop on a vertex v then take v into the solution set, and reduce to the instance $(G - v, k - 1)$.

Rule 4: Multiedge Reduction. If there are more than two edges between u and v then delete all but two of these. The parameter k is unchanged.

The soundness of these reduction rules is trivial. In linear time we can determine if any of the above reduction rules can be applied to a problem instance. An instance (G, k) is *reduced* if none of the reduction rules can be applied.

We will always assume that an instance we are working with is reduced with regards to Reduction Rules (1-4).

Step One
The first step of our algorithm employs the polynomial-time 2-approximation algorithm of Bafna, et al. [BBF99] to compute an approximate solution. If the approximate solution S that is produced is too big, $|S| > 2k$, then we can decide that (G, k) is a no-instance, and we are done.

Thus we can assume as the first part of our *structural map*, a feedback vertex set S for G, where $|S| \leq 2k$. Let F denote the forest $G - S$.

Step Two
In the second step, we greedily compute a maximal set $\mathcal{P}$ of pairwise internally vertex-disjoint paths (really, cycles) ρ that satisfy:

(1) ρ begins at a vertex v of S.
(2) If ρ begins at v then ρ also ends at v (thus forming a cycle).
(3) All internal vertices of ρ belong to $V - S$.

The collection of paths $\mathcal{P}$ is the second part of our structural map. Note that by our assumption that (G, k) is reduced relative to Reduction Rules (1-4), every path $\rho \in \mathcal{P}$ must have at least one internal vertex, because a reduced (G, k) does not have any loops. The following reduction rule is defined relative to this structure.

Rule 5: Flower Reduction. If there is a vertex $v \in S$ such that $\mathcal{P}$ contains at least $k + 1$ paths that begin and end at v, then reduce (G, k) to $(G - v, k - 1)$.

The soundness of Rule 5 is immediately apparent: any k-element feedback vertex set V' must contain v, since otherwise at least one of the $k + 1$ internally vertex-disjoint "loops on v" provided by $\mathcal{P}$ would fail to contain a vertex in V', as there are too many of them, and any vertex in V' can only hit one of them.

Step Three (Repeated)
This is the main loop of our algorithm. Our algorithm maintains a *structural map* for the reduced instance (G, k), developed in the first two steps, consisting of:

(1) the feedback set S, where $|S| \leq 2k$, and
(2) the collection of paths $\mathcal{P}$.

We repeatedly either:

• Improve the *quality* of this structural map, according to a list of priorities detailed below, or

- Discover an opportunity to apply a reduction rule, or
- Output a kernelized instance.

There is one more reduction rule that is defined relative to our structural map, but first we need some definitions concerning the overall structural picture provided by our structural map information.

If ρ is a path in $\mathcal{P}$, then let ρ' denote the ("internal") subpath formed by the vertices of ρ in $V - S$. Let $\mathcal{P}'$ denote the collection of such subpaths ρ' of paths ρ in $\mathcal{P}$. Let F' denote the subforest of F that results by deleting from F any vertex that belongs to a path in $\mathcal{P}'$. The vertices of F can be viewed as partitioned into:

(1) the paths ρ' in $\mathcal{P}'$, and
(2) the trees in F'.

Let $\mathcal{C}$ denote the set of vertices that either belong to S, or belong to a path $\rho' \in \mathcal{P}'$.

In order to distinguish between different kinds of trees in F', we consider a graph, the *F-model graph*, that models this situation. In this model graph, there is one (*red*) vertex for each tree in F', and one (*blue*) vertex for each path ρ' in $\mathcal{P}'$. Each vertex x in the model represents a set of vertices $V(x)$ of F. (If x is a red vertex, then it represents the vertices in a tree of F'; if x is a blue vertex, then it represents the vertices of a path ρ' in $\mathcal{P}'$.)

In the F-model graph, a vertex x is adjacent to a vertex x' if and only if there is a vertex $u \in V(x)$ that is adjacent to a vertex $u' \in V(x')$. Note that the F-model graph is acyclic, since otherwise S would fail to be a feedback vertex set in G. If T is a tree in F', then $v(T)$ denotes the red vertex of the F-model graph corresponding to T. Similarly, if ρ' is a path in $\mathcal{P}'$, then $v(\rho')$ denotes the blue vertex of the F-model graph corresponding to ρ'.

Let $\mathcal{T}_0$ denote the set of trees of F' whose corresponding vertices in the F-model graph have degree 0. Similarly, let $\mathcal{T}_1$ denote the set of trees of F' whose corresponding vertices in the F-model graph have degree 1, and let $\mathcal{T}_2$ denote the set of trees of F' whose corresponding vertices in the F-model graph have degree at least 2.

Rule 6: Tree Elimination. Suppose (G, k) is an instance with structural map $(S, \mathcal{P})$ produced at the end of Step Two, reduced with respect to Reduction Rules (1-5), and suppose that there is a tree T of F' such that for every pair of distinct vertices s, t in $\mathcal{C}$ where T is adjacent to both s and t in G, there are at least $k + 2$ different trees T' in F' where each T' is also adjacent to both s and t in G. Then reduce (G, k) to $(G - T, k)$.

It is easy to see that determining whether Reduction Rule 6 applies in the situation described, can be accomplished in polynomial time. Less obvious is the soundness of the reduction rule.

Lemma 1. Reduction Rule 6 is sound.

Proof. It is obvious that if (G, k) is a yes-instance, then so is the reduced instance, since the yes-instances are hereditary under deletion. In the converse direction, suppose A is a feedback set of size k in the reduced instance. We argue that A must also be a solution for (G, k). If not, then there must be a cycle C in G that avoids A. Because A is a solution for the reduced graph, the cycle C must pass through T (perhaps more than once).

Any vertex v of C is joined to T by at most one edge. There are two cases to consider to justify this assertion. If v belongs to S and is joined by two edges to T, then $\mathcal{P}$ is not maximal. If v belongs to a path $\rho' \in \mathcal{P}'$ and is joined by two edges to T, then S fails to be a feedback vertex set.

It follows that the intersection of C with T consists of a number of disjoint paths ρ_i, $i = 1, ..., m$, where for each i, a traversal of C enters ρ_i from a vertex $s_i \in H$ and exits ρ_i to a vertex $t_i \in H$ with $s_i \neq t_i$. It also follows that for $i \neq j$, the pair of vertices $\{s_i, t_i\}$ is disjoint from the pair of vertices $\{s_j, t_j\}$. Suppose that the ρ_i are indexed in the order of a traversal of C. The cycle C' that consists of (1) a path in T from s_1 to t_m, and (2) the path in C through the reduced graph from t_m to s_1, also avoids A. In particular, s_1 and t_m do not belong to A. The pre-conditions for the reduction rule include that there are at least $k + 2$ trees T' different from T in F', each of these connected to both s_1 and t_m, and thus there are at least $k + 2$ internally vertex-disjoint paths from s_1 to t_m in the reduced instance. But then, at least two of these must avoid A (since A has only k vertices) and form (together with s_1 and t_m) a cycle in the reduced graph that avoids A, a contradiction. Therefore A must also be a solution for (G, k). □

In the next series of lemmas we advance some claims about the situation where (G, k) is an instance with structural map $(S, \mathcal{P})$ (with $\mathcal{P}$ maximal, as computed in Step Two), that is reduced with respect to Reduction Rules (1-5).

Lemma 2. $\mathcal{P}$ contains at most $2k^2$ paths.

Proof. If there were more paths in $\mathcal{P}$, then the Flower Rule would apply.   □

Lemma 3. The number of trees in $\mathcal{T}_2$ is bounded by $2k^2 - 1$.

Proof. This follows from Lemma 2, since F is a forest.   □

Lemma 4. If T is any tree in the forest F', then $|T| \leq 2k^2 + 2k - 2$.

Proof. We first argue that for any tree T on m vertices, there must be at least $m + 2$ edges connecting T to the rest of G, that is, to $G - T$. First note that any leaf of T must be connected to the rest of G by at least two edges, since otherwise either the Degree One Reduction Rule or the Degree Two Reduction Rule would apply. If T consists of only one vertex, then similarly it must be connected to the rest of G by at least three edges. If u is an internal vertex of T of degree 2 relative to T, then at least one edge must join u to the rest of G, since otherwise the Degree Two Rule would apply. Let l denote the number of leaves of T, let j denote the number of internal vertices of T of degree 2, and let b denote the number of internal vertices of T of degree greater than 2. Then $m = l + j + b$. By the above observations, the number of edges joining T to the rest of G is at least $c = 2l + j$. The inequality we seek, $c \geq m + 2$, follows from the elmentary fact that $l \geq b + 2$.

Lemma 4 follows, since if the bound stated in the lemma did not hold, then T would be joined by two edges either: (1) to a path $\rho' \in \mathcal{P}'$, or (2) to a vertex of S. Case (1) contradicts that S is a feedback vertex set for G, and case (2) contradicts that $\mathcal{P}$ is maximal.   □

The Quality of the Structural Map. The *quality* of the structural map may be improved according to the following list of priorities:

(1) The size of $\mathcal{P}$ should be maximized.
(2) The sum of the lengths of the paths in $\mathcal{P}'$

$$\sum_{\rho' \in \mathcal{P}'} |\rho'|$$

should be minimized.

Lemma 5. Suppose (G, k) is an instance with structural map $(S, \mathcal{P})$ (with $\mathcal{P}$ maximal, as computed in Step Two), that is reduced with respect to Reduction Rules (1-5). Then either every path $\rho' \in \mathcal{P}'$ has length at most $6k + 2(2k^2 - 1)$, or in polynomial time we can improve the quality of the the structural map.

Proof. Consider a path $\rho' \in \mathcal{P}'$. We will argue about the length of a subpath "in the middle" of ρ'. Suppose the vertices of ρ' in the order of a traversal are:

$$\rho' = (v_1, v_2, ..., v_l)$$

Let ρ'' denote the subpath

$$\rho'' = (v_{2k+1}, v_{2k+2}, ..., v_{l-2k-1}, v_{l-2k})$$

In other words, ρ'' is the "middle" subpath excluding both the first $2k$ vertices of ρ' and the last $2k$ vertices of ρ'.

We argue that if $|\rho''| > 2k + 2(2k^2 - 1)$ then we can improve the quality of $\mathcal{P}$ with respect to the second priority by replacing ρ with a shorter path. The vertices of ρ'' do not have degree 2 in G, since otherwise the Degree Two Reduction Rule would apply. Therefore each vertex x of ρ'' is adjacent to some vertex x' that is not one of its two neighbors with respect to the path ρ'. Since F is a forest, $x' \notin \rho'$. There are four possibilities: (1) $x' \in S$, (2) x' is a vertex of a tree in $\mathcal{T}_2$, (3) x' is a vertex of a different path σ' in $\mathcal{P}'$, or (4) x' is a vertex of a tree T in $\mathcal{T}_1$.

In case (4), there must be a path ρ_s from x through T to a vertex $s \in S$. Any such path is disjoint from the other paths in $\mathcal{P}'$. By the argument in the proof of Lemma 4, if $|T| = m$, then T must be joined by at least $m + 1$ edges to S, in view of the definition of $\mathcal{T}_1$. Also, T cannot be joined to $s \in S$ by two different edges, otherwise $\mathcal{P}$ is not maximal. It follows that $|T| \leq 2k - 1$, and that the number of internal vertices of ρ_s is bounded by $2k - 1$.

Two distinct vertices x, y of ρ'' cannot both be adjacent to a path $\sigma' \neq \rho'$, else S fails to be a feedback vertex set. Similarly, they cannot both be adjacent to the same tree in $\mathcal{T}_2$. If there are distinct vertices x, y of ρ'' that both have paths to $s \in S$ (via trees in $\mathcal{T}_1$), then the second priority can be improved.

Since S contains at most $2k$ vertices, there are at most $2k^2 - 1$ trees in $\mathcal{T}_2$, and there are at most $2k^2 - 1$ paths in $\mathcal{P}'$ other than ρ', the lemma follows, by the Pigeonhole Principle.   □

Theorem 1. There is a kernelization bound $g(k) = O(k^{11})$ such that for an instance (G, k) either:

(1) $|G| \leq g(k)$, that is, the instance is kernelized, or
(2) In polynomial time, we can compute a structural map $(S, \mathcal{P})$ for (G, k) and either discover an opportunity to apply a reduction rule, or discover an opportunity to improve the structural map according to the priorities listed above.

Proof. The forest F' consists of trees of size $O(k^2)$, by Lemma 4. The total number of vertices that either belong to S, or belong to a path in $\mathcal{P}'$, by Lemmas 2 and 5, is bounded by $z = O(k^4)$. Both of these statements hold under the assumption that the instance is reduced (in polynomial time) under Reduction Rules (1-5). Now consider going through the list of trees in F', checking, for each tree encountered on the list, if there is an opportunity to apply Reduction Rule 6, the Tree Elimination Rule. We can keep a count, for each pair of vertices s, t in $\mathcal{C}$, of the number of trees on the list, up to that point, that are attached to both s and t. Each tree T on the list must either increase the count for at least one such pair (up to a limit of $k + 2$), otherwise T qualifies for the Tree Elimination Rule. There are at most $\binom{z}{2} = O(k^8)$ such pairs in $\mathcal{C}$, and therefore $O(k^9)$ trees in F', else the Tree Elimination Rule is triggered. By Lemma 4, the total number of vertices in F' is $O(k^{11})$.   □

Corollary. In polynomial time, we can kernelize an instance of UFVS to a kernel of size $O(k^{11})$.

Proof. This follows from the fact that the priorities can be improved only a polynomial number of times.   □

3   Summary and Open Problems

In this paper, we have addressed a notable concrete open problem about FPT kernelization, and we have shown that the fixed-parameter tractable UNDIRECTED FEEDBACK VERTEX SET problem has an $O(k^{11})$ kernel. Might this problem have a linear-size kernel? We think this may well be so, despite our rather weak result here, in view of the fact that the problem admits a constant factor P-time approximation algorithm (a different, but seemingly related issue). Polynomial-time "crown-type" reductions may potentially play a role in showing that UFVS is in $Lin(k)$ [CFJ04, ACFLSS04, DFRS04, LS05, Pr05, Slo05].

The subject of parameterized complexity has been dogged from the beginning by concerns that the definition of FPT is "too lax" to capture a really meaningful framework for "coping with intractability" in the sense of [GJ79]. (The early entanglement of the subject with graph minors theory perhaps did not help in this regard.) In particular, the usual definition of FPT calls for solvability in time $f(k)n^c$ where $f(k)$ is a completely unrestricted function. Despite the fact that many of the most important and natural FPT parameterized problems seem to admit FPT algorithms with relatively (surprisingly?) benign $f(k)$, there has been motivation to explore subclasses of FPT with stronger definitional claims on practical parametric feasibility [FGW04, Wey04, FG06]. Certainly, considering FPT from the point of view of polynomial-time data reduction gives a very natural in-road to more sensitive distinctions concerning parameterized tractability.

Pre-processing is a humble strategy for coping with hard problems, almost universally employed. It has become clear, however, that far from being trivial and uninteresting, that pre-processing has unexpected practical power for real-world input distributions [Wei98,Wei00,BKV05], and is mathematically a much deeper subject than has generally been understood. It is almost impossible to talk about pre-processing in the classical complexity framework in any sensible and interesting way, and the historical neglect of this vital subject may be related to this fact.

Here is the difficulty. If my problem Π is NP-hard, then probably there is no P-time algorithm to *solve* the problem, that is, to completely dispose of the input. If you suggested that perhaps I should settle for a P-time algorithm that instead of completely disposing of the input, at least simplifies it by getting rid of, or reducing away, the easy parts — then this would seem a highly compelling suggestion. But how can this be formalized? The obvious first shot is to ask for a P-time algorithm that reduces the input I to an input I' where $|I'| < |I|$ in a way that loses no essential information (i.e., trades the original input for smaller input, which can be called *data reduction*). The difficulty with this "obvious" formalization of the compelling suggestion is that if you had such a P-time data reduction algorithm, then by repeatedly applying it, you could dispose of the entire input in polynomial time, and this is impossible, since Π is NP-hard. Thus, in the classical framework, an effort to formulate a mathematically interesting program to explore polynomial-time preprocessing immediately crashes.

In the parameterized complexity framework, however, such a program can be formulated in an absolutely interesting and productive way. The effectiveness of P-time processing is measured against the structure represented by the parameter. This is precisely the intellectual location of this paper. You might reasonably call the subject of FPT kernelization the *Lost Continent of Polynomial Time*.

Kernelization as we have considered it here is a polynomial time many:1 transformation of a parameterized problem Π to itself. One can also consider Turing kernelizations. A nontrivial example of a $2k$ Turing kernelization, based on iterative compression, for the VERTEX COVER problem is described by Dehne, et al. in [DFRS04]. Of course, in the case of VERTEX COVER, this is matched by a $2k$ many:1 kernelization. Nevertheless, Turing kernelization may in general have greater power than many:1 kernelization. It makes sense to ask if the UNDIRECTED FEEDBACK VERTEX SET problem might admit a linear-size Turing kernelization.

There seems to be much interesting work to do in exploring kernelizability, and its limits, for the many parameterized problems now known to be in FPT.

References

[ACFLSS04] F. N. Abu-Khzam, R. L. Collins, M. R. Fellows, M. A. Langston, W. H. Suters and C. T. Symons. Kernelization algorithms for the vertex cover problem: theory and experiments. *Proceedings of the 6th Workshop on Algorithm Engineering and Experiments (ALENEX)*, New Orleans, January, 2004, ACM/SIAM, *Proc. Applied Mathematics 115*, L. Arge, G. Italiano and R. Sedgewick, eds.

[AFN04] J. Alber, M. Fellows and R. Niedermeier. Polynomial time data reduction for dominating set. *Journal of the ACM* 51 (2004), 363–384.

[BBF99] V. Bafna, P. Berman and T. Fujito. A 2-approximation algorithm for the undirected feedback vertex set problem. *SIAM Journal on Discrete Mathematics* 12 (1999), 289–297.

[BBG00] A. Becker, R. Bar-Yehuda and D. Geiger. Random algorithms for the loop cutset problem. *Journal of Artificial Intelligence Research* 12 (2000), 219–234.

[BKV05] H. Bodlaender, A. Koster and F. van den Eijkhof. Preprocessing rules for triangulation of probabilistic networks. *Computational Intelligence* 21 (2005), 286–305.

[Bod94] H. Bodlaender. On disjoint cycles. *International Journal of Foundations of Computer Science* 5 (1994), 59–68.

[CFJ04] B. Chor, M. Fellows and D. Juedes. Linear kernels in linear time, or how to save k colors in $O(n^2)$ steps. *Proceedings WG 2004*, Springer-Verlag, *Lecture Notes in Computer Science* 3353 (2004), 257–269.

[DF92] R. Downey and M. Fellows. Fixed-parameter tractability and completeness. *Congressus Numerantium* 87 (1992), 161–187.

[DF99] R. G. Downey and M. R. Fellows. *Parameterized Complexity*. Springer-Verlag, 1999.

[DFLRS05] F. Dehne, M. Fellows, M. Langston, F. Rosamond and K. Stevens. An $O^*(2^{O(k)})$ FPT algorithm for the undirected feedback vertex set problem. In: *Proceedings COCOON 2005*, Springer-Verlag, *Lecture Notes in Computer Science* 3595 (2005), 859–869.

[DFRS04] F. Dehne, M. Fellows, F. Rosamond and P. Shaw. Greedy localization, iterative compression and modeled crown reductions: new FPT techniques, an improved algorithm for set splitting and a novel $2k$ kernelization for vertex cover. *Proceedings of the First International Workshop on Parameterized and Exact Computation*, Springer-Verlag, *Lecture Notes in Computer Science* vol. 3162 (2004), 271–280.

[FG06] *Parameterized Complexity Theory*, J. Flum and M. Grohe, Springer-Verlag, 2006.

[FGW04] J. Flum, M. Grohe and M. Weyer. Bounded fixed-parameter tractability and $\log^2 n$ nondeterministic bits. *Proceedings of the 31st ICALP*, Springer-Verlag, *Lecture Notes in Computer Science* 3142 (2004), 555–567.

[FPR99] P. Festa, P. M. Pardalos and M. G. C. Resende. Feedback set problems. In: D. Z. Du, P. M. Pardalos (Eds.), *Handbook of Combinatorial Optimization*, Vol. A, Kluwer (1999), 209–258.

[GGHNW05] J. Guo, J. Gramm, F. Hueffner, R. Niedermeier, S. Wernicke. Improved fixed-parameter algorithms for two feedback set problems. *Proceedings of WADS 2005*, Springer-Verlag, *Lecture Notes in Computer Science* 3608 (2005), 158–169.

[GJ79]   M. R. Garey and D. S. Johnson. *Computers and Intractability: A Guide to the Theory of NP-Completeness.* W.H. Freeman, 1979.

[Guo06]   J. Guo. *Algorithm design techniques for parameterized problems.* Ph.D. Thesis, Friedrich-Schiller-Universität, Jena, 2006.

[KPS04]   I. Kanj, M. Pelsmajer and M. Schaefer. Parameterized algorithms for feedback vertex set. *Proceedings of the First International Workshop on Parameterized and Exact Computation,* Springer-Verlag, *Lecture Notes in Computer Science* vol. 3162 (2004), 235–247.

[LS05]   D. Lokshtanov and C. Sloper. Fixed-parameter set-splitting, linear kernel and improved running time. In: *Algorithms and Complexity in Durham 2005: Proceedings of the First ACiD Workshop,* King's College Press, *Texts in Algorithmics* 4 (2005), 105–113.

[Nie02]   R. Niedermeier. *Invitation to fixed-parameter algorithms,* Habilitationschrift, University of Tubingen, 2002. (Electronic file available from R. Niedermeier.)

[Nie06]   R. Niedermeier. *Invitation to Fixed Parameter Algorithms.* Oxford University Press, 2006.

[NT75]   G. L. Nemhauser and L. E. Trotter. Vertex packings: structural properties and algorithms. *Mathematical Programming* 8 (1975), 232–248.

[Pr05]   E. Prieto-Rodriguez. *Systematic kernelization in FPT algorithm design.* Ph.D. Thesis, School of EE&CS, University of Newcastle, Australia, 2005.

[RSS02]   V. Raman, S. Saurabh and C. Subramanian. Faster fixed-parameter tractable algorithms for undirected feedback vertex set. In *Proceedings of the 13th Annual International Symposium on Algorithms and Computation,* Springer, *Lecture Notes in Computer Science* vol. 2518 (2002), 241–248.

[RSS05]   V. Raman, S. Saurabh and C.R. Subramanian. Faster algorithms for feedback vertex set. In: *Proceedings of the 2nd Brazilian Symposium on Graphs, Algorithms and Combinatorics,* GRACO 2005, April 27-29, 2005, Angra dos Reis (Rio de Janeiro), Brazil. Elsevier, *Electronic Notes in Discrete Mathematics* (2005).

[Slo05]   C. Sloper. *Techniques in parameterized algorithm design.* Ph.D. Thesis, Department of Informatics, University of Bergen, Norway, 2005.

[Wei98]   K. Weihe. Covering trains by stations, or the power of data reduction. *Proc. ALEX'98* (1998), 1–8.

[Wei00]   K. Weihe, "On the Differences Between 'Practical' and 'Applied' (invited paper)," *Proc. WAE 2000,* Springer-Verlag, *Lecture Notes in Computer Science* 1982 (2001), 1–10.

[Wey04]   M. Weyer. Bounded fixed-parameter tractability: the case of $2^{poly(k)}$. *Proceedings of 1st IWPEC,* Springer-Verlag, *Lecture Notes in Computer Science* 3164 (2004), 49–60.

Fixed-Parameter Tractability Results for Full-Degree Spanning Tree and Its Dual

Jiong Guo[*], Rolf Niedermeier, and Sebastian Wernicke[**]

Institut für Informatik, Friedrich-Schiller-Universität Jena
Ernst-Abbe-Platz 2, D-07743 Jena, Germany
{guo, niedermr, wernicke}@minet.uni-jena.de

Abstract. We provide first-time fixed-parameter tractability results for the NP-complete problems MAXIMUM FULL-DEGREE SPANNING TREE and MINIMUM-VERTEX FEEDBACK EDGE SET. These problems are dual to each other: In MAXIMUM FULL-DEGREE SPANNING TREE, the task is to find a spanning tree for a given graph that maximizes the number of vertices that preserve their degree. For MINIMUM-VERTEX FEEDBACK EDGE SET the task is to minimize the number of vertices that end up with a reduced degree. Parameterized by the solution size, we exhibit that MINIMUM-VERTEX FEEDBACK EDGE SET is fixed-parameter tractable and has a problem kernel with the number of vertices linearly depending on the parameter k. Our main contribution for MAXIMUM FULL-DEGREE SPANNING TREE, which is W[1]-hard, is a linear-size problem kernel when restricted to planar graphs. Moreover, we present subexponential-time algorithms in the case of planar graphs.

1 Introduction

The NP-complete MAXIMUM FULL-DEGREE SPANNING TREE (FDST) problem is defined as follows.

> **Input**: An undirected graph $G = (V, E)$ and an integer $k \geq 0$.
> **Task**: Find a spanning tree T of G (called *solution tree*) in which at least k vertices have the same degree as in G.

Referring to vertices that maintain their degree as *full-degree* vertices and to the remaining ones as *reduced-degree* vertices, the task basically is to maximize the number of full-degree vertices. FDST is motivated by applications in water distribution and electrical networks [3,4,13]. It is a notoriously hard problem and, as such, not polynomial-time approximable within a factor of $O(n^{1/2-\epsilon})$ for any $\epsilon > 0$ unless NP-complete problems have randomized polynomial-time algorithms [3]. The approximability bound is almost tight in that Bhatia et al. [3] provide an algorithm with an approximation ratio of $\Theta(n^{1/2})$. FDST remains NP-complete in planar graphs; however, polynomial-time approximation schemes

---

[*] Supported by the Deutsche Forschungsgemeinschaft, Emmy Noether research group PIAF (fixed-parameter algorithms), NI 369/4.
[**] Supported by the Deutsche Telekom Stiftung.

H.L. Bodlaender and M.A. Langston (Eds.): IWPEC 2006, LNCS 4169, pp. 203–214, 2006.
© Springer-Verlag Berlin Heidelberg 2006

(PTAS) are known here [3,4]. Broersma et al. [4] present further tractability and intractability results for various special graph classes. By way of contrast, the parameterized complexity [7,10,14] of FDST has so far been unexplored.

The complement (dual) problem of FDST, called MINIMUM-VERTEX FEED-BACK EDGE SET (VFES), is to find a feedback edge set[1] in a given graph such that this set is incident to as few vertices as possible. In other words, we want to minimize the number of reduced-degree vertices. VFES is motivated by an application of placing pressure-meters in fluid networks [12,15,16]. Khuller et al. [12] show that VFES is APX-hard and present a polynomial-time approximation algorithm with ratio $(2 + \epsilon)$ for any fixed $\epsilon > 0$. Moreover, they develop a PTAS for VFES in planar graphs. As with FDST, the parameterized complexity of VFES has so far not been investigated.

Parameterized by the respective solution size k, this work provides first-time parameterized complexity results for FDST and its dual VFES. Somewhat analogously to the study of approximability properties, we observe that FDST seems to be the harder problem when compared to its dual VFES: Whereas VFES is fixed-parameter tractable, FDST is W[1]-hard.[2] More specifically, our findings are as follows:

- VFES has a problem kernel with less than $4k$ vertices and it can be solved in $O(4^k \cdot k^2 + m)$ time for an m-edges graph.
- FDST becomes fixed-parameter tractable in the case of planar graphs. In particular, as the main technical contribution of this paper, we prove a linear-size problem kernel for FDST when restricted to planar graphs.
- For planar graphs, both VFES and FDST are solvable in subexponential time. More specifically, in n-vertices planar graphs VFES is solvable in $O(2^{O(\sqrt{k} \log k)} + k^5 + n)$ time and FDST in $O(2^{O(\sqrt{k} \log k)} + k^5 + n^3)$ time. Herein, we make use of tree decomposition-based dynamic programming.

We remark that, when restricted to planar graphs, this work amends the so far few examples where both a problem and its dual possess linear-size problem kernels. Other examples we are aware of (again restricted to planar graphs) are VERTEX COVER and its dual INDEPENDENT SET and DOMINATING SET and its dual NONBLOCKER [1,5,6].

We omit most proofs in this extended abstract due to a lack of space.

2   Minimum-Vertex Feedback Edge Set

MINIMUM-VERTEX FEEDBACK EDGE SET (VFES), the dual of MAXIMUM FULL-DEGREE SPANNING TREE (FDST), appears to be better tractable from a parameterized point of view than FDST. We subsequently present a simple problem kernelization with a kernel graph whose number of vertices linearly depends on the parameter (Sect. 2.1) and, based on this, an efficient fixed-parameter algorithm (Sect. 2.2) for VFES.

---

[1] A feedback edge set is a set of edges whose deletion destroys all cycles in a graph.
[2] The reduction is from INDEPENDENT SET and the same as the one in [3,12].

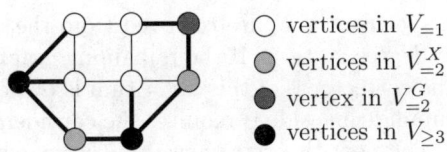

Fig. 1. Given a spanning tree for a graph (bold edges), the proof of the linear kernel for VFES in Theorem 1 partitions them into four disjoint subsets as illustrated above (see proof for details)

2.1   Data Reduction and Problem Kernel

To reduce a given instance of VFES to a problem kernel, we make use of two very simple data reduction rules already used by Khuller et al. [12].

Rule 1. Remove all degree-one vertices.

Rule 2. For any two neighboring degree-two vertices that do not have a common neighbor, contract the edge between them.

The correctness and the linear running time of these two rules are easy to verify; we call an instance of VFES *reduced* if the reduction rules cannot be applied any more.

Theorem 1. *A reduced instance* $(G = (V, E), k)$ *of* VFES *only has a solution tree if* $|V| < 4k$.

Proof. Assume that G has a solution tree T. Let X denote the set of the reduced-degree vertices in T. We partition the vertices in V into three disjoint subsets according to their degree in T, namely $V_{=1}$ contains all degree-1 vertices, $V_{=2}$ contains all degree-2 vertices, and $V_{\geq 3}$ contains all vertices of degree higher than 2. Furthermore, let $V_{=2}^X := V_{=2} \cap X$ and $V_{=2}^G := V_{=2} \setminus V_{=2}^X$. The partition is illustrated in Figure 1. Since G does not contain any degree-1 vertices by Rule 1, every degree-1 vertex in T is a reduced-degree vertex. Hence, T can have at most k leaves and $|V_{=1}| \leq |X| \leq k$. Since T is a tree, this directly implies $|V_{\geq 3}| \leq k - 2$.

As for $V_{=2}$, the vertices in $V_{=1} \cup V_{\geq 3}$ are either directly connected to each other or via a path P consisting of vertices from $V_{=2}$. Because Rule 2 contracts edges between two degree-2 vertices that have no common neighbor in the input graph, at least one of every two neighboring vertices of P has to be a reduced-degree vertex. Clearly, $V_{=2}^X \cup V_1 \subseteq X$. Since T is a tree, it follows that $|V_{=2}^G| \leq |V_{=1} \cup V_{\geq 3} \cup V_{=2}^X| - 1 \leq 2k - 3$.

Overall, this shows that $|V| = |V_{=1} \cup V_{=2}^X \cup V_{=2}^G \cup V_{\geq 3}| < 4k$ as claimed. $\square$

2.2   A Fixed-Parameter Algorithm

The problem kernel obtained in Theorem 1 suggests a simple fixed-parameter algorithm for MINIMUM-VERTEX FEEDBACK EDGE SET: For $i = 1, \ldots, k$, we consider all $\binom{4k}{i}$ size-i subsets X of kernel vertices. For each of these subsets, all edges between the vertices in X are removed from the input graph, that is, they become reduced-degree vertices. If the remaining graph is a forest, we have found a solution.[3] The correctness of this algorithm is obvious by its exhaustive nature, but on the running-time side it requires the consideration of $\sum_{i=1}^{k} \binom{4k}{i} > 9.45^k$ vertex subsets. The next theorem shows that we can do better because an exhaustive approach does not need to consider all vertices of the kernel but only those with degree at least three.

Theorem 2. *For an m-edge instance $(G = (V, E), k)$ of VFES, it can be decided in $O(4^k \cdot k^2 + m)$ time whether it has a solution tree.*

Proof. Given an instance $(G = (V, E), k)$ of VFES, we first perform the kernelization which needs $O(m)$ time. By Theorem 1, we know that the remaining graph only has a solution tree if it contains fewer than $4k$ vertices. We now partition the vertices in V according to their degree: vertices in $V_{=2}$ have degree two and $V_{\geq 3}$ contains all vertices with degree at least three. For every size-i subset $X_{\geq 3} \subseteq V_{\geq 3}$, $1 \leq i \leq k$, the following two steps are performed:

> **Step 1.** Remove all edges between vertices in $X_{\geq 3}$. Call the resulting graph $G' = (V, E')$.
>
> **Step 2.** For each edge $e \in E'$, assign it a weight $w(e) = m + 1$ if it is incident to a vertex in $V_{\geq 3} \setminus X_{\geq 3}$ and a weight of 1, otherwise. Then, find a maximum-weight spanning tree for every connected component of G'. If the total weight of edges that are *not* in a spanning tree is at most $k - i$, then the original VFES instance $(G = (V, E), k)$ has a solution tree and the algorithm terminates; otherwise, the next subset $X_{\geq 3}$ is tried.

To justify Step 2, observe that the edges incident to vertices in $V_{\geq 3} \setminus X_{\geq 3}$ have to be in the solution tree because these vertices preserve their degree. Therefore, if a component in G' contains cycles we can only destroy these by removing edges between a vertex in $X_{\geq 3}$ and a vertex in $V_{=2}$. Moreover, removing such an edge results in exactly one additional reduced-degree vertex. Thus, searching for a maximum-weight spanning tree in the second step leads to a solution tree (the components can easily be reconnected by edges between vertices in $X_{\geq 3}$).

The running time of the algorithm is composed of the linear preprocessing time, the number of subsets that need to be considered, and the time needed for Steps 1 and 2. By Theorem 1, a reduced graph contains less than $4k$ vertices and, hence, $O(k^2)$ edges, upper-bounding the time needed for Steps 1 and 2 by $O(k^2)$. From the proof of Theorem 1, we know that $|V_{\geq 3}| \leq 2k$, and hence the overall running time is bounded above by $O(m + \sum_{1 \leq i \leq k} \binom{2k}{i} k^2) = O(4^k \cdot k^2 + m)$. $\square$

3 Maximum Full-Degree Spanning Tree in Planar Graphs

The reduction from INDEPENDENT SET to MAXIMUM FULL-DEGREE SPANNING TREE (FDST) that is given in [3] already shows that FDST is W[1]-hard with

---

[3] Since the input graph must be connected, we can add some edges from G that are between the vertices in X in order to reconnect connected components with each other and thus obtain a spanning tree.

respect to the number k of full-degree vertices.[4] For *planar* graphs, however, this does not hold; in this section, we show that it is even possible to achieve a linear-size problem kernel for FDST in planar graphs.

While the proof of the linear size upper bound is very involved and technical, the actual computation of the problem kernel is based on several straightforward data reduction rules on a given instance of FDST.[5]

Reduction Rules. Let $N(v)$ denote the neighborhood of a vertex v. For two vertices $v \neq w$ with $|N(v) \cap N(w)| \geq 3$, perform the following reductions:
1. Let u_1, u_2, u_3 be three vertices in $N(v) \cap N(w)$.
    1.1 If $N(u_1) = \{v, w\}$ and additionally either $N(u_2) = \{v, w\}$ or $N(u_2) = \{u_3, v, w\}$, then remove u_2.
    1.2 If $N(u_1) = \{u_2, v, w\}$, $N(u_2) = \{u_1, u_3, v, w\}$, and $N(u_3) = \{u_2, v, w\}$, then remove u_3.
2. Let u_1, u_2, u_3, u_4 be four vertices in $N(v) \cap N(w)$.
    2.1 If $N(u_1) = \{u_2, v, w\}$, $N(u_2) = \{u_1, v, w\}$, $N(u_3) = \{u_4, v, w\}$, and $N(u_4) = \{u_3, v, w\}$, then remove u_3 and u_4.
    2.2 If $N(u_2) = \{u_1, u_3, v, w\}$, $N(u_3) = \{u_2, u_4, v, w\}$, and there is no edge $\{u_1, u_4\}$, then remove the edge $\{u_2, u_3\}$.
3. Let u_1, u_2, u_3, u_4, u_5 be five vertices in $N(v) \cap N(w)$.
    If $N(u_2) = \{u_1, u_3, v, w\}$, $N(u_3) = \{u_2, v, w\}$, $N(u_4) = \{u_5, v, w\}$, and $N(u_5) = \{u_4, v, w\}$, then remove u_3.

The five subcases are illustrated in Figure 2. Omitting a formal proof here, exhaustively applying these rules yields a graph that has a spanning tree with k full-degree vertices iff the original graph has a spanning tree with k full-degree vertices, that is, the reduction rules are "correct."

As we shall show in the remainder of this section, a planar graph that is reduced with respect to the reduction rules is a linear-size problem kernel for MAX-IMUM FULL-DEGREE SPANNING TREE in planar graphs:

Theorem 3. *For a given planar n-vertex graph G, let k be the maximum number of full-degree vertices in any spanning tree for G. If G is reduced with respect to the given reduction rules, then $n = O(k)$, that is, we have a linear-size problem kernel for FDST on G that can be computed in $O(n^3)$ time.*

The proof of this theorem is quite involved. We basically achieve it by contradiction, that is, we assume that we are given an optimal solution tree to FDST on G and show that either $|V| = O(k)$ must hold or this optimality is contradicted.[6] Throughout this section, we denote the set of full-degree vertices in the optimal solution tree by F (note that $|F| = k$).

---

[4] W[1]-hardness stands for (presumable) parameterized intractability. Refer to [7,10,14] for details.

[5] Note that the reduction rule applies to an arbitrary graph but the "linear-size kernel" performance-guarantee is only shown for planar graphs.

[6] This sort of proof strategy has first been used in work dealing with the MAX LEAF SPANNING TREE problem [8,9].

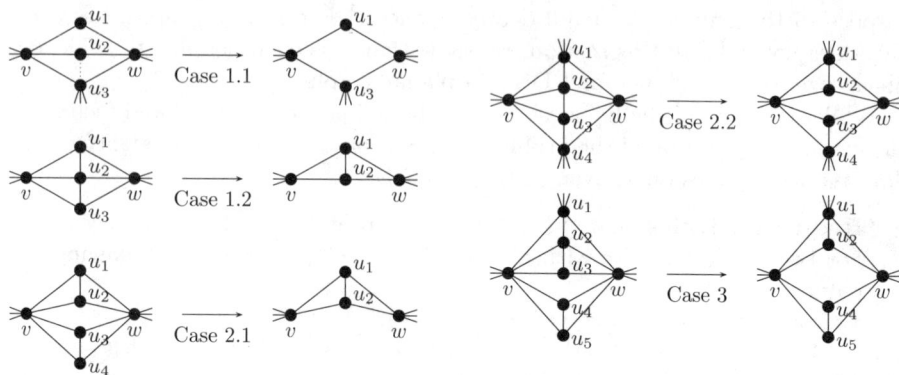

Fig. 2. The five subcases of the reduction rule for planar FDST that yields a linear-size problem kernel

To this end, the following lemma is very helpful.

Lemma 4. *Every vertex in G has distance at most 2 to a vertex in F.*   □

Using this lemma, our strategy to prove Theorem 3 is for every vertex in F to partition the set of vertices that have distance at most 2 to it into two sets and separately upper-bound their size. More specifically, Section 3.1 introduces the concept of *region decompositions* which use the set F to divide the input graph into small areas.[7]

In Section 3.2, it is shown that the number of regions in a region decomposition is bounded above by $O(|F|)$ (Lemma 9) and that in the reduced graph, every region contains only a constant number of vertices (Lemma 13). Overall, this shows that there are at most $O(|F|)$ vertices lying inside regions (Proposition 14). Essentially following the same strategy in Section 3.3, we upper-bound the number of vertices that do not lie inside regions by $O(|F|)$ (Proposition 18). The linear kernel claimed in Theorem 3 directly follows from the $O(|F|)$ upper bounds on the number of vertices in regions and outside regions.

3.1   Neighborhood Partition and Region Decomposition

This section prepares the proof of the size-$O(|F|)$ upper bound of a reduced graph by introducing two partitions of the vertices in $V \setminus F$. One partition is "local" in that for every vertex in F, it concerns vertices within distance at most 2 to it; we partition this *2-neighborhood* into so-called "exit vertices" and "prison vertices." The other partition, called *region decomposition*, is somewhat more "global" in that it concerns the union of 2-neighborhoods for some pairs of vertices in F. These partitions are subsequently used in Sections 3.2 and 3.3 to show the desired linear size of a reduced planar graph.

---

[7] The proof concept is, in this sense, similar to the one for the linear-size problem kernel for DOMINATING SET in planar graphs due to Alber et al. [1].

As to the notation used, for a set $V' \subseteq V$ the subgraph of G *induced* by V' is denoted by $G[V'] = (V', E')$. For a vertex $v \in V$, we denote the vertices having distance 2 to v by $N_2(v)$. By $N_{1,2}(v)$, we denote $N(v) \cup N_2(v)$. Furthermore, we set $N^1(v) := N(v) \setminus F$ and let $N^2(v)$ denote all vertices that have distance exactly 2 to v in the induced graph $G[(V \setminus F) \cup \{v\}]$. Finally, $N^{1,2}(v) := N^1(v) \cup N^2(v)$. This section and the following always consider some fixed embedding of G in the plane (hence, we call G plane instead of planar).

As already mentioned, Lemma 4 shows that every vertex in G has distance at most 2 to a full-degree vertex. Thus, if we can upper-bound the number of vertices in $\bigcup_{v \in F} N^{1,2}(v)$ by $O(|F|)$, the linear problem kernel follows as claimed. In order to do this, we partition the vertices in $N^{1,2}(v)$ into two subsets, separately upper-bounding their sizes in Sections 3.2 and 3.3:

$$N^{1,2}_{exit}(v) := \{u \in N^{1,2}(v) \mid u \in N(w) \text{ for a vertex } w \notin N^{1,2}(v)\},$$
$$N^{1,2}_{prison}(v) := N^{1,2}(v) \setminus N^{1,2}_{exit}(v).$$

For a subset $V' \subseteq V$, we use $N^{1,2}_{exit}(V')$ to denote $\bigcup_{v \in V'} N^{1,2}_{exit}(v)$. The intuition of the partition is that the "exit" vertices have edges to vertices that lie outside of $N^{1,2}(v)$ whereas the "prison" vertices have no edge to a vertex different from $v \cup N^{1,2}(v)$.[8] As an example, the following illustration shows the partition of a neighborhood $N^{1,2}(v)$ into $N^{1,2}_{exit}(v)$ and $N^{1,2}_{prison}(v)$:

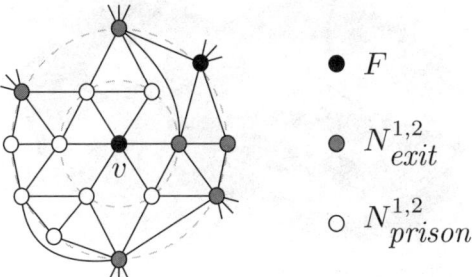

As it turns out, every vertex in $N^{1,2}_{exit}(v)$ is "caught" between two vertices in F, that is, for these vertices a stronger variant of Lemma 4 holds.

Lemma 5. *Given a vertex $v \in F$, every vertex $u \in N^{1,2}_{exit}(v)$ lies on a length-at-most-five path between v and an other vertex $w \in F$ (where $w \neq v$).*   □

Lemma 5 can be used to eventually upper-bound the size of $N^{1,2}_{exit}(F)$ because the length-at-most-five paths form "bounded areas" between vertices from F called "regions."

Definition 6. *A region $R(v, w)$ between two vertices $v, w \in F$ is a closed subset of the plane with the following properties:*

1. *The boundary of $R(v, w)$ is formed by two paths between v and w; the length of each path is at most five.*

---

[8] See [1] for a similar notion in the context of DOMINATING SET.

2. *All vertices which lie* strictly *inside of $R(v, w)$—that is, they do not lie on the boundary—are from $N^{1,2}(v) \cup N^{1,2}(w)$. (Observe that no vertex from F lies strictly inside of a region.)*

To denote the vertices that lie in $R(v, w)$, we use $V(R(v, w))$, that is, $V(R(v, w))$ is the union of the boundary vertices and the vertices lying strictly inside of $R(v, w)$.

Using this definition, a plane graph can be partitioned into a number of regions by a so-called *region decomposition*.

Definition 7. *An F-region decomposition of G is a set $\mathcal{R}$ of regions $R(v, w)$ with $v, w \in F$ such that the following holds:*

1. *Except for v and w, no vertex from $V(R(v, w))$ belongs to F.*
2. *There is no vertex that lies strictly inside of more than one region from $\mathcal{R}$. (The boundaries of regions may touch each other.)*

For an F-region decomposition $\mathcal{R}$, we let $V(\mathcal{R}) := \bigcup_{R \in \mathcal{R}} V(R)$. An F-region decomposition $\mathcal{R}$ is called maximal *if there is no region $R \notin \mathcal{R}$ such that $\mathcal{R}' := \mathcal{R} \cup \{R\}$ is an F-region decomposition with $V(\mathcal{R}) \subsetneq V(\mathcal{R}')$.*

An example of a (maximal) F-region decomposition is the following (the full-degree vertices in F are colored black, the shaded areas are the regions of the decomposition):

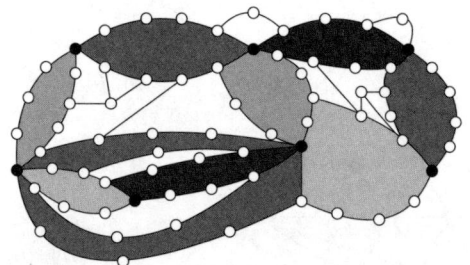

Our notions of "region" and "region decomposition" are similar to the technique used by Alber et al. [1] for proving a linear-size kernel for DOMINATING SET in planar graphs. However, the problem structure of FDST makes the proof somewhat more involved: Whereas Alber et al. were able to bound their regions by length-3 paths, the FDST problem appears to require longer bounds (that is, length-5 paths in our proofs). The reason for this is that in DOMINATING SET, a vertex affects only its direct neighborhood whereas the full-degree property affects vertices in the 2-neighborhood.

The following lemma shows that it is possible to construct a maximal F-region decomposition $\mathcal{R}$ satisfying $N_{exit}^{1,2}(F) \subseteq V(\mathcal{R})$. By subsequently bounding the number of vertices that lie in regions in Section 3.2, this allows us to upper-bound the number of vertices in $N_{exit}^{1,2}(F)$ so that we only have to deal with vertices from $N_{prison}^{1,2}(F)$ in Section 3.3.

Lemma 8. *For a plane graph $G = (V, E)$, there exists a maximal F-region decomposition $\mathcal{R}$ of G such that $N_{exit}^{1,2}(F) \subseteq V(\mathcal{R})$.*   □

3.2 Bounding the Number of Vertices in Regions

In this section, we show that the F-region decomposition $\mathcal{R}$ from Lemma 8 has $O(|F|)$ vertices lying inside of regions (Proposition 14). The proof of this is achieved in several steps: First, Lemma 9 shows that the total number of regions is $O(|F|)$. Then, Proposition 12 upper-bounds the number of length-2 paths that can occur between two vertices in the reduced graph; this proposition is heavily used to prove that every region contains at most $O(1)$ vertices in Lemma 13. Finally, Proposition 14 follows from Lemmas 9 and 13.

Lemma 9. *For $|F| \geq 3$, the maximal F-region decomposition $\mathcal{R}$ in Lemma 8 consists of at most $6|F| - 12$ regions.* □

To show the constant number of vertices in each region, we make use of the following structure:

Definition 10. *Let v and w be two distinct vertices in a plane graph G. A diamond $D(v, w)$ is a closed area of the plane that is bounded by two length-2 paths between v and w such that every vertex that lies inside of the closed area is a neighbor of both v and w. If i vertices lie strictly inside of a diamond, then the diamond is said to have $(i + 1)$ facets (a facet is an area enclosed by two length-2 paths).*

Lemma 11. *A reduced plane graph does not contain a diamond with more than 5 facets.* □

The only two possible 5-facet diamonds (the worst-case diamonds, so to say) that might remain after the data reduction are the following:

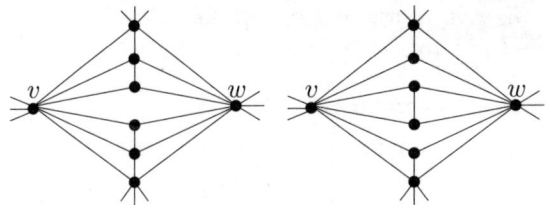

The absence of diamonds with too many facets is central to showing the linear size of the problem kernel for FDST: At most 2 vertices of any diamond can be full-degree, and hence the possibility of arbitrarily large diamonds would prohibit a provable upper bound of the reduced graph. For the same reason, it is important that there cannot be arbitrarily many length-2 paths between two vertices of the input graph. Therefore, Lemma 11 is generalized in order to obtain a upper bound on the number of length-2 paths between two vertices.

Proposition 12. *Let v and w be two vertices in a reduced plane graph G such that an area $A(v, w)$ of the plane is enclosed by two length-2 paths between v and w. If neither the middle vertices of the enclosing paths nor any vertex inside of the area are contained in F, then the following holds:*

1. If $v, w \notin F$, at most eight length-2 paths from v to w lie inside of $A(v, w)$.
2. If $v \notin F$ or $w \notin F$, at most sixteen length-2 paths from v to w lie inside of $A(v, w)$. □

Using the upper bound on the number of length-2 paths between two vertices that Proposition 12 establishes, it is possible to bound above the number of vertices that can lie inside of a region.

Lemma 13. *Every region $R(v, w) \in \mathcal{R}$ contains $O(1)$ vertices.* □

In conjunction with the $O(|F|)$ upper bound on the number of regions that was established in Lemma 9, this directly allows us to bound above the number of vertices that lie inside of regions.

Proposition 14. *The total number of vertices lying inside of regions of $\mathcal{R}$ is $O(|F|)$.* □

3.3 Bounding the Number of Vertices Outside of Regions

In the last section, we have bounded above the number of vertices that lie inside of regions of a maximal F-region decomposition $\mathcal{R}$. It remains to bound above the vertices that do not lie in regions. By Lemma 8, every one of these remaining vertices is in $N_{prison}^{1,2}(v)$ for some vertex $v \in F$. To bound the number of these vertices by $O(|F|)$, we essentially follow the same strategy as in the last section: We show that each vertex in $N_{prison}^{1,2}(v)$ lies in one of $O(|F|)$ so-called "prison areas" and that each such prison area contains a constant number of vertices.

Definition 15. *Given a maximal F-region decomposition, a* prison area *for a vertex $v \in F$ is a closed area of the plane with the following properties:*

1. *All vertices that lie strictly inside of the area are from $N_{prison}^{1,2}(v)$.*
2. *The area cannot be extended to include any vertex from $N_{prison}^{1,2}(v)$ without violating the first condition.*

Analogously to the preceding section, we first show that the number of prison areas is upper-bounded by $O(|F|)$ and then show that each area contains a constant number of vertices.

Lemma 16. *Given a maximal F-region decomposition $\mathcal{R}$, the vertices that do not lie inside of any region of $\mathcal{R}$ form at most $12|F| - 24$ prison areas (again, we assume $|F| \geq 3$).* □

Lemma 17. *Every prison area contains $O(1)$ vertices.* □

Proposition 18. *The number of vertices lying outside of regions of $\mathcal{R}$ is bounded above by $O(|F|)$.* □

4   A Tree Decomposition-Based Algorithm

There exists a tree decomposition-based algorithm that solves both MINIMUM-VERTEX FEEDBACK EDGE SET and MAXIMUM FULL-DEGREE SPANNING TREE in $O((2\omega)^{3\omega} \cdot \omega \cdot n)$ time on an n-vertex graph with a given tree decomposition of width ω.[9] We omit its description here due to lack of space.

Theorem 19. *For an n-vertex graph G with a given width-ω tree decomposition (T, X), both* MINIMUM-VERTEX FEEDBACK EDGE SET *and* MAXIMUM FULL-DEGREE SPANNING TREE *can be solved in* $O((2\omega)^{3\omega} \cdot \omega \cdot n)$ *time.*   □

Using the linear problem kernels for VFES and FDST that were established in Theorems 1 and 3, Theorem 19 implies subexponential-time algorithms for these problems when restricted to planar graphs:

Theorem 20. *In n-vertex planar graphs,* MINIMUM-VERTEX FEEDBACK EDGE SET *is solvable in* $O(2^{O(\sqrt{k}\log k)} + k^5 + n)$ *time and* MAXIMUM FULL-DEGREE SPANNING TREE *in* $O(2^{O(\sqrt{k}\log k)} + k^5 + n^3)$ *time, where k is the number of degree-reduced vertices or full-degree vertices, respectively.*

Proof. Both problems have linear-size kernels in planar graphs, that is, the reduced graphs have at most $O(k)$ vertices after performing the respective kernelization. The claim follows from Theorem 19 and the facts that planar $O(k)$-vertex graphs have treewidth $\omega = O(\sqrt{k})$ and that a corresponding tree decomposition of width $3\omega/2$ can be computed in $O(k^5)$ time [17].   □

5   Conclusion

Our main technical contribution is the proof of the linear-size problem kernel for FDST in planar graphs. It is easily conceivable that there is room for significantly improving the involved worst-case constant factors—the situation is comparable (but perhaps even more technical) with analogous results for DOMINATING SET in planar graphs [1,5]. Having obtained linear problem kernel sizes for a problem and its dual, the way for applying the lower bound technique for kernel size due to Chen et al. [5] now seems open.

Another interesting line of future research might be to investigate whether our problem kernel result for planar graphs can be lifted to superclasses of planar graphs—corresponding results for DOMINATING SET in this direction are reported in [11]. Perhaps even more importantly, it would be interesting to pursue experimental studies (similar to Bhatia et al. [3]) with real-world data in order to explore the practical usefulness of our algorithms and problem kernelizations.

---

[9] Broersma et al. [4] have already shown that both VFES and FDST are solvable in linear time for graphs with bounded treewidth by expressing the problems in monadic second-order logic. The linear-time solvability follows then from a result of Arnborg et al. [2]. However, this approach does not provide an exact dependency of the running time on the treewidth. Therefore, their result cannot imply a subexponential-time algorithm.

References

1. J. Alber, M. R. Fellows, and R. Niedermeier. Polynomial-time data reduction for Dominating Set. *Journal of the ACM*, 51:363–384, 2004.
2. S. Arnborg, J. Lagergren, and D. Seese. Easy problems for tree-decomposable graphs. *Journal of Algorithms*, 12:308–340, 1991.
3. R. Bhatia, S. Khuller, R. Pless, and Y. Sussmann. The full degree spanning tree problem. *Networks*, 36:203–209, 2000.
4. H. J. Broersma, A. Huck, T. Kloks, O. Koppius, D. Kratsch, H. Müller, and H. Tuinstra. Degree-preserving trees. *Networks*, 35:26–39, 2000.
5. J. Chen, H. Fernau, I. A. Kanj, and G. Xia. Parametric duality and kernelization: Lower bounds and upper bounds on kernel size. In *Proc. of 22nd STACS*, volume 3404 of *LNCS*, pages 269–280. Springer, 2005.
6. F. K. H. A. Dehne, M. R. Fellows, H. Fernau, E. Prieto, and F. A. Rosamond. Nonblocker: Parameterized algorithmics for minimum dominating set. In *Proc. of 32nd SOFSEM*, volume 3831 of *LNCS*, pages 237–245. Springer, 2006.
7. R. G. Downey and M. R. Fellows. *Parameterized Complexity*. Springer-Verlag, Berlin, 1999.
8. V. Estivill-Castro, M. R. Fellows, M. A. Langston, and F. A. Rosamond. FPT is P-time extremal structure I. In *Proc. 1st ACiD*, pages 1–41, 2005.
9. M. R. Fellows, C. McCartin, F. A. Rosamond, and U. Stege. Coordinatized kernels and catalytic reductions: An improved FPT algorithm for Max Leaf Spanning Tree and other problems. In *Proc. of 20th FSTTCS*, volume 1974 of *LNCS*, pages 240–251. Springer, 2000.
10. J. Flum and M. Grohe. *Parameterized Complexity Theory*. Springer-Verlag, 2006.
11. F. V. Fomin and D. M. Thilikos. Fast parameterized algorithms for graphs on surfaces: Linear kernel and exponential speed-up. In *Proc. 31st ICALP*, volume 3142 of *LNCS*, pages 581–592. Springer, 2004.
12. S. Khuller, R. Bhatia, and R. Pless. On local search and placement of meters in networks. *SIAM Journal on Computing*, 32(2):470–487, 2003.
13. M. Lewinter. Interpolation theorem for the number of degree-preserving vertices of spanning trees. *IEEE Transactions on Circuits Systems I: Fundamental Theory and Applications*, 34(2):205–205, 1987.
14. R. Niedermeier. *Invitation to Fixed-Parameter Algorithms*. Oxford University Press, 2006.
15. L. E. Ormsbee. Implicit network calibration. *Journal of Water Resources Planning Management*, 115:243–257, 1989.
16. L. E. Ormsbee and D. J. Wood. Explicit pipe network calibration. *Journal of Water Resources Planning Management*, 112:116–182, 1986.
17. P. D. Seymour and R. Thomas. Call routing and the ratcatcher. *Combinatorica*, 14(2):217–241, 1994.

On the Effective Enumerability of NP Problems

Jianer Chen[1,*], Iyad A. Kanj[2,**], Jie Meng[1], Ge Xia[3,*], and Fenghui Zhang[1,*]

[1] Department of Computer Science, Texas A&M University,
College Station, TX 77843, USA
{chen, jmeng, fhzhang}@cs.tamu.edu
[2] School of CTI, DePaul University, 243 S. Wabash Avenue,
Chicago, IL 60604, USA
ikanj@cs.depaul.edu
[3] Department of Computer Science, Lafayette College,
Easton, PA 18042, USA
gexia@cs.lafayette.edu

Abstract. In the field of computational optimization, it is often the case that we are given an instance of an NP problem and asked to enumerate the first few "best" solutions to the instance. Motivated by this, we propose in this paper a new framework to measure the effective enumerability of NP optimization problems. More specifically, given an instance of an NP problem, we consider the parameterized problem of enumerating a given number of best solutions to the instance, and study its average complexity in terms of the number of solutions. Our framework is different from the previously-proposed ones. For example, although it is known that counting the number of k-paths in a graph is $\#W[1]$-complete, we present a fixed-parameter enumeration algorithm for the problem. We show that most algorithmic techniques for fixed-parameter tractable problems, such as search trees, color coding, and bounded treewidth, can be used for parameterized enumerations. In addition, we design elegant and new enumeration techniques and show how to generate small-size structures and enumerate solutions efficiently.

1 Introduction

Most computational problems are concerned with finding a single solution for a problem instance. For example, decision problems ask for the existence of a solution (to a given instance) that satisfies certain properties [24]. On the other hand, many computational problems in practice seek a number of good solutions rather than a single one. Examples include seeking significant sub-structures in biological networks [20,27], studying sequence Motifs and alignments [25], and constructing a list of codewords in list decoding [18]. Moreover, because of the proneness of computation to errors, a computed optimal solution may not be the "real" optimal one. Therefore, it becomes desirable to generate several "best" solutions rather than a single one.

* Supported in part by NSF Grants CCR-0311590 and CCF-4030683.
** Supported in part by DePaul University Competitive Research Grant.

H.L. Bodlaender and M.A. Langston (Eds.): IWPEC 2006, LNCS 4169, pp. 215–226, 2006.

Several approaches trying to meet this need have been proposed. The most notable one is the study of the counting complexity of a given problem, which is the computational hardness of counting all the solutions to a given instance. Since its initialization by Valiant [29], significant work has been done in the study of counting complexity. Most of this work has focused on the negative side, i.e., proving the intractability of certain counting problems. For example, Valiant [29] proved that counting the number of perfect matchings in a bipartite graph is #P-complete. Flum and Grohe [17] studied the parameterized complexity of counting problems and, in particular, proved that the problem of counting the number of k-paths in a graph is #W[1]-complete.

Another approach along this line of research studies the complexity of enumerating *all* solutions to a given problem instance. Tomita, Tanaka, and Takahashi [28] presented an exponential time algorithm that enumerates all maximal cliques in a graph. Fernau [16] considered a number of enumeration paradigms and studied their respective complexities.

None of the above approaches, however, has perfectly met the practical needs. The study of counting complexity does not provide hints on how to generate solutions. Even worse, the counting complexity of a problem can be significantly different from that of generating a single solution (e.g., perfect matchings in bipartite graphs [29], k-paths in a graph [17]). The enumeration approach (i.e., enumerating all solutions to a given instance) may easily become computationally infeasible, simply because the number of solutions is too large.

Motivated by this, we propose a new framework to study the effective enumerability of NP optimization problems. we will be mainly interested in NP optimization problems that have efficient algorithms for generating a single solution. We will also be seeking solutions of small size k, and study the enumerability of problems whose best solution can be generated in time $f(k)n^{O(1)}$, where f is a recursive function. We associate each solution to the problem with a "weight" that indicates the quality/ranking of the solution. We say that an NP optimization problem is *fixed-parameter enumerable* if there is an algorithm that, for a given problem instance (x, k) and an integer K, generates the K best (in terms of the solution weight) solutions of size k to x in time $f(k)n^{O(1)}K^{O(1)}$.

By setting $K = 1$, we can see that fixed-parameter enumerable problems are also fixed-parameter tractable. It will be interesting to know whether these two notions are equivalent. Along this line, we examine the most popular techniques used in developing parameterized algorithms, including bounded search-tree method, color coding schemes, and bounded tree-width algorithms. We show that these most popular algorithm-design techniques can be non-trivially transformed into techniques for fixed-parameter enumerable problems.

There has been some research in the literature that is related to this research. For example, Chegireddy and Hamacher [8] developed algorithms for finding the K largest perfect matchings in a weighted graph, and Eppstein considered the problem of enumerating the K shortest paths in a digraph [14]. However, to the authors' knowledge, all these researches dealt with problems in P. On the other hand, the current paper mainly focuses on NP-hard optimization problems, and

on developing a systematical approach for effective enumeration of a larger class
of such problems.

2   Definitions and Preliminaries

Recall that a *parameterized problem* consists of instances of the form (x, k), where
$x \in \Sigma^*$ for a finite alphabet Σ, and k is a non-negative integer. A parameterized
problem Q is *fixed parameter tractable* if there is an algorithm A which on input
(x, k) decides if (x, k) is a yes-instance of Q in time $f(k)n^{O(1)}$, where f is a
recursive function independent of $n = |x|$. We extend the standard definition of
NP optimization problems [4] to encompass their parameterized versions.

Definition 1. A *parameterized NP optimization problem* Q is a 4-tuple
(I_Q, S_Q, f_Q, opt_Q) where:

1. I_Q is the set of input instances of the form (x, k), with $x \in \Sigma^*$ for a fixed
   finite alphabet Σ, and k is a non-negative integer called the *parameter*.
2. For each instance (x, k) in I_Q, $S_Q(x, k)$ is the set of *feasible solutions* for
   (x, k), which is defined by a polynomial p and a polynomial time computable
   predicate Φ as $S_Q(x, k) = \{y : |y| \leq p(|x|) \ \& \ \Phi(x, k, y)\}$.
3. $f_Q(x, k, y)$ is a polynomial-time computable function mapping (x, k, y) to a
   real number, where $(x, k) \in I_Q$ and $y \in S_Q(x, k)$.
4. $opt_Q \in \{\max, \min\}$.

Note that since the length of a solution to an instance (x, k) in Q is bounded by
a polynomial of $|x|$, the number of solutions to the instance (x, k) is bounded by
$2^{q(|x|)}$ for a fixed polynomial q. Therefore, the weights of the solutions in the set
$S_Q(x, k)$ can be given in a finite sorted list $L = [f_Q(x, k, y_1), f_Q(x, k, y_2), \ldots]$, in
a non-decreasing order when $opt_Q = \min$, and in a non-increasing order when
$opt_Q = \max$. We say that a set $\{y_1', \ldots, y_K'\}$ of K solutions in $S_Q(x, k)$ are the K
best solutions for the instance (x, k), if the values $f_Q(x, k, y_1')$, $\ldots$, $f_Q(x, k, y_K')$,
when sorted accordingly, are identical to the first K values in the list L.

Definition 2. A parameterized NP optimization problem Q is *fixed-parameter
enumerable* if there are two algorithms A_1 and A_2 such that the following holds.

1. Given an instance (x, k) of Q, the algorithm A_1 generates a structure $\tau_{x,k}$ in
   time $f(k)n^{O(1)}$, where f is a recursive function independent of $n = |x|$.
2. Given the structure $\tau_{x,k}$ and an integer $K \geq 0$, the algorithm A_2 generates
   the K best solutions to the instance (x, k) in time $O(|\tau_{x,k}|^{O(1)} K^{O(1)})$.[1]

The algorithm A_1 will be called the *structure algorithm*, and the algorithm A_2
will be called the *enumeration algorithm*. We say that the problem Q is *linearly
fixed-parameter enumerable* if the running time of the enumeration algorithm A_2
is $O(|\tau_{x,k}|^{O(1)} K)$.

---

[1] Note that it is possible that the total number $|S_Q(x, k)|$ of solutions is smaller than
K. To avoid repeatedly distinguishing the two possible cases, we will simply use K
to refer to the value $K_0 = \min\{K, |S_Q(x, k)|\}$.

We comment on the above definition. Since the algorithm A_1 runs in time $f(k)n^{O(1)}$, the size $|\tau_{x,k}|$ of the structure $\tau_{x,k}$ is bounded by $f(k)n^{O(1)}$. In consequence, the running time of the enumeration algorithm A_2 is bounded by $f_1(k)n^{O(1)}K^{O(1)}$, where f_1 is a recursive function independent of n. Moreover, we require that for each input instance (x, k), the fixed-parameter enumerable problem Q have a small structure $\tau_{x,k}$ whose size is independent of K. Thus, the following theorem holds directly from the above definitions.

Theorem 1. *If a parameterized NP optimization problem Q is fixed-parameter enumerable, then it is fixed-parameter tractable.*

3   Effective Enumerations Based on Branch-and-Search

The branch-and-search method based on bounded search-trees has been a very popular and powerful technique in the development of efficient exact and parameterized algorithms [13]. The reader is referred to [13] for more information about the bounded search tree technique and its analysis.

We discuss how this technique can be employed in designing algorithms for the structure-generation phase of enumeration algorithms for parameterized NP optimization problems. As a running example, we describe the algorithm with VERTEX COVER as the underlying problem. Recall that a vertex set C in a graph G is a *vertex cover* for G, if each edge in G has at least one end in C. A vertex cover of k vertices will be called a *k-vertex cover*. The VERTEX COVER problem is a well-known fixed-parameter tractable problem, and parameterized algorithms for it have been extensively studied (e.g., [6]). The best parameterized algorithm for VERTEX COVER problem uses polynomial space and $O(1.2738^k + kn)$ running time. Moreover, the counting complexity (i.e., counting the number of solutions to a given instance) and the complexity of enumerating all solutions to an instance have also been examined. Moelle, Richter and Rossmanith[30] showed that counting the number of vertex cover of size k can be done in time $O^*(1.3803^k)$. The complexity of enumerating all k-vertex covers, however, depends on whether k is the size of a minimum vertex cover of the graph or not. Fernau [16] showed that if k is equal to the size of a minimum vertex cover, then enumerating all k-vertex covers can be done in time $O(2^k k^2 + kn)$, while if k is not equal to the size of a minimum vertex cover, then no algorithm of running time $f(k)n^{O(1)}$, for any recursive function f, can enumerate all k-vertex covers.

We investigate the fixed parameter enumerability of the problem. We assume that the input graph G is weighted, and each vertex is associated with a real number (the *vertex weight*). The *weight* of a vertex cover C is the sum of the weights of the vertices in C. Therefore, a vertex cover C_1 is *smaller than* a vertex cover C_2 if the weight of C_1 is smaller than the weight of C_2.

> WEIGHTED VERTEX COVER: Given a weighted graph G, and non-negative integers k and K, generate the K smallest k-vertex covers in G.

3.1   The Structure Algorithm

Let (G, k, K) be an instance of the WEIGHTED VERTEX COVER problem, where G is a graph of n vertices. Since a vertex of degree larger than k must be in every k-vertex cover of G, we can first remove all vertices of degree larger than k in the graph and then work on the remaining graph. This pre-processing can be done in linear time. The resulting graph G' has $O(n)$ vertices and $O(kn)$ edges, in which we consider k'-vertex covers for an integer $k' \leq k$. So without loss of generality, we will assume that the input graph G has n vertices and $O(kn)$ edges.

The structure algorithm for WEIGHTED VERTEX COVER is a recursive algorithm based on the branch-and-search method, which on an input instance (G, k) returns a collection $\mathcal{L}(G, k)$ of triples (I, O, R), where each (I, O, R) is a partition of the vertex set of the graph G, representing the set of all k-vertex covers that include all vertices in I and exclude all vertices in O. Moreover, we require that in the subgraph induced by the vertex set R, all the vertices have degree bounded by 2. The structure algorithm is given in Figure 1.

Algorithm **structure-vc**

INPUT: *G: a weighted graph;   k: an integer;*

1. **if** $(k < 0)$ or $(k = 0$ but the edge set of G is not empty)
   then return $\mathcal{L}(G, k) = \emptyset$;
2. **if** there is no vertex of degree larger than 2 in G
   then return $\mathcal{L}(G, k) = \{(\emptyset, \emptyset, V(G))\}$;
3. pick any vertex v of degree $d \geq 3$;
4. let $G_1 = G - v$ and $G_2 = G - (v \cup N(v))$, where $N(v)$ is the set of neighbors of v;
5. recursively call **structure-vc**$(G_1, k - 1)$ and **structure-vc**$(G_2, k - d)$;
   let the returned collections be $\mathcal{L}(G_1, k - 1)$ and $\mathcal{L}(G_2, k - d)$, respectively;
6. $\mathcal{L}(G, k) = \emptyset$;
7. **for** each triple (I_1, O_1, R_1) in $\mathcal{L}(G_1, k - 1)$
   do   add $(I_1 \cup \{v\}, O_1, R_1)$ to $\mathcal{L}(G, k)$;
8. **for** each triple (I_2, O_2, R_2) in $\mathcal{L}(G_2, k - d)$
   do   add $(I_2 \cup N(v), O_2 \cup \{v\}, R_2)$ to $\mathcal{L}(G, k)$;

Fig. 1. The structure algorithm for WEIGHTED VERTEX COVER

Theorem 2. *On an input (G, k), the algorithm **structure-vc** runs in time $O(1.47^k n)$ and returns a collection $\mathcal{L}(G, k)$ of at most 1.466^k triples.*

We say that a vertex cover C of the graph G is *consistent with* a partition (I, O, R) of the vertex set of G if C contains all vertices in I and excludes all vertices in O. The following lemma can be proved by simple induction.

Lemma 1. *Let $\mathcal{L}(G, k)$ be the collection returned by the algorithm* **structure-vc** *on input (G, k). Then every k-vertex cover of G is consistent with exactly one triple in $\mathcal{L}(G, k)$.*

The collection $\mathcal{L}(G, k)$ forms the structure $\tau_{G,k}$ for the instance (G, k) of the WEIGHTED VERTEX COVER problem. By Theorem 2, the structure $\tau_{G,k}$ can be constructed in time $O(1.47^k n)$.

3.2   The Enumeration Algorithm

Let $\mathcal{L}(G, k)$ be the structure returned by the algorithm **structure-vc** on the input (G, k). By Lemma 1, every k-vertex cover C of G is consistent with exactly one triple (I, O, R) in $\mathcal{L}(G, k)$, and contains all the vertices in I and excludes all the vertices in O. Thus, the k-vertex cover C must consist of the vertex set I, plus a vertex cover of $k - |I|$ vertices for the subgraph $G(R)$ induced by the vertex set R. Therefore, the K smallest k-vertex covers for the graph G that are consistent with the triple (I, O, R) can be generated by generating the K smallest $(k - |I|)$-vertex covers for the induced subgraph $G(R)$. Finally, the K smallest k-vertex covers for the original graph G can be obtained by performing the above process on all the triples in the structure $\mathcal{L}(G, k)$, and then picking the K smallest $k-$vertex covers among all the generated k-vertex covers.

   By looking at the algorithm **structure-vc**, we observe that all the vertices in the induced subgraph $G(R)$ have degree bounded by 2. Therefore, we first discuss how we deal with such graphs.

Lemma 2. *Let G be a graph of n vertices in which all vertices have degree bounded by 2. Then the K smallest k-vertex covers of G can be generated in time $O(Kkn)$.*

Proof. Since all the vertices in G have degree bounded by 2, every connected component of G is either an isolated vertex, a simple path, or a simple cycle. Order the vertices of G to form a list $W = [v_1, v_2, \ldots, v_n]$ such that the vertices of each connected component of G appear consecutively in W. In particular, the vertices of a simple path appear in W in the order by which we traverse the path from an arbitrary end to the other end, and the vertices of a simple cycle appear in W in the order by which we traverse the entire cycle starting from an arbitrary vertex in the cycle. A vertex $v_i \in G$ is a *type-1 vertex* if it has degree 0, a *type-2 vertex* if it is in a simple path of length at least 1, and a *type-3 vertex* if it is in a simple cycle.

   For each i, $1 \le i \le n$, let G_i be the subgraph of G induced by the vertex set $\{v_1, v_2, \ldots, v_i\}$. For each induced subgraph G_i, we build a list $L_i = [S_{i,0}, S_{i,1}, \ldots, S_{i,k}]$, $S_{i,j}$ is a set of j-vertex covers for G_i, defined as follows:

   (1) If v_i is of type-1, then $S_{i,j}$ is the set of the K smallest j-vertex covers for G_i (recall that by this we really mean "the K smallest j-vertex covers or all the j-vertex covers if the total number of j-vertex covers is smaller than K"— this remark also applies to the following discussion);

(2) If v_i is of type-2, then $S_{i,j}$ consists of two sets $S'_{i,j}$ and $S''_{i,j}$, where $S'_{i,j}$ contains the K smallest j-vertex covers of G_i that contain v_i, and $S''_{i,j}$ contains the K smallest j-vertex covers of G_i that do not contain v_i;

(3) If v_i is of type-3 and appears in a simple cycle $[v_h, \ldots, v_i, \ldots, v_t]$ in G, then $S_{i,j}$ consists of four sets $S'_{i,j}$, $S''_{i,j}$, $S'''_{i,j}$ and $S''''_{i,j}$, where $S'_{i,j}$ is the set of the K smallest j-vertex covers of G_i that contain both v_h and v_i, $S''_{i,j}$ is the set of the K smallest j-vertex covers of G_i that contain v_h but not v_i, $S'''_{i,j}$ is the set of the K smallest j-vertex covers of G_i that contain v_i but not v_h, and $S''''_{i,j}$ is the set of the K smallest j-vertex covers of G_i that contain neither v_h nor v_i.

Note that since each set $S_{i,j}$ contains at most $4K$ j-vertex covers, the set $S^0_{i,j}$ consisting of the K smallest j-vertex covers of the graph G_i can be constructed from $S_{i,j}$ in time $O(K)$.

The list L_1 can be trivially constructed: (1) if v_1 is of type-1, then all the sets $S_{1,j}$ are empty except $S_{1,0} = \{\emptyset\}$ and $S_{1,1} = \{(v_1)\}$; (2) if v_i is of type-2, then all the sets $S'_{1,j}$ and $S''_{1,j}$ are empty except $S''_{1,0} = \{\emptyset\}$ and $S'_{1,1} = \{(v_1)\}$; and (3) if v_i is of type-3, then all the sets $S'_{1,j}$, $S''_{1,j}$, $S'''_{1,j}$, and $S''''_{1,j}$ are empty except $S''''_{1,0} = \{\emptyset\}$ and $S'_{1,1} = \{(v_1)\}$.

Inductively, suppose that we have built the list L_{i-1}. To build the list L_i, we distinguish the following cases based on the type of the vertex v_i.

Case 1. The vertex v_i is of type-1. Then the graph G_i is the graph G_{i-1} plus an isolated vertex v_i. For each j, $0 \leq j \leq k$, let $S^0_{i-1,j}$ be the set of the K smallest j-vertex covers of the graph G_{i-1}, which can be constructed in time $O(K)$. Since each vertex cover of G_i is either a vertex cover of G_{i-1}, or a vertex cover of G_{i-1} plus the vertex v_i, the set $S_{i,j}$ in L_i can be constructed as follows: take each $(j-1)$-vertex cover of G_{i-1} from $S^0_{i-1,j-1}$ and add the vertex v_i to it to make a j-vertex cover of G_i. This gives a set F of K j-vertex covers for G_i. It is clear that the K smallest j-vertex covers of G_i must be contained in the set $F \cup S^0_{i-1,j}$, which is a set of $2K$ j-vertex covers for G_i. Thus, the K smallest j-vertex covers in the set $F \cup S^0_{i-1,j}$ make the set $S_{i,j}$. Each set $S_{i,j}$ can be constructed in time $O(K)$, and the list L_i can be constructed from the list L_{i-1} in time $O(Kk)$.

Case 2. The vertex v_i is of type-2. Then v_i is on a simple path $[v_h, \ldots, v_i, \ldots, v_t]$ in G of length at least 1. As in Case 1, for each j, let $S^0_{i-1,j}$ be the set of the K smallest j-vertex covers for G_{i-1}.

If $v_i = v_h$ is the first vertex on the path, then the graph G_i is the graph G_{i-1} plus an isolated vertex v_i. Thus, the set $S'_{i,j}$ can be obtained from $S^0_{i-1,j-1}$ by adding the vertex v_i to each $(j-1)$-vertex cover of G_{i-1} in $S^0_{i-1,j-1}$. The set $S''_{i,j}$ is equal to the set $S^0_{i-1,j}$.

If $h < i$ and v_i is not the first vertex on the path, then the graph G_i is the graph G_{i-1} plus the vertex v_i and the edge $[v_{i-1}, v_i]$. Therefore, each vertex cover of G_i is either a vertex cover of G_{i-1} plus v_i, or a vertex cover of G_{i-1} that contains v_{i-1}. Thus, the set $S'_{i,j}$ is again obtained from $S^0_{i-1,j-1}$ by adding the vertex v_i to each $(j-1)$-vertex cover of G_{i-1} in $S^0_{i-1,j-1}$. On the other hand, now the set $S''_{i,j}$ is equal to the set $S'_{i-1,j}$.

In this case, the list L_i can be constructed from the list L_{i-1} in time $O(Kk)$.

Case 3. The vertex v_i is of type-3. v_i is on a simple cycle $[v_h, \ldots, v_i, \ldots, v_t]$ of G. For each j, let $S^0_{i-1,j}$ be the set of the K smallest j-vertex covers of G_{i-1}.

If $v_i = v_h$ is the first vertex on the cycle, then the graph G_i is the graph G_{i-1} plus an isolated vertex v_i. Thus, the set $S'_{i,j}$ can be obtained from $S^0_{i-1,j-1}$ by adding the vertex v_i to each $(j-1)$-vertex cover of G_{i-1} in $S^0_{i-1,j-1}$, and the set $S'''''_{i,j}$ is equal to the set $S^0_{i-1,j}$. By the definition, the sets $S''_{i,j}$ and $S'''_{i,j}$ are empty.

If $h < i < t$, then the graph G_i is the graph G_{i-1} plus the vertex v_i and the edge $[v_{i-1}, v_i]$. Therefore, the set $S'_{i,j}$ can be obtained by adding the vertex v_i to each $(j-1)$-vertex cover in the union $S'_{i-1,j-1} \cup S''_{i-1,j-1}$ then selecting the K smallest ones; the set $S''_{i,j}$ is equal to the set $S'_{i-1,j}$; the set $S'''_{i,j}$ is obtained by adding the vertex v_i to each $(j-1)$-vertex cover in the union $S'''_{i-1,j-1} \cup S''''_{i-1,j-1}$ then selecting the K smallest ones; and the set $S'''''_{i,j}$ is equal to the set $S'''_{i-1,j}$.

If $v_i = v_t$ is the last vertex on the cycle, then the graph G_i is the graph G_{i-1} plus the vertex v_i and two edges $[v_h, v_i]$ and $[v_{i-1}, v_i]$. In this case, the set $S'_{i,j}$ can be obtained by adding the vertex v_i to each $(j-1)$-vertex cover in the union $S'_{i-1,j-1} \cup S''_{i-1,j-1}$ then selecting the K smallest ones; the set $S''_{i,j}$ is equal to the set $S'_{i-1,j}$; the set $S'''_{i,j}$ is obtained by adding the vertex v_i to each $(j-1)$-vertex cover in the union $S'''_{i-1,j-1} \cup S'''''_{i-1,j-1}$ then selecting the K smallest ones; and the set $S'''''_{i,j}$ is empty because $[v_h, v_i]$ is an edge in G_i.

The correctness of the above constructions can be easily verified using the definitions of the sets $S'_{i,j}$, $S''_{i,j}$, $S'''_{i,j}$, and $S'''''_{i,j}$. Moreover, it is also easy to see that the list L_i can be constructed from the list L_{i-1} in time $O(Kk)$.

Summarizing all the above, we conclude that the list L_n can be constructed in time $O(Kkn)$. Now the K smallest k-vertex covers of the graph $G = G_n$ can be easily obtained in time $O(K)$ from the set $S_{n,k}$ in the list L_n. This completes the proof of the lemma.   □

Now it should be obvious in principle how we can generate the K smallest k-vertex covers for the graph G: they can be obtained by first generating the K smallest consistent k-vertex covers, for each triple in $\mathcal{L}(G, k)$. However, by applying some enumeration tricks, we can significantly speedup this enumeration process, as shown in the following theorem.

Theorem 3. *Let (G, k) be an instance of the* WEIGHTED VERTEX COVER *problem, and let $\mathcal{L}(G, k)$ be the structure returned by the algorithm* **structure-vc** *on (G, k). Then the K smallest k-vertex covers of the graph G can be generated in time $O(1.47^k n + 1.22^k Kn)$.*

Corollary 1. *The* WEIGHTED VERTEX COVER *problem is linearly fixed-parameter enumerable. More specifically, given an instance (G, k) and a nonnegative integer K, the K smallest k-vertex covers of the graph G can be generated in time $O(1.47^k n + 1.22^k Kn)$, where n is the number of vertices in the graph.*

4   Effective Enumeration Based on Color Coding

The *color coding* technique [2] is very powerful and useful in the development of efficient parameterized algorithms. In this section, we show that the color coding technique is also very helpful in developing effective algorithms for the structure-generation phase of enumeration algorithms for parameterized NP optimization problems. We will illustrate this fact by presenting an enumeration algorithm for the k-PATH problem. A simple path in a graph G is a k-*path* if it contains exactly k vertices. The *weight* of a path in a weighted graph is the sum of the weights of the vertices in the path. The problem can be formally defined as follows.

> WEIGHTED k-PATH: given a weighted graph G and integers k and K, generate the K largest k-paths in G.

4.1   The Structure Algorithm

A k-*coloring* of a set S is a function from S to $\{1, 2, \ldots, k\}$. A collection $\mathcal{F}$ of k-colorings of S is a k-*color coding scheme* for S if for any subset W of k elements in S, there is a k-coloring f_W in $\mathcal{F}$ such that no two elements in W are assigned the same color by f_W. The *size* of the k-color coding scheme $\mathcal{F}$ is equal to the number of k-colorings in $\mathcal{F}$. Alon, Yuster, and Zwick [2] showed that there is a k-color coding scheme of size $2^{O(k)}n$ for a set of n elements. This bound has been improved recently to $O(6.4^k n)$ [7]. In the following discussion, we will assume a k-color coding scheme $\mathcal{F}$ of size $O(6.4^k n)$ for a set of n elements.

On a given instance (G, k) of the WEIGHTED k-PATH problem, where G is a graph of n vertices, the structure algorithm for WEIGHTED k-PATH produces $h = O(6.4^k n)$ copies $\{G_1, G_2, \ldots, G_h\}$ of the graph G, where each copy G_i is colored by a k-coloring in the k-color coding scheme $\mathcal{F}$. Note that by the definition of k-color coding schemes, every k-path in the graph G has all its vertices colored with different colors in at least one of these copies of the graph G. The list $\tau_{G,k} = \{G_1, G_2, \ldots, G_h\}$ is the structure returned by the structure algorithm for the WEIGHTED k-PATH problem, whose running time is $O(6.4^k n^2)$.

4.2   The Enumeration Algorithm

The enumeration algorithm for WEIGHTED k-PATH is a careful and non-trivial generalization of the dynamic programming algorithm described in [2] that finds a k-path in a k-colored graph. We first discuss how we deal with each copy G_i of the colored graphs in the list $\tau_{G,k}$. We say that a k-path in a k-colored graph is *properly colored* if no two vertices on the path are colored with the same color. Consider the algorithm given in Figure 2, where $c(w)$ denotes the color assigned to the vertex w in the k-colored graph G. Inductively, before the j-th execution of the loop in steps 2.1-2.5 of the algorithm, we assume that each vertex w is associated with a collection $\mathcal{C}_j(w)$ of pairs (C, P), where C is a subset of j colors in the k-color set, and P is the set of up to K largest properly colored j-paths ending at w that use exactly the colors in C. Then the j-th execution of steps 2.1-2.5 will produce a similar collection $\mathcal{C}_{j+1}(w)$ for $(j + 1)$-paths in G based on the collection $\mathcal{C}_j(w)$ of j-paths.

Note that at the end of the algorithm **enumerate-path**(G, k, K), for each vertex w in the k-colored graph G, the collection $\mathcal{C}_k(w)$ is either empty, or contains a single pair (C, P), where C is the set of all k colors and P is a set of properly colored k-paths ending at w in G.

Algorithm **enumerate-path**(G, k, K)

INPUT: *a k-colored graph G, and integers k and K*

1. **for** each vertex w in G **do** $C_1(w) = [(\{c(w)\}; \{w\})]$;
2. **for** $j = 1$ **to** $k - 1$ **do**
   2.1. **for** each edge $[v, w]$ in G **do**
   2.2.   **for** each pair (C, P) in $C_j(v)$ **do**
   2.3.   **if** $(c(w) \notin C)$ **then**
   2.4.   construct $|P|$ $(j + 1)$-paths ending at w by extending each paht in P to the vertex w;
   2.5.   add these $(j + 1)$-paths to P' in the pair $(C \cup \{c(w)\}, P')$ in $C_{j+1}(w)$ and only keep the K largest $(j + 1)$-paths in P';

Fig. 2. The enumeration algorithm for WEIGHTED k-PATH

Lemma 3. *For each vertex w in the k-colored graph G, the pair (C, P) in the collection $C_k(w)$ returned by the algorithm **enumerate-path**(G, k, K) contains the K largest properly colored k-paths ending at w. The running time of the algorithm **enumerate-path**(G, k, K) is $O(2^k k^2 n^2 K)$.*

Theorem 4. *Given the structure $\tau_{G,k}$ and an integer K, the K largest k-paths in the graph G can be generated in time $O(12.8^k + 6.4^k k^2 n^3 K)$. In consequence, the WEIGHTED k-PATH problem is linearly fixed-parameter enumerable.*

5   Effective Enumeration Based on Tree Decomposition

The concept of the tree decomposition of a graph has played an important role in the study of algorithmic graph theory [5] and in developing efficient exact and parameterized algorithms for graph problems on planar graphs (see [1]). In this section, we discuss how this approach can be used to develop algorithms for the structure-generation phase of enumeration algorithms.

A vertex set D in a graph G is a *dominating set* if each vertex in $G - D$ is adjacent to a vertex in D. A dominating set of k vertices is called a *k-dominating set*. Given a weighted graph, the *weight* of a dominating set D is the sum of the weights of the vertices in D. The following problem is our running example.

WEIGHTED PLANAR DOMINATING SET. Given (G, k, K), where G is a weighted planar graph, and k and K are nonnegative integers, generate the K smallest k-dominating sets in the graph G.

5.1   The Structure Algorithm

We omit the terminology and proofs of theorems in this section, interested readers are referred to [31].

Theorem 5. *There is an $O(\sqrt{k}n)$ time algorithm that given a planar graph G on n vertices and a positive integer k, either constructs a nice tree decomposition $(\mathcal{V}, \mathcal{T})$ for G of width $O(\sqrt{k})$ and $O(n)$ nodes, or reports that no dominating set for G of size bounded by k exists.*

The structure $\tau_{G,k}$ is simply the nice tree decomposition $(\mathcal{V}, \mathcal{T})$ obtained by the above theorem.

5.2   The Enumeration Algorithm

Given the nice tree decomposition $\tau_{G,k} = (\mathcal{V}, \mathcal{T})$, we can generate the K smallest k-dominating sets in the graph G using dynamic programming.

Theorem 6. *Given a planar graph G on n vertices and two nonnegative integers k and K, the K smallest k-dominating sets in G can be generated in time $O(2^{O(\sqrt{k})}nK \log K)$.*

6   Final Remarks

We have introduced the concept of effective enumerability, or more precisely, fixed-parameter enumerability of NP optimization problems. Our objective is to solve enumeration problems that have an increasing demand in computational science. Further investigation on the relationship between fixed-parameter tractability and fixed-parameter enumerability may open up an interesting research direction, which seems interesting and important, from both the theoretical and the practical points of view.

References

1. J. ALBER, H. BODLAENDER, H. FERNAU, T. KLOKS, AND R. NIEDERMEIER, Fixed parameter algorithms for dominating set and related problems on planar graphs, *Algorithmica* **33**, pp. 461-493, (2002).
2. N. ALON, R. YUSTER, AND U. ZWICK, Color-coding, *Journal of the ACM* **42**, pp. 844-856, (1995).
3. V. ARVIND AND V. RAMAN, Approximation algorithms for some parameterized counting problems, ISAAC'02, pp. 453-464, (2002).
4. G. AUSIELLO, P. CRESCENZI, G. GAMBOSI, V. KANN, A. MARCHETTI-SPACCAMELA, M. PROTASI, *Complexity and Approximation, Combinatorial optimization problems and their approximability properties*, Springer-Verlag, 1999.
5. H. BODLAENDER, Treewidth: algorithmic techniques and results, *Lecture Notes in Computer Science* **1295**, pp. 19-36, (1997).
6. J. CHEN, I. A. KANJ, AND W. JIA, Vertex cover: further observations and further improvements, *Journal of Algorithms* **41**, pp. 280-301, (2001).
7. J. CHEN, S. LU, S.-H. SZE, AND F. ZHANG, Improved algorithms for the k-path problem, *Manuscript*, (2005).
8. C. CHEGIREDDY AND H. HAMACHER, Algorithms for finding K-best perfect matchings, *Discrete Applied Mathematics 18*, pp. 155-165, (1987).

9. S. CHIEN, A determinant-based algorithm for counting perfect matching in a general graph, SODA'04, pp. 728-735, (2004).

10. T. CORMEN, C. LEISERSON, R. RIVEST, AND C. STEIN, *Introduction to Algorithms*, 2nd Edition, McGraw-Hill Book Company, Boston, MA, 2001.

11. V. DAHLLOF AND P. JONSSON, An algorithm for counting maximum weighted independent sets and its applications, SODA'02, pp. 292-298, (2002).

12. M. DYER, Approximate counting by dynamic programming, STOC'03, pp. 693-699, (2003).

13. R.G. DOWNEY AND M.R. FELLOWS, *Parameterized Complexity*, Springer-Verlag, 1999.

14. D. EPPSTEIN, Finding the k shortest paths, *SIAM J. Computing* **28-2**, pp. 652-673, (1998).

15. M. FELLOWS, C. KNAUER, N. NISHIMURA, P. RAGDE, F. ROSAMOND, U. STEGE, D. THILIKOS, AND S. WHITESIDES, Faster fixed-parameter tractable algorithms for matching and packing problems, *Lecture Notes in Computer Science 3221*, (2004), pp. 311-322.

16. H. FERNAU, On parameterized enumeration, COCOON'02, pp. 564-573, (2002).

17. J. FLUM AND M. GROHE, The parameterized complexity of counting problems, *SIAM Journal on Computing 33*, pp. 892-922, (2004).

18. V. GURUSWAMI, *List decoding of error-correcting codes, Lecture Notes in Computer Science* **3282**, 2005.

19. H. HUNT III, M. MARATHE, V. RADHAKRISHNAN, AND R. STEARNS, The complexity of planar counting problems, *SIAM Journal on Computing 27*, pp. 1142-1167, (1998).

20. B. KELLEY, R. SHARAN, R. KARP, T. SITTLER, D. ROOT, B. STOCKWELL, AND T. IDEKER, Conserved pathways within bacteria and yeast as revealed by global protein network alignment, *Proc. Natl. Acad. Sci. USA* **100**, pp. 11394-11399, (2003).

21. T. KLOKS, Treewidth, computations and approximations, *Lecture Notes in Computer Science* **842**, (1994).

22. I. KOUTIS, A faster parameterized algorithm for set packing, *Information Processing Letters 94*, (2005), pp. 7-9.

23. S. NAKANO, Efficient generation of triconnected plane triangulations, COCOON'01, pp. 131-141, (2001).

24. C. H. PAPADIMITRIOU, *Computational Complexity*, Addison-Wesley, 1994.

25. P. PEVZNER AND S.-H. SZE, Combinatorial approaches to finding subtle signals in DNA sequences, ISMB'2000, pp. 269-278, (2000).

26. S. RAVI AND H. HUNT III, An application of the planar separator theorem to counting problems, *Information Processing Letters 25*, pp. 317-321, (1987).

27. J. SCOTT, T. IDEKER, R. KARP, AND R. SHARAN, Efficient algorithms for detecting signaling pathways in protein interaction networks, RECOMB 2005, to appear.

28. E. TOMITA, A. TANAKA, AND H. TAKAHASHI, The worst-case time complexity for generating all maximal cliques, COCOON'04, pp. 161-170, (2004).

29. L. VALIANT, The complexity of computing the permanent, *Theoretical Computer Science 8*, pp. 189-201, (1979).

30. D. MOELLE, S. RICHTER, P. ROSSMANITH, Enumerate and Expand: New Runtime Bound for Vertex Cover Variants, COCOON 06, to appear

31. JIANER CHEN, IYAD A. KANJ, JIE MENG, GE XIA, FENGHUI ZHANG, On the Effective Enumerability of NP Problems, Technique Report tr2006-5-2, Department of Computer Science, Texas A&M University.

The Parameterized Complexity of Enumerating Frequent Itemsets[*]

Matthew Hamilton[1], Rhonda Chaytor[2], and Todd Wareham[2]

[1] Department of Computing Science, University of Alberta, Edmonton, AB, Canada
hamilton@cs.ualberta.ca
[2] Department of Computer Science, Memorial University of Newfoundland,
St. John's, NL, Canada
{rchaytor, harold}@cs.mun.ca

Abstract. A core problem in data mining is enumerating frequently-occurring itemsets in a given set of transactions. The search and enumeration versions of this problem have recently been proven NP- and #P-hard, respectively (Gunopulos *et al*, 2003) and known algorithms all have running times whose exponential terms are functions of either the size of the largest transaction in the input and/or the largest itemset in the output. In this paper, we analyze the complexity of the size-k frequent itemset enumeration problem relative to a variety of parameterizations. Many of our hardness results are proved using a recent extension of parameterized complexity to solution-counting problems (McCartin, 2002). These results include hardness for versions of this problem based on restricted transaction-set structure. We also derive a collection of fixed-parameter algorithms using off-the-shelf parameterized algorithm design techniques, several of which suggest new algorithmic directions for the frequent itemset enumeration problem.

1 Introduction

A core problem in data mining is frequent itemset enumeration – that is, given a database of transactions T over a set I, what are the frequent itemsets, *i.e.*, itemsets that occur in at least t transactions in T for some specified threshold t? The task of mining frequent itemsets plays a critical part in many knowledge discovery tasks, from association rule mining [2,3] to finding frequent subgraphs in biological networks [4].

One of the main challenges facing the data mining field is discovering ways to scale knowledge discovery algorithms to larger database systems while meeting the demand for a quick response to data mining requests [5]. It is clear that along with the necessary advances in hardware, efficient algorithms to solve fundamental problems such as frequent itemset discovery must be developed in order to meet this challenge.

From the parameterized complexity perspective, current algorithms for this problem exploit a single parameter. Many algorithms put forth to enumerate frequent itemsets (see [6,7,8] and references) are incremental improvements on the

---

[*] This paper is an extension of results that first appeared in [1].

H.L. Bodlaender and M.A. Langston (Eds.): IWPEC 2006, LNCS 4169, pp. 227–238, 2006.

Apriori algorithm introduced in [9], and the exponential terms in the time complexities of these algorithms are all functions based on the sizes of the largest transaction in the input and the largest frequent itemset in the output [10] (note that the former is an upper bound on the latter). It has been reported [11,8] that these algorithms perform well on sparse databases such as market basket data containing only small frequent itemsets with small average transaction size. However, their performance on dense databases suffers as the transaction and frequent itemset sizes no longer occupy such a narrow range of values. One attempt to get around such problems has been to focus on enumerating maximal or closed[1] frequent itemsets, which are much fewer in number than frequent itemsets of specified sizes and from which size-specific frequent itemsets can be reconstructed; however, to date, algorithms for enumerating closed and maximal itemsets have been plagued by the same limitations listed above (see [12,6,13] and references).

Given all this, it is worthwhile to investigate other algorithmic possibilities for frequent itemset enumeration relative to different parameterizations. To this end, we describe in this paper the results of a systematic parameterized complexity analysis [14] of the following problem:

FREQUENT ITEMSET ENUMERATION (FIE)
Input: A set $I = \{i_1, i_2, \ldots, i_m\}$ of items, a set $T = \{t_1, t_2, \ldots, t_n\}$ of subsets of I whose largest subset is of size $l \leq m$, an integer k, $1 \leq k \leq m$, and an integer t, $1 \leq t \leq n$.
Output: All t-frequent k-itemsets, *i.e.*, all subsets s of I of size k such that s occurs in at least t members of T.

As the decision version of this problem is NP-complete ([6, Theorem 6] and [3]), it is unlikely that there is an efficient algorithm for finding even a single maximum-sized frequent itemset, and counting or enumerating all of them will be more difficult still. Indeed, it has been shown that the counting version of the problem is $\#P$-complete [6, Theorem 7] even when we are only required to count the maximal or closed frequent itemsets [13, Theorem 14].

Our analysis includes the obvious problem-aspects k, t, l, m, and n. Aspects l and k are of practical interest as data mining frequently involves small transaction sizes and consequent small frequent itemset sizes [15]; moreover, as such mining also frequently requires large frequent itemsets with small support and small frequent items sets with large support [15], the dual versions[2] of k and t (denoted k_d and t_d, respectively) are also of great interest.

In our analysis, we delimit the algorithmic possibilities for the FIE problem relative to parameters composed of all subsets of size two of the aspects in Table 1 by deriving hardness results and fixed-parameter algorithms relative to the

---

[1] A t-frequent itemset x is **maximal** if no single item i can be added to it without lowering the occurrence of $x \cup \{i\}$ below t in T. A t-frequent itemset x is **closed** if the largest subset of items in I that is common to all itemsets in T that contain at least one item in x is x itself.

[2] In these dual cases, the FIE problem is parameterized such that we ask for a frequent itemset of size $m - k_d$ and of support $n - t_d$, respectively.

Table 1. Aspects of the Frequent Itemset Enumeration Problem

k	size of frequent itemset	k_d	dual of size
t	support threshold	t_d	dual of support threshold
l	maximum transaction size		
m	number of items		
n	number of transactions		

corresponding parameterized problems. Our hardness results are atypical in that we will derive lower bounds on FIE via hardness of the corresponding counting problem FREQUENT ITEMSET COUNTING (FIC) (which outputs the number of t-frequent k-itemsets) rather than the corresponding decision problem.

The paper is organized as follows. In Section 2, we review some parameterized complexity definitions particular to analyzing counting and enumeration problems. In Section 3, we give parameterized hardness results for the frequent itemset enumeration problem as well as versions of this problem in which the structure of the transaction-set is restricted. In Section 4, we give fixed-parameter tractability results. For reasons of space, several of the proofs of lemmas and theorems in these sections are omitted. Finally, in Section 5 we summarize our results and give some promising directions for future research.

2   Parameterized Complexity Analysis of Counting and Enumeration Problems

The majority of parameterized complexity results derived over the last 15 years have focused on decision and search problems. Recently, work has been done in the area of parameterized counting and enumeration complexity [16,17,18,19,20]. This is of great interest in applications such as data mining, which are more interested in enumerating solutions than just finding one solution that is in some sense optimal. It also provides an additional route for deriving enumeration problem hardness results – even if the associated decision problem cannot be shown intractable or is in FPT, the enumeration problem may yet be shown intractable either directly or indirectly via a hardness result for the associated counting problem. In the remainder of this paper, denote the version of problem Π parameterized by k as $\langle k \rangle$-Π and an instance of such a problem by $\langle x, k \rangle$.

To show enumeration intractability results, we will use the parameterized counting complexity framework developed in [20]. Let $FFPT$ be the class of parameterized counting problems $\langle k \rangle$-Π that can be solved by an algorithm A that runs in time $f(k)|x|^c$ for an arbitrary function f and a constant c. Note that $FFPT$ bears the same relation to FP, the class of function-computing problems that can be solved in polynomial time, that FPT has to P. The standard parameterized reducibility can be easily modified along the lines in [21] such that it preserves solution-number between the parameterized counting problems involved. In addition, we define a new class $\#XP$ which is the class of parameterized counting problems $\langle k \rangle$-Π that can be solved by an algorithm A that runs in time $f(k)|x|^{g(k)}$ for arbitrary functions f and g.

We will state parameterized enumeration tractability results within the framework defined in [17,18]. This framework defines different types of tractability relative to different types of solution-sets that can be enumerated. In particular, given a parameterized enumeration problem $\langle k \rangle$-Π, an algorithm that runs in time $f(k)|x|^c$ for an arbitrary function f and a constant c may enumerate all, all optimal (minimum / maximum), or all minimal / maximal solutions. In [17,18], such a problem is said to be fixed-parameter enumerable, optimally fixed-parameter enumerable and inclusion-minimally fixed-parameter enumerable, respectively. In this paper, we define the classes of problems with these three types of algorithms as FPE, $OFPE$, and $IMFPE$, respectively.

The following lemmas are fairly trivial (the proofs are similar to those for the analogous lemmas given in [14] and are left to the reader) but are nonetheless useful for extending given results to fill out the final result-table in Section 5. Where the type of the problem is clear from context, we will say that all problems in FPT, $FFPT$, FPE, $OFPE$, or $IMFPE$ are fixed-parameter tractable.

Lemma 1. *Given a set S of aspects of a counting problem Π, if Π is #P-hard when the value of every aspect $s \in S$ is fixed, then the parameterized counting problem $\langle S \rangle$-Π is not in #XP unless FP=#P.*

Lemma 2. *Given sets $S \subseteq S'$ of aspects of a problem Π, if parameterized problem $\langle S \rangle$-Π is fixed-parameter tractable then parameterized problem $\langle S' \rangle$-Π is fixed-parameter tractable.*

Lemma 3. *Given sets $S \subseteq S'$ of aspects of a problem Π, if parameterized problem $\langle S' \rangle$-Π is not fixed-parameter tractable unless **X** for some conjecture **X** then parameterized problem $\langle S \rangle$-Π is not fixed-parameter tractable unless **X**.*

3   Negative Results: Hardness

In this section, we show fixed-parameter intractability of FIC (and hence FIE) relative to various sets of aspects by giving parameterized counting reductions from known intractable counting problems. This gives us two types of results:

1. There exists no algorithm that runs in time $O(f(k)|x|^\alpha)$ for problem $\langle k \rangle$-X ($X \notin FFPT$ unless $W[2] = FPT$ or $W[1] = FPT$).
2. There exists no algorithm that runs in time $O(f(k)|x|^{g(k)})$ for problem $\langle k \rangle$-X ($X \notin \#XP$ unless $FP = \#P$).

In the subsections below, we further break down our results into those involving the general FIE problem and those in which transaction-set structure is restricted.

Our reductions are based on both classical and parameterized versions of the problems WEIGHTED MONOTONE CNF SATISFIABILITY (WCS$^+$) and WEIGHTED ANTIMONOTONE CNF SATISFIABILITY (WCS$^+$).

Let the versions of these problems parameterized by the weight k of the satisfying solutions be $\langle k \rangle$-WCS$^+$ and $\langle k \rangle$-WCS$^-$, respectively. Both WCS$^+$ and WCS$^-$ are NP-complete [22, Problems LO1 and LO2], and $\langle k \rangle$-WCS$^+$ is $W[2]$-complete [23, Theorem 2.1] while $\langle k \rangle$-WCS$^-$ is $W[1]$-complete [24, Theorem 12.6]. We also need the counting versions of WCS$^+$ and WCS$^-$, denoted #WCS$^+$ and #WCS$^-$ respectively, which output the number of truth assignments of weight k that satisfy F. Problem #WCS$^+$ is known to be #P-hard [21, Section 4, Problem 7].

3.1   General FIE

We first give a reduction from #WCS$^+$ to FIC. This reduction is similar to the one used in [6] to prove the #P-hardness of FIC.

Lemma 4. *#WCS$^+$ reduces to FIC*

Theorem 5. *Unless $FPT = W[2]$, $\langle k, t \rangle$-FIC $\notin FFPT$.*

Proof: Lemma 4 gives a parameterized counting reduction from $\langle k \rangle$-#WCS$^+$ to $\langle k, t \rangle$-FIC. Hence, $\langle k, t \rangle$-FIC $\in FFPT$ implies $\langle k \rangle$-#WCS$^+$ $\in FFPT$. This in turn implies that $\langle k \rangle$-WCS$^+$ $\in FPT$, which is not true unless $FPT = W[2]$ as $\langle k \rangle$-WCS$^+$ is $W[2]$-complete. ∎

Theorem 6. *FIC is #P-hard when $t = 1$.*

Proof: Follows from the #P-hardness of WCS$^+$ and the reduction in Lemma 4 in which $t = 1$ in the created instance of FIC. ∎

Corollary 7. *$\langle t \rangle - FIC \notin \#XP$ unless $FP = \#P$.*

Proof: Follows from Theorem 5 and Lemma 1. ∎

We can derive further results by showing the hardness of FIC is with respect to t and k.

Theorem 8. *$\langle t \rangle$-FIC reduces to $\langle k \rangle$-FIC and $\langle k \rangle$-FIC reduces to $\langle t \rangle$-FIC.*

Corollary 9. *$\langle k \rangle - FIC \notin \#XP$ unless $FP = \#P$.*

Proof: Follows from Corollary 9, Theorem 8, and Lemma 1. ∎

We now consider parameters containing dual aspects. The following reduction from #WCS$^-$ to FIC is similar to that in Lemma 4.

Lemma 10. *#WCS$^-$ reduces to FIC*

Proof: Given an instance $\langle X = \{X_1, \ldots, X_n\}, C = \{C_1, \ldots, C_m\}, k \rangle$ of #WCS$^-$, construct an instance $\langle I, T = \{T_1, \ldots, T_m\}, k', t \rangle$ of FIC such that $I = X$, $T_i =$

$\{x \in X | x \notin C_i\}$ for $1 \leq i \leq m$, $k' = n - k$, and $t = 1$. This construction can be done in polynomial time. We now prove correctness.

Claim: There exists a weight k assignment of variables that falsifies F iff there is a corresponding 1-frequent itemset of size $n - k$ in T.

Proof of Claim: ($\leftarrow$) Suppose we have $X_s = \{x_{s_1}, ..., x_{s_k}\} \subset X$, a weight k truth assignment that falsifies the given formula F. Then there must be some C_i that is false under assignment X_s; that is such that all of its literals are in X_s (an antimonotone clause C_i is false iff all its literals have value true). Hence, $T_i \subset X_s^c$, the literals in X set false under the assignment X_s, which is of size $n - k$. Therefore T_i contains a 1-frequent $n - k$-itemset.

($\rightarrow$) Suppose we have a 1-frequent $n - k$-itemset. Then some T_i contains at least $n - k$ items. Then the weight k truth assignment where the k literals not in T_i are assigned to be true falsifies clause C_i. Hence there exists a weight k falsifying truth assignment. ∎

Let w and v be witness functions for $\#WCS^-$ and FIC, respectively. By the claim above, the number of falsifying weight-k assignments of F is equal to $v(\langle I, T, n - k, 1 \rangle)$. As there are $\binom{n}{k}$ weight k truth assignments, the number of weight-k satisfying assignments of F is $\binom{n}{k} - v(\langle I, T, n - k, 1 \rangle)$. Set $\tau(v(\langle I, T, n - k, 1 \rangle)) = \binom{n}{k} - v(\langle I, T, n - k, 1 \rangle)$. Hence, $w(\langle X, C, k \rangle) = \tau(v(\langle I, T, n - k, 1 \rangle)$. ∎

Theorem 11. *Unless $FPT = W[1]$, $\langle k_d, t \rangle$-FIC $\notin FFPT$.*

Proof: Lemma 10 gives a parameterized counting reduction from $\langle k \rangle$-$\#WCS^-$ to $\langle k_d, t \rangle$-FIC. Hence, $\langle k_d, t \rangle$-FIC $\in FFPT$ implies $\langle k \rangle$-$\#WCS^- \in FFPT$. This in turn implies that $\langle k \rangle$-$WCS^- \in FPT$, which is not true unless $FPT = W[1]$ as $\langle k \rangle$-WCS^- is $W[1]$-complete. ∎

Corollary 12. *Unless $FPT = W[1]$, $\langle k, t_d \rangle$-FIC $\notin FFPT$.*

Proof: Follows from Theorems 11 and 8. ∎

3.2   Restricted FIE

In light of the bad news in the last section, it might make sense to consider parameterized versions of the FIE problem where the underlying structure of the transaction-set T is restricted. One way of visualizing such restrictions is to consider each itemset in T as specifying a hyperedge in a hypergraph based on vertex-set I. If a hypergraph constructed in such a fashion from a collection C of sets is acyclic, C is said to be **tree-like** [25]. The effects of this restriction can be dramatic, *e.g.*, for k the maximum subset size, $\langle k \rangle$-SET COVER is not in XP in general but fixed-parameter tractable when restricted to tree-like collections of sets [25], and it would be interesting to know how it affects the FIE problem.

Let FTIE be the version of FIE in which T is tree-like and R be the tree in the hypergraph underlying T; we will refer to R as the **subset-tree** associated with T.

Lemma 13. *FIE reduces to FTIE.*

Proof: Given an instance $\langle I, T, k, t \rangle$ of FIE, construct an instance $\langle I', T', k', t' \rangle$ of FTIE such that $I' = I$, $T' = \{T_a\} \cup T$ where $T_a = \{i \mid i \in I\}$, *i.e.*, a transaction-set containing every item, $k' = k$ and $t' = t + 1$.

Claim: T' is a tree-like collection of subsets of I'.

Proof of Claim: The subset-tree R consists of a root node with the remaining nodes being leaf nodes. Specifically T_a is the root with every other $T_i \in T$ mapped to a leaf node. Since for all $T_i \in T$, $T_i \subseteq T_a$ and every subset of nodes of R clearly is a subtree of R, for each $i \in I'$ all nodes that correspond to subsets of T' containing i must induce a subtree. ∎

This construction can be done in polynomial time. We now prove correctness.

Let $M \subseteq I$ be a t-frequent itemset in T of size k. Then $M \subset T_i$ for each $T_i' \in T' \setminus T_a$, since these are sets of T' from T. Hence, M is at least t-frequent in T'. Also clearly $M \subseteq T_a$, hence M occurs $t' = t + 1$ times in T', as we require.

Let $M \subseteq I'$ be a t'-frequent itemset in T' of size k. It is clear $M \subseteq T_a$, hence M must appear $t' - 1 = t$ times in $T' \setminus T_a$. It follow that M occurs t times in T, as we require. ∎

Theorem 14. *FTIE is NP-complete.*

Proof: FTIE is in NP since we know FIE is in NP. By Lemma 13, FTIE is NP-hard since FIE is NP-hard. Thus, FTIE is NP-complete. ∎

The hypergraph underlying T can also be characterized in terms of parameters capturing notions of bounded cyclicity, *e.g.*, hypertreewidth [26]. However, as the acyclic case is the most basic case (often having a value of one) in each such measure, the corresponding parameterized decision versions of FIE for these measures are not even in XP unless $P = NP$, which bodes even worse for the associated counting and enumeration problems. The news is not appreciably better when we consider FTIE relative to the aspects defined in Section 1.

Theorem 15. *All hardness results for parameterizations of FIE relative to aspect-subsets of k, t, k_d, and t_d also hold for FTIE.*

Proof: In the reduction described in the proof of Lemma 13, k and t in the constructed instances of FTIE are functions of k and t respectively in the given instances of FIE; hence, all parameterizations of FTIE involving aspect-subsets containing k, t, k_d, and t_d are as hard as the corresponding parameterizations of FIE. ∎

Despite the above, there remain some avenues of promise. In the special case that the subset-tree is a path (or equivalently, the associated set-collection has the consecutive ones property), FTIE is solvable in polynomial time [27]. The method of distance from triviality [28] might then be applicable if we can find a parameter that captures some notion of distance from this trivial case, especially if we combine this with some other restriction on the problem. This begs the questions of whether typical instances of frequent itemset mining would actually fall into a range of small distances from such a trivial case and whether the derived algorithm is of use in practice; however, given the current paucity of algorithmic options for solving frequent itemset enumeration problems at present, *cf.* Section 4, it is worth following up.

4   Positive Results: Fixed-Parameter Algorithms

In this section, we give various fixed-parameter algorithms for solving the FIE problem. Many of these algorithms are based on naive brute-force enumeration that is inefficient in practice; however, they all serve to establish fixed-parameter tractability relative to various parameters and hence pave the way for future algorithms research.

Theorem 16. $\langle l \rangle$-, $\langle m \rangle$-, and $\langle n \rangle$-FIE are fixed-parameter tractable.

We can derive other algorithms by exploiting a connection between FIE and the following problem:

CONSTRAINT BIPARTITE VERTEX COVER ENUMERATION (CBVCE)
Input: A bipartite graph $G = (V_1, V_2, E)$ and nonnegative integers k_1 and k_2.
Output: All subsets $C_1 \subseteq V_1$ and $C_2 \subseteq V_2$ of sizes $|C_1| \leq k_1$ and $|C_2| \leq k_2$ respectively such that each edge in E has at least one endpoint in $C_1 \cup C_2$.

Theorem 17. $\langle t_d, k_d \rangle$-FIE reduces to $\langle k_1, k_2 \rangle$-CBVCE and $\langle k_1, k_2 \rangle$-CBVCE reduces to $\langle t_d, k_d \rangle$-FIE.

Essentially, what we do in this reduction is consider the bipartite graph G_t induced by the transaction-itemset inclusion relation and delete bipartite vertex covers with bipartite sets of size k_d and t_d from the bipartite complement graph G_t^c of G_t. In G_t', the subgraph of G_t remaining after these vertices are deleted, there remains a $(n - t_d)$-frequent itemset of size $(m - k_d)$. We can delete all such bipartite vertex covers, allowing us to enumerate all the frequent itemsets of the size we desire. Likewise in the other direction, deleting a frequent itemset and then edge complementing the resultant bipartite graph leaves a remaining constrained bipartite vertex cover

Though the $\langle k_1, k_2 \rangle$-CBVC decision problem can be solved in $O(1.3999^{k_1+k_2} + (k_1 + k_2)(|V_1| + |V_2|))$ time [29,30], as $\langle k_1, k_2 \rangle$-CBVCE is not in FPE [18], Theorem 17 implies that $\langle t_d, k_d \rangle$-FIE is not in FPE either. However, other types of enumeration are still fixed-parameter tractable. In [17], it was shown that though enumerating all k-sized vertex covers is not in FPE, enumerating all minimal k-sized vertex covers is fixed-parameter tractable. It is easy to see the same is true for CBVCE (and hence $\langle t_d, k_d \rangle$-FIE) if we restrict to enumerating minimal signatures (maximal signatures) as defined in [18] for multi-aspect parameters.

Corollary 18. $\langle t_d, k_d \rangle$-FIE is in $IMFPE$.

Proof: The reduction in Theorem 17 can be composed with the algorithm given in [29] for solving the decision version of CBVC, yielding a $O(2^{t_d+k_d}(|I| + |T|))$ time algorithm for $\langle t_d, k_d \rangle$-FIE. ∎

This fairly naive algorithm can be readily improved with a full kernel [17]. The kernelization described in [29] for CBVC works here since we are only required to enumerate minimal signatures, hence allowing for a $O((t_d + k_d)2^{t_d+k_d} + (t_d + k_d)(|I| + |T|))$ time algorithm. Note that the 2 in the base of the exponent is necessary and this algorithm is in a sense optimal – if we consider the bipartite graph with $k_1 + k_2$ vertices in each of its bipartite sets with a single edge joining pairs of vertices across these sets, there are $2^{k_1+k_2}$ minimal signatures that must be processed.

Does further relaxing the succinctness of the enumerated sets of frequent itemsets yield still better algorithms? It is remarked in [18] that the best decision problem algorithm for CBVC can be used to enumerate a representative solution for each minimal signature, allowing the 2 in the base of the algorithm runtime exponent to be improved to 1.39. Other such improvements arise when CBVC is restricted to enumerating minimum solutions [31,30]. Though such restrictions are probably inadequate for most frequent itemset mining applications, they are suggestive of directions for future research.

An obvious objection to a CBVC-based approach to FIE is that the corresponding parameterization seems to correspond to impractical mining queries. That is, small values for these parameters correspond to large itemsets of high support, which is not what is typically desired. However, we can look to [32] where we find an analogous situation. In this case, cliques are found in graphs using a reduction to VERTEX COVER. Due to good kernelization, fairly large vertex covers can be found in practice despite a theoretical worst-case complexity bound that suggests otherwise. One of the key lessons of the paper was that good preprocessing can often lead to problem instances where input parameter values are much smaller than in the given problem instances. This is yet another direction for future research.

It is very important to note that the approaches sketched above are fundamentally different from those used in current algorithms for FIE. In combination with attention to implementation details such as the use of space-efficient data structures, these approaches may yield algorithms which perform much better than current FIE algorithms for certain types of frequent itemset mining tasks.

5   Conclusions and Future Directions

All results derived in Sections 3 and 4 as well as other results derived by applying Lemmas 2 and 3 are summarized in Table 2. These results specify the frontier of parameterized intractability with respect to the problem-aspects we consider here, at least for the most basic kinds of frequent itemset enumeration; moreover, these results (in particular, those in Section 4) suggest several new algorithmic directions for this problem. That being said, many open questions remain to be answered, some of which are briefly discussed below.

Table 2. Summary of Parameterized Results for FIE and FTIE. Note that entry "$\notin FPE[\mathbf{X}]$ in this table relative to a parameter p means that $\langle p \rangle$-FIE and $\langle p \rangle$-FTIE are not fixed-parameter enumerable because $\langle p \rangle$-FIC is not in $\mathbf{X}$ unless certain unlikely class-collapse results detailed in Section 3 occur.

	$-$	t	t_d	n
$-$	$NP\text{-}C$	$\notin FPE[\#XP]$	$\notin FPE[FFPT]$	FPE
k	$\notin FPE[\#XP]$	$\notin FPE[FFPT]$	$\notin FPE[FFPT]$	FPE
k_d	$\notin FPE[FFPT]$	$\notin FPE[FFPT]$	$\notin FPE, \in IMFPE$	FPE
l	FPE	FPE	FPE	FPE
m	FPE	FPE	FPE	FPE

One such question is the parameterized complexity of the decision version of FIE relative to the parameters for which we derived counting hardness results. Quite aside from the theoretical interest of such results, they may be useful in investigating certain restrictions on frequent itemset enumeration that cannot be formulated in counting problems. For example, consider relaxing the requirement to enumerate *all* frequent itemsets to enumerating only s such sets for a given number s. A hard decision problem would automatically imply that such a variant is hard; however, FPT results for the decision problem may give some clues on how to enumerate s frequent itemsets efficiently.

In [17] there is some discussion about an approach to parameterized enumeration that involves computing a succinct representation of the solutions for a problem from which the solutions can quickly be explicitly enumerated. Several such succinct representations (namely, closed and maximal frequent itemsets) are already being used in this manner in data mining (see [33] and references). Building on preliminary results given in Section 4, it would be interesting to know whether FIE relative to some of the hard parameters described in this paper becomes fixed-parameter tractable if these or other succinct representations are used.

Acknowledgments. This work has been supported by an NSERC USRA award (MH) and NSERC Grant 228104 (RC, TW).

References

1. Hamilton, M.: B.Sc.h Dissertation, The Parameterized Complexity of Enumerating Frequent Itemsets (2005)
2. Agrawal, R., Imielinski, T., Swami, A.N.: Mining Association Rules between Sets of Items in Large Databases. In Buneman, P., Jajodia, S., eds.: Proceedings of the 1993 ACM SIGMOD International Conference on Management of Data. (1993) 207–216
3. Zaki, M.J., Ogihara, M.: Theoretical Foundations of Association Rules. In: In Proceedings of 3rd SIGMOD'98 Workshop on Research Issues in Data Mining and Knowledge Discovery (DMKD'98). (1998)

4. Koyutürk, M., Grama, A., Szpankowski, W.: An Efficient Algorithm for Detecting Frequent Subgraphs in Biological Networks. In: ISMB/ECCB (Supplement of Bioinformatics). (2004) 200–207
5. Grossman, R., Kasif, S., Moore, R., Rocke, D., Ullman, J.: Data mining research: Opportunities and challenges. A Report of three NSF Workshops on Mining Large, Massive, and Distributed Data (1998)
6. Gunopulos, D., Khardon, R., Mannila, H., Saluja, S., Toivonen, H., Sharm, R.S.: Discovering All Most Specific Sentences. ACM Transactions on Database Systems **28**(2) (2003) 140–174
7. Lin, D.I., Kedem, Z.M.: Pincer Search: A New Algorithm for Discovering the Maximum Frequent Set. In: Proceedings of the 6th International Conference on Extending Database Technology (EDBT), Lecture Notes in Computer Science. Volume 1377. (1998) 105–119
8. Zaki, M.J., Hsiao, C.J.: Efficient Algorithms for Mining Closed Itemsets and Their Lattice Structure. IEEE Transactions on Knowlege and Data Engineering **17**(4) (2005) 462–478
9. Agrawal, R., Srikant, R.: Fast Algorithms for Mining Association Rules. In Bocca, J.B., Jarke, M., Zaniolo, C., eds.: Proceedings of the 20th International Conference on Very Large Data Bases (VLDB), Morgan Kaufmann (1994) 487–499
10. Goethals, B.: Survey on Frequent Pattern Mining (2003)
11. Zaki, M.J., Gouda, K.: Fast Vertical Mining Using Diffsets. In: Proceedings of the Ninth ACM SIGKDD International Conference on Knowledge Discovery and Data Mining, New York, NY, USA, ACM Press (2003) 326–335
12. Goethals, B., Zaki, M.J.: Advances in frequent itemset mining implementations: Introduction to fimi03. In: Proceedings of the ICDM 2003 Workshop on Frequent Itemset Mining Implementations. (2003)
13. Yang, G.: The Complexity of Mining Maximal Frequent Itemsets and Maximal Frequent Patterns. In: Proceedings of the Tenth ACM SIGKDD International Conference on Knowledge Discovery and Data Mining. (2004) 344–353
14. Wareham, H.T.: Systematic Parameterized Complexity Analysis in Computational Phonology. PhD thesis, University of Victoria (1999)
15. Seno, M., Karypis, G.: SLPMiner: An Algorithm for Finding Frequent Sequential Patterns Using Length-Decreasing Support Constraint. In: Proceedings of the 2nd IEEE International Conference on Data Mining (ICDM). (2002) 418–425
16. Arvind, V., Raman, V.: Approximation Algorithms for Some Parameterized Counting Problems. In: Proceedings of the 13th International Symposium on Algorithms and Computation (ISAAC), Lecture Notes in Computer Science. Volume 2518. (2002) 453–464
17. Damaschke, P.: Parameterized Enumeration, Transversals, and Imperfect Phylogeny Reconstruction. In: Proceedings of 1st International Workshop on Parameterized and Exact Computation (IWPEC'2004), Lecture Notes in Computer Science. Volume 3162. (2004) 1–12
18. Fernau, H.: On Parameterized Enumeration. In: Proceedings of the 8th Annual International Conference on Computing and Combinatorics (COCOON), Lecture Notes In Computer Science. Volume 2387. (2002) 564–573
19. Flum, J., Grohe, M.: The Parameterized Complexity of Counting Problems. SIAM Journal on Computing **33**(4) (2004) 892–922
20. McCartin, C.: Parameterized Counting Problems. In: Mathematical Foundations of Computer Science 2002, 27th International Symposium (MFCS 2002), Lecture Notes in Computer Science. Volume 2420. (2002) 556–567

21. Valiant, L.G.: The Complexity of Enumeration and Reliability Problems. SIAM Journal on Computing **8**(3) (1979) 410–421
22. Garey, M.R., Johnson, D.S.: Computers and Intractablity: A Guide to the Theory of NP-Completeness. Freeman (1979)
23. Downey, R.G., Fellows, M.R.: Fixed-Parameter Tractability and Completeness I: Basic Results. SIAM Journal on Computing **24**(4) (1995) 873–921
24. Downey, R.G., Fellows, M.R.: Parameterized Complexity. Springer-Verlag (1999)
25. Guo, J., Niedermeier, R.: Exact algorithms and applications for tree-like weighted set cover. Journal of Discrete Algorithms (Accepted June 2005 (to appear))
26. Gottlob, G., Grohe, M., Musliu, N., Samer, M., Scarcello, F.: Hypertree decompositions: Structure, algorithms, and applications. In: Proceedings of the 31st International Workshop on Graph-Theoretic Concepts in Computer Science (WG), Lecture Notes in Computer Science. Volume 3787. (2005) 1–15
27. Alexe, G., Alexe, S., Crama, Y., Foldes, S., Hammer, P.L., Simeone, B.: Consensus algorithms for the generation of all maximal bicliques. Discrete Applied Mathematics **145**(1) (2004) 11–21
28. Jiong G., Huffner, F., Niedermeier, R.: A Structural View on Parameterizing Problems: Distance from Triviality. In: Proceedings of the 1st International Workshop on Parameterized and Exact Computation (IWPEC 2004), Lecture Notes in Computer Science. Volume 3162. (2004) 162–173
29. Fernau, H., Niedermeier, R.: An Efficient Exact Algorithm for Constraint Bipartite Vertex Cover. In Ambos-Spies, K., et al., eds.: 38. Workshop über Komplexitätstheorie, Datenstrukturen und effiziente Algorithmen. Volume 44. (1999) 8
30. Chen, J., Kanj, I.A.: Constrained Minimum Vertex Cover in Bipartite Graphs: Complexity and Parameterized Algorithms. Journal of Computer and System Sciences **67**(4) (2003) 833–847
31. Chlebík, M., Chlebíková, J.: Improvement of Nemhauser-Trotter Theorem and Its Applications in Parametrized Complexity. In: Proceedings of the 9th Scandinavian Workshop on Algorithm Theory (SWAT), Lecture Notes in Computer Science. Volume 3111. (2004) 174–186
32. Abu-Khzam, F.N., Collins, R.L., Fellows, M.R., Langston, M.A., Suters, W.H., Symons, C.T.: Kernelization Algorithms for the Vertex Cover Problem: Theory and Experiments. In: Proceedings of the Sixth Workshop on Algorithm Engineering and Experiments and the First Workshop on Analytic Algorithmics and Combinatorics (ALENEX/ANALC). (2004) 62–69
33. Afrati, F.N., Gionis, A., Mannila, H.: Approximating a collection of frequent sets. In: Proceedings of the Tenth ACM SIGKDD International Conference on Knowledge Discovery and Data Mining. (2004) 12–19

Random Separation: A New Method for Solving Fixed-Cardinality Optimization Problems

Leizhen Cai*, Siu Man Chan, and Siu On Chan

Department of Computer Science and Engineering
The Chinese University of Hong Kong
Shatin, Hong Kong SAR, China
{lcai, smchan2, sochan2}@cse.cuhk.edu.hk

Abstract. We develop a new randomized method, *random separation*, for solving fixed-cardinality optimization problems on graphs, i.e., problems concerning solutions with exactly a fixed number k of elements (e.g., k vertices V') that optimize solution values (e.g., the number of edges covered by V'). The key idea of the method is to partition the vertex set of a graph randomly into two disjoint sets to separate a solution from the rest of the graph into connected components, and then select appropriate components to form a solution. We can use universal sets to derandomize algorithms obtained from this method.

This new method is versatile and powerful as it can be used to solve a wide range of fixed-cardinality optimization problems for degree-bounded graphs, graphs of bounded degeneracy (a large family of graphs that contains degree-bounded graphs, planar graphs, graphs of bounded tree-width, and nontrivial minor-closed families of graphs), and even general graphs.

1 Introduction

1.1 Motivations and Related Work

Many NP-hard problems, when some part of input I is taken as a fixed parameter k to form *fixed-parameter problems*, can be solved by algorithms that run in *uniformly polynomial time*, i.e., $f(k)|I|^{O(1)}$ time for some function $f(k)$. Prominent and influential examples of such algorithms include $O(n^3)$ algorithms of Robertson and Seymour [12] for solving the subgraph homomorphism and minor containment problems, an $O(n)$ algorithm of Bodlaender [3] for finding tree-decompositions of tree-width k, $O(n)$ algorithms of Courcelle [7] and Arnborg *et al.* [2] for solving problems expressible in monadic second-order logic on graphs of tree-width k, and an $O(kn + 1.286^k)$ algorithm of Chen, Kanj and Jia [6] for finding a vertex cover of k vertices. Downey and Fellows [8] have established a general framework for studying the complexity of fixed-parameter problems.

In this paper, we develop a new randomized method, called *random separation*, for designing uniformly polynomial-time algorithms to solve fixed-parameter

---

* Partially supported by a Direct Research Grant from the Chinese University of Hong Kong.

H.L. Bodlaender and M.A. Langston (Eds.): IWPEC 2006, LNCS 4169, pp. 239–250, 2006.

problems on graphs, especially fixed-cardinality optimization problems. A *fixed-cardinality optimization problem* is a problem that asks for a solution with exactly a fixed number k of elements (e.g., k vertices V') to optimize the solution value (e.g., the number of edges covered by V'), and recently Cai [4] has initiated a systematic study of fixed-cardinality optimization problems from the parameterized complexity point of view. The key idea in our random separation method is to partition the vertex set of a graph randomly into two disjoint sets to separate a solution from the rest of the graph into connected components, and then select appropriate components to form a solution. We can use universal sets to derandomize algorithms obtained from this method.

Our initial inspiration for developing the random separation method came from the colour-coding method of Alon, Yuster and Zwick [1] for solving fixed-parameter problems on graphs. The basic idea of their method is to colour vertices randomly in k colours and then try to find a *colourful k-solution*, i.e., a solution consisting of k vertices in distinct colours. To derandomize the algorithm, they use perfect hash functions. They have used colour-coding to find, for each fixed k, a k-path in $O(n)$ expected time and $O(n \log n)$ worst-case time, a k-cycle in $O(n^\alpha)$ expected time and $O(n^\alpha \log n)$ worst-cast time, where $\alpha < 2.376$, and a subgraph isomorphic to a k-vertex graph H of tree-width w in $O(n^{w+1})$ expected time and $O(n^{w+1} \log n)$ worst-case time. Unfortunately, for most fixed-parameter problems, it seems very difficult to find colourful k-solutions, which greatly limits the applicability of colour-coding.

On the other hand, it is much easier to deal with connected components, which enables us to use random separation to solve a wide range of fixed-parameter problems, especially when the input graph has bounded degree or degeneracy. In fact, it is rather surprising that we can use random separation to obtain uniformly polynomial-time algorithms for classes of fixed-parameter problems, especially fixed-cardinality optimization problems, on graphs of bounded degree.

For derandomization, our main tools are universal sets and perfect hash functions. A collection of binary vectors of length n is (n, t)-*universal* if for every subset of size t of the indices, all 2^t configurations appear. Naor, Schulman and Srinivasan [11] have a deterministic construction for (n, t)-universal sets of size $2^t t^{O(\log t)} \log n$ that can be listed in linear time. A family $\mathcal{F}$ of functions mapping a domain of size n into a range of size k is an (n, k)-*family of perfect hash functions* if for every subset S of size k from the domain there is a function in $\mathcal{F}$ that is 1-to-1 on S. Based on work of Schmidt and Siegel [13] and pointed out by Moni Naor [1], an (n, k)-family of perfect hash functions of size $2^{O(k)} \log n$ can be deterministically constructed in linear time.

Following the framework of Downey and Fellows [8], we say that a fixed-parameter problem is *fixed-parameter tractable* if it has a uniformly polynomial-time algorithm, and *fixed-parameter intractable* if it is $W[i]$-hard for some $W[i]$ in the W-hierarchy. We note that a $W[i]$-hard problem cannot be solved in uniformly polynomial time unless all problems in $W[i]$ can be solved in uniformly polynomial time.

1.2   Main Results

We focus on graphs of bounded degree or degeneracy as many fixed-parameter problems are fixed-parameter intractable for general graphs. A *degree-bounded graph* is a graph whose maximum degree is bounded by a constant d. A graph G is *d-degenerate* if every induced subgraph of G has a vertex of degree at most d. It is easy to see that every d-degenerate graph admits an acyclic orientation such that the outdegree of each vertex is at most d. Many interesting families of graphs are d-degenerate for some fixed constant d. For example, graphs embeddable on some fixed surface (planar graphs are 5-degenerate), degree-bounded graphs, graphs of bounded tree-width, and nontrivial minor-closed families of graphs.

In this paper we will demonstrate the power of our random separation method by the following results:

1. For degree bounded graphs G, we obtain uniformly polynomial-time algorithms for a wide range of fixed-parameter problems that ask us to find k vertices (edges) S to optimize a value $\phi(S)$ defined by an objective function ϕ that is computable in uniformly polynomial time (Section 3.2).
2. For every degree bounded graph G and every k-vertex graph H, we can find a maximum (minimum) weight induced (partial) H-subgraph in G in $O(n \log n)$ time for each fixed k (Section 2.2).
3. For each fixed k, we can find a subset of vertices in a general graph to cover exactly k edges in $O(m + n \log n)$ time (Section 2.4).
4. For every k-vertex tree (forest) H and fixed k and d, we can find an induced H-subgraph in a d-degenerate graph that contains one in $O(n)$ expected time and $O(n \log^2 n)$ worst-case time (Section 4).
5. For fixed k and d, we can find an induced k-cycle in a d-degenerate graph that contains one in $O(n^2)$ expected time and $O(n^2 \log^2 n)$ worst-case time (Section 4).

Furthermore, we can also use random separation to solve fixed-parameter problems on satisfiability, integer programming, set packing and covering, and many others when the input obeys some "degree" constraints, which will be discussed in the full paper.

1.3   Notation and Organization

We use $G = (V, E)$ to denote the input graph (or digraph) with n vertices and m edges. For a graph H, its vertex (edge) set is denoted by $V(H)$ (respectively, $E(H)$). For a subset V' of vertices, $N_G(V')$ denotes the *neighbourhood* of V', i.e., the set of vertices not in V' that are adjacent to some vertices in V', and $N_G^+(V')$ the *out-neighbourhood* of V', i.e., vertices not in V' that are heads of edges connected with some vertices in V'. For a subgraph H, we use $N_G(H)$ as a shorthand for $N_G(V(H))$ and $N_G^+(H)$ for $N_G^+(V(H))$. We use $d_G(v)$ to denote the degree of vertex v in G.

A subgraph in G that is isomorphic to a given graph H is an *H-subgraph*. For two vertices u and v, $d_G(u, v)$ denotes their distance in G; and for two subgraphs

H_1 and H_2 of G, $d_G(H_1, H_2)$ denotes their distance in G, i.e., $d_G(H_1, H_2) = \min\{d_G(u, v) \mid u \in V(H_1), v \in V(H_2)\}$.

In Section 2, we introduce our basic random separation method together with several working examples. In Section 3, we extend the method to solve a wide range of fixed-cardinality optimization problems for degree-bounded graphs, and in Section 4 we combine random separation with colour-coding to find induced k-vertex trees (forests) and induced k-cycles for d-degenerate graphs. We conclude the paper with a brief summary and some open problems in Section 5.

2   Random Separation

The basic idea of our random separation method is to use a random partition of the vertex set V of a graph $G = (V, E)$ to separate a solution from the rest of G into connected components and then select appropriate components to form a solution. To be precise, we colour each vertex randomly and independently by either green or red to define a random partition of V into green vertices V_g and red vertices V_r, which forms the *green subgraph* $G_g = G[V_g]$ and the *red subgraph* $G_r = G[V_r]$. For a solution S with k vertices, there is $2^{-(k+|N_G(S)|)}$ chance that a random partition has the property that S is entirely in the green subgraph G_g and its neighbourhood $N_G(S)$ is entirely in the red subgraph G_r, i.e., S consists of a collection of connected components of G_g, referred to as *green components*. For such a partition, we can usually find an appropriate collection of green components to form a required k-solution by using the standard dynamic programming algorithm for the 0-1 knapsack problem (see, for instance, the textbook of Kleinberg and Tardos [10]). Therefore, with probability $2^{-(k+|N_G(S)|)}$, we can find a required k-solution from a random partition. To derandomize the algorithm, we use (n, t)-universal sets for $t = k + |N_G(S)|$. The total time of the whole algorithm is uniformly polynomial when t is bounded by a function of k. We give several examples to illustrate this method in the rest of this section.

2.1   Dense k-Vertex Subgraphs in Degree-Bounded Graphs

Let us start with the problem of finding an induced subgraph on k vertices that contains the maximum number of edges. Let d be a fixed constant and $G = (V, E)$ a graph of maximum degree d. First we randomly colour each vertex of G by either green or red to form a random partition (V_g, V_r) of V. Let G' be a maximum k-vertex induced subgraph of G. A partition of V is a "good partition" for G' if all vertices in G' are green and all vertices in its neighbourhood $N_G(G')$ are red. Note that $N_G(G')$ has at most dk vertices as $d_G(v) \leq d$ for each vertex v. Therefore the probability that a random partition is a good partition for G' is at least $2^{-(d+1)k}$ and thus, with at least such a probability, G' is the union of some green components.

To find a maximum k-vertex induced subgraph for a good partition of G', we need only find a collection $\mathcal{H}'$ of green components such that the total number of vertices in $\mathcal{H}'$ is k and the total number of edges in $\mathcal{H}'$ is maximized. For this purpose, we first compute in $O(dn)$ time the number n_i of vertices and the number m_i of edges inside each green component H_i. Then we find a collection $\mathcal{H}'$ of green components that maximizes

$$\sum_{H_i \in \mathcal{H}'} m_i$$

subject to $\sum_{H_i \in \mathcal{H}'} n_i = k$. This can be solved in $O(kn)$ time by using the standard dynamic programming algorithm for the 0-1 knapsack problem (see [10]). Therefore, with probability at least $2^{-(d+1)k}$, we can find a maximum k-vertex induced subgraph of G in $O((d+k)n)$ time.

To derandomize the algorithm, we need a family of partitions such that for every partition Π of any $(d+1)k$ vertices into k vertices and dk vertices, there is a partition in the family that is consistent with Π. Clearly, any family of $(n, (d+1)k)$-universal sets can be used as the required family of partitions. Using a construction of Naor, Schulman and Srinivasan [11], we obtain a required family of partitions of size $2^{(d+1)k}(dk+k)^{O(\log(dk+k))} \log n$ that can be listed in linear time. Therefore we obtain a deterministic algorithm that runs in $O(f(k,d)n \log n)$ time where

$$f(k,d) = 2^{(d+1)k}(dk+k)^{O(\log(dk+k))}(d+k),$$

which is $O(n \log n)$ for fixed k and d, and uniformly polynomial for parameter k and fixed d.

2.2   Subgraph Isomorphism for Degree-Bounded Graphs

Although subgraph isomorphism problems are W[1]-hard for general graphs [8], we can use random separation to solve them easily for degree-bounded graphs, even for the weighted case. We note that the following theorem for the unweighted case also follows from a result of Seese [14] and a more general result of Frick and Grohe [9].

Theorem 1. *Let $G = (V, E; w)$, where $w : V \bigcup E \to R$, be a weighted graph (digraph) whose maximum degree is d, and H an arbitrary k-vertex graph (digraph). If G contains an induced (a partial) H-subgraph, then for fixed k and d, it takes $O(n \log n)$ time to find a maximum (minimum) weight induced (partial) H-subgraph in G.*

Proof.   We consider induced subgraph first. Let H' be a maximum (minimum) weight induced H-subgraph in G. Generate a random partition (V_g, V_r) of V. With probability at least $2^{-(d+1)k}$, each connected component of H' is a green component. For each connected component H_i of H, we find a maximum (minimum) weight H_i-subgraph H_i^* in G_g, which takes $O(k!kdn)$ time. Then with probability at least $2^{-(d+1)k}$, $\cup_{H_i \in H} H_i^*$ is a maximum (minimum) weight induced H-subgraph in G, and therefore we can solve the problem in $O(n)$ expected time for fixed k and d. We derandomize the algorithm by using $(n, (d+1)k)$-universal sets to obtain a deterministic algorithm that runs in $O(n \log n)$ time for fixed k and d.

For the partial subgraph case, we use a random partition to partition edges E into green and red edges (E_g, E_r). Let H' be a maximum (minimum) weight partial H-subgraph in G. With probability at least 2^{-kd} (note that k vertices are incident with at most kd edges), edges in H' are green and all other edges adjacent to edges in H' are red, i.e., each connected component of H' is a green component in $G[E_g]$. The rest of the arguments is the same as the induced subgraph case and is omitted.   □

2.3   Weighted Independent k-Sets in d-Degenerate Graphs

Let d be a fixed constant, and $G = (V, E; w)$ a weighted d-degenerate graph with $w : V \to R$. Consider the problem of finding a maximum weight independent k-set in G, i.e., a set of k mutually nonadjacent vertices of maximum total weight. Although the problem is W[1]-hard for general graphs [8], it is trivially solvable for d-degenerate graphs by using random separation.

First we orient edges of G so that the outdegree of each vertex is at most d, which is easily done in $O(dn)$ time. Then we generate a random partition (V_g, V_r) of V. The probability that a maximum weight independent k-set V' is entirely inside G_g and the out-neighbourhood of each vertex of V' is entirely in G_r is at least $2^{-(d+1)k}$. Therefore, with probability at least $2^{-(d+1)k}$, V' consists of sinks of G_g and thus k sinks of largest weights in G_g. We can easily find such a V' in $O(kn)$ time, and thus, with probability at least $2^{-(d+1)k}$, we can find a maximum weight independent k-set in $O((d + k)n)$ time. Again, we can derandomize the algorithm by using $(n, (d+1)k)$-universal sets to obtain a deterministic algorithm that runs in $O(n \log n)$ time for fixed k and d.

Remark 2. For fixed constants k and d, we can also use random separation to find a maximum weight induced k-matching in d-degenerate graphs in $O(n \log n)$ time, and in Section 4 we will combine with colour coding to solve several other induced subgraph isomorphism problems for d-degenerate graphs.

2.4   Covering Exactly k Edges in General Graphs

We now consider the problem of finding a set of vertices to cover exactly k edges in a general graph G. W.l.o.g., we may assume that G has no isolated vertices. Let V' be a set of vertices that cover exactly k edges. Clearly $|V'| \le k$ and thus every vertex in V' has degree at most k. Let V_k be the set of vertices of degree at most k in G and $F = G[V_k]$.

A random partition of V_k is a "good partition" for V' if all vertices in V' are green and all vertices in $N_F(V')$ are red. Since $N_F(V')$ has at most k vertices, the probability that a random partition is a good partition for V' is at least 2^{-2k}. Given a good partition for V', the problem of finding a subset of vertices to cover exactly k edges is equivalent to the problem of finding a collection $\mathcal{H}'$ of green components such that the total number of edges in G covered by vertices in $\mathcal{H}'$ is exactly k. To find such an $\mathcal{H}'$, we compute, for each green component H_i, the number e_i of edges in G covered by vertices in H_i. Since for any two green components H_i and H_j, the number of edges covered by vertices in $H_i \cup H_j$ equals $e_i + e_j$, we can obtain such a collection $\mathcal{H}'$ in $O(kn)$ time by using the standard dynamic programming algorithm for the subset sum problem (see [10]). Therefore we can solve the problem in $O(m + 4^k kn)$ expected time when G contains such a vertex cover and, using $(n, 2k)$-universal sets for derandomization, $O(m + 4^k (2k)^{O(\log k)} kn \log n)$ worst-case time, which is $O(m + n \log n)$ for each fixed k.

Remark 3. This example illustrates that random separation is also useful for finding a solution in a general graph if there is a required solution such that the total number of elements in the solution and its neighbourhood is bounded by a function of k. In the full paper, we will solve several problems with such a property, in particular, some fixed-cardinality optimization problems studied in [4].

3   Extended Random Separation

We have seen in the previous section that random separation is quite useful for solving fixed-parameter problems. In this section, we will extend our basic method to solve a large class of fixed-cardinality optimization problems on degree-bounded graphs.

A random partition (V_g, V_r) is i-*separating*, $i \geq 1$, if there is a solution S such that all vertices in S are green and all other vertices at most distance i away from S are red. We note that our basic idea in random separation is to use a 1-separating partition to separate a solution.

3.1   Maximum Dominating k-Sets for Degree-Bounded Graphs

Let us consider the problem of finding k vertices V' in a graph G of maximum degree d to dominate the maximum number of vertices, i.e., to maximize $|N_G(V')|$. First we note that the dynamic programming approach based on a 1-separating partition does not work as a red vertex can be simultaneously dominated by vertices in several green components. In fact, it seems that no i-separating partition, $i \geq 1$, enables us to use dynamic programming directly based on the information of each green component.

To solve the problem, we use 2-separating partitions together with the new idea of merging green components into clusters. Let (V_g, V_r) be a 2-separating partition of G, and V' a maximum dominating k-set such that all vertices in V' are green and all other vertices that are at most distance 2 away from V' are red. Observe that for any two green components H_1 and H_2, if $d_G(H_1, H_2) \leq 2$ then $V(H_1) \subseteq V'$ iff $V(H_2) \subseteq V'$. Therefore we can merge green components into clusters so that all vertices in a cluster are either all or none in V'.

Let G_H be the graph whose vertices are green components and whose edges correspond to pairs of green components with distance at most 2 in G. Then each connected component of G_H corresponds to a *cluster of green components*, called 2-*cgc*, whose vertices must be either all or none in V'. Furthermore, the distance in G between any two 2-cgcs is at least 3 and thus a red vertex is dominated by at most one 2-cgc. Therefore V' consists of a collection of 2-cgcs.

We can find all 2-cgcs in $O(dn)$ time by using breadth-first search. For each 2-cgc C, we compute the number $\phi(C)$ of vertices dominated by C. Then $\phi(C_1 \cup C_2) = \phi(C_1) + \phi(C_2)$ hold for any two 2-cgcs C_1 and C_2 as no vertex is simultaneously dominated by both C_1 and C_2. Therefore, given a 2-separating partition, we can use the standard dynamic programming algorithm for the (0,1)-knapsack problem based on 2-cgcs to find a maximum dominating k-set in $O(kn)$ time.

Since the probability that a random partition is 2-separating is at least $2^{-k(1+d^2)}$, our algorithm finds, with at least this probability, a maximum dominating k-set in $O((d + k)n)$ time. We can use $(n, k(1 + d^2))$-universal sets to derandomize the algorithm so that it runs in $O(n \log n)$ time for fixed k and d.

3.2   Fixed-Cardinality Optimization Problems on Degree-Bounded Graphs

The idea for solving the maximum dominating k-set problem can be generalized to form the base of our *extended random separation* method. We use a random partition (V_g, V_r) of V to obtain, with some probability p, an i-separating partition for some $i \geq 1$ that separates a solution S from the rest of the graph by i layers of red vertices.

Let G_H be the graph whose vertices are green components and whose edges correspond to pairs of green components with distance at most i in G. Then each connected component of G_H corresponds to a *cluster of green components*, called i-cgc, whose vertices must be either all or none in S. Furthermore, the distance in G between any two i-cgcs is at least $i + 1$.

We merge green components into i-cgcs. Then solution S consists of a collection of i-cgcs. We use dynamic programming based on information of i-cgcs to find an appropriate collection of i-cgcs to produce a solution and derandomize the algorithm by using $(n, t(k, d))$-universal sets for some integer $t(k, d)$.

The extended random separation method is quite powerful and can be used to solve classes of fixed-cardinality optimization problems on degree-bounded graphs that require us to find k vertices (edges) to optimize a value ϕ defined on vertices (edges).[1]

Theorem 4. *Let $k, d \in N$ be fixed constants and $G = (V, E)$ a graph of maximum degree d. Let $\phi : 2^V \to R \cup \{-\infty, +\infty\}$ be an objective function to be optimized. Then it takes $O(n^{\max\{c', c+1\}} \log n)$ time to find k vertices V' in G that optimizes $\phi(V')$ if the following conditions are satisfied:*

1. *For all $V' \subseteq V$ with $|V'| \leq k$, $\phi(V')$ can be computed in $O(g(k, d)n^c)$ time for some function $g(k, d)$ and constant $c > 0$.*
2. *There is a positive integer i computable in $O(h(k, d)n^{c'})$ time for some function $h(k, d)$ and constant $c' > 0$ such that for all $V_1, V_2 \subseteq V$ with $|V_1| + |V_2| \leq k$, if $d_G(V_1, V_2) > i$ then $\phi(V_1 \cup V_2) = \phi(V_1) + \phi(V_2)$.*

---

[1] We use $\phi(S) = -\infty$ for maximization problems and $\phi(S) = +\infty$ for minimization problems to indicate that S is not a feasible solution.

Proof. (sketch) First we generate a random partition (V_g, V_r) of V. Let V' be an optimal k-solution. Then the probability that (V_g, V_r) is an i-separating partition for V' is at least $2^{-t(k,d)}$ for $t(k,d) = k + kd \sum_{j=1}^{i} (d-1)^{j-1}$.

It takes $O(h(k,d)n^{c'})$ time to compute i, and $O(dn)$ time to find all i-cgcs of an i-separating partition. For each i-cgc C, we delete it if it contains more than k vertices. Otherwise we determine the number of vertices in C and compute $\phi(V(C))$ in $O(g(k,d)n^c)$ time. For any two i-cgcs C_1 and C_2, their distance in G is at least $i + 1$ and thus $\phi(V(C_1 \cup C_2)) = \phi(V(C_1)) + \phi(V(C_2))$ when $|V(C_1)| + |V(C_2)| \le k$. This enables us to use the standard dynamic programming algorithm for 0-1 knapsack problem to find an optimal k-solution in $O(kn)$ time. We can derandomize the algorithm by $(n, t(k,d))$-universal sets to make it run in $O(n^{\max\{c',c+1\}} \log n)$ time. $\qquad\square$

Remark 5. Theorem 4 is easily adapted for ϕ being a property of V'. Furthermore, the theorem can be generalized to problems of selecting k disjoint (induced or partial) subgraphs $S_1, S_2, \ldots, S_k$ from a degree-bounded graph (digraph) to optimize the value of an objective function defined on them, provided that the total number of vertices in all S_i's is bounded by a function of k and S_i's are *homogeneous*, for instance, all S_i's are edges, triangles, trees, or planar graphs. Further generalizations to hypergraphs are also possible, and we will discuss these issues in the full paper.

4   Combining with Colour-Coding

As demonstrated in the previous two sections, random separation is very effective in solving fixed-parameter problems on degree-bounded graphs. However, it is much more difficult to solve fixed-parameter problems on graphs of bounded degeneracy. In fact, we can show that several problems that are fixed-parameter tractable for degree-bounded graphs, including the (induced) subgraph isomorphism problem, are W[1]-hard even for 2-degenerate graphs [5].

In this section, we will combine random separation with colour-coding to solve some induced subgraph isomorphism problems for d-degenerate graphs. Let G be a d-degenerate graph whose edges have been oriented so that the outdegree of each vertex is at most d. To solve a fixed-parameter problem for G, we use random separation first to separate a k-solution from its out-neighbours and then use colour-coding for the green subgraph to locate a solution. In other words, we use $k + 1$ colours, instead of k colours in colour coding, to form a base of our randomized algorithm, where the role of the extra colour is the same as red in random separation. Surprisingly, this extra colour allows us to solve problems that seem not manageable by colour-coding or random separation alone. We can use universal sets and perfect hash functions together to derandomize the algorithm.

The following simple observation is one of the main reasons that our combined approach works for finding certain isomorphic induced subgraphs in d-degenerate graphs.

Lemma 1. *Let $\vec{G}$ be an arbitrary orientation of a graph G and H a subgraph of G. If $N_{\vec{G}}^+(v) \cap V(H) \subseteq N_H(v)$ for every vertex v in H, then H is an induced subgraph of G.*

Proof. Let $\vec{uv}$ be an edge in $\vec{G}$ between two arbitrary vertices u and v in H. Then $v \in N_{\vec{G}}^+(u) \cap V(H)$ and thus $v \in N_H(u)$. Therefore uv is an edge of H and hence H is an induced subgraph of G.   □

We start with the isomorphic induced subtree problem.

Theorem 6. *Let k and d be fixed constants, T a tree on k vertices, and $G = (V, E)$ a d-degenerate graph that contains an induced T-subgraph. Then we can find an induced T-subgraph in G in $O(n)$ expected time and $O(n \log^2 n)$ worst-case time.*

Proof. Arbitrarily choose a leaf of T as the root and define a post-order traversal of T. For convenience, we assume that vertices of T are $1, 2, \cdots, k$ following the post-order traversal. For each vertex i, let T_i denote the subtree of T rooted at i, and $p(i)$ the parent of i in T. Then $i < p(i)$ and k is the root of T.

We orient edges of G so that the outdegree of each vertex is at most d, which is easily done in $O(dn)$ time. For a $(k+1)$-colouring $c : V \to \{0, 1, 2, \ldots, k\}$ of G, an induced T-subgraph T' in G is "well-coloured" if the vertex in T' corresponding to vertex i in T has colour i and every vertex in the out-neighbourhood $N_G^+(T')$ of T' has colour 0.

Given a $(k+1)$-colouring c of V, the following algorithm finds a well-coloured induced T-subgraph T' in G if such a T' exists. To do so, we process vertices v of colour i for each i from 1 to k: roughly speaking, when there is a T_i-subgraph rooted at v, we mark v and add appropriate edges to E'.

Algorithm Iso-Tree

Step 1. Generate a $(k + 1)$-colouring $c : V \to \{0, 1, 2, \ldots, k\}$ of G as follows: produce a random partition (V_g, V_r) of V, colour all red vertices V_r by colour 0 and then randomly colour all green vertices V_g by colours in $\{1, 2, \ldots, k\}$.

Step 2. For each vertex v of colour 1, mark it if at most one vertex in $N_G^+(v)$ has colour $p(1)$, and all other vertices in $N_G^+(v)$ have colour 0.

Step 3. For each i from 2 to k, process vertices of colour i as follows. For each vertex v of colour i, mark it if the following conditions are satisfied:

   1. For each child j of vertex i in T, there is a marked vertex $u_j \in N_G(v)$ of colour j.
   2. If $i \neq k$ then at most one vertex in $N_G^+(v)$ has colour $p(i)$.
   3. All other vertices in $N_G^+(v)$ have colour 0.

   Add edge vu_j to E' when v is marked.

Step 4. If there is a marked vertex v of colour k, then E' contains the edges of an induced T-subgraph.

The correctness of the algorithm can be established by Lemma 1 and induction on i. For the running time, it is easy to see that the algorithm takes $O(dn)$ time for a given $k + 1$ colouring. The probability that an induced T-subgraph in G is well-coloured is at least $2^{-k(d+1)}k^{-k}$, and thus the expected time of the algorithm is $O(2^{k(d+1)}k^k dn)$. To derandomize the algorithm, we use $(n, k(d+1))$-universal sets and then (n, k)-perfect hash functions. Therefore the derandomized algorithm runs in $O(n \log^2 n)$ time for fixed k and d.   □

A slight modification to **Iso-Tree** enables us to extend the above theorem to T being an arbitrary k-vertex forest. We can also use the ideas in **Iso-Tree** to find induced k-cycles in graphs of bounded degeneracy.

Theorem 7. *For fixed constants k and d, we can find an induced k-cycle, if it exists, in a d-degenerate graph G in $O(n^2)$ expected time and $O(n^2 \log^2 n)$ worst-case time.*

Proof. (sketch) Basically, we use algorithm **Iso-Tree**. The idea is to use it to find well-coloured induced k-paths with the following modification: mark a vertex v of colour 1 if there are at most two vertices $x, y \in N_G^+(v)$ with $c(x) = 2$ and $c(y) = k$ and all other vertices in $N_G^+(v)$ have colour 0.

Then for each marked vertex v of colour 1, independently do a round of the marking process to find marked vertices $M_k(v)$ of colour k. If there is an edge between v and some vertex in $M_k(v)$, then we find an induced k-cycle in G.

Since we need to do $O(n)$ rounds of independent marking, the algorithm takes $O(n^2)$ expected time and thus $O(n^2 \log^2 n)$ worst-case time after derandomization.   □

Note that it is W[1]-hard to find an induced k-path (k-cycle) in a general graph [5].

Remark 8. We can easily extend Theorem 6 and Theorem 7 to deal with weighted graphs. We can also use the idea in **Iso-Tree** to find in $O(n \log^2 n)$ time an induced k-vertex tree (forest) in a d-degenerate graph, which is W[1]-hard for general graphs [5]. Furthermore, it seems possible that we can generalize the idea to find, for any k-vertex graph H of tree-width w, an induced H-subgraph in a d-degenerate graph in $O(n^w \log^2 n)$ time.

5   Concluding Remarks

We have introduced the innovative random separation method for solving fixed-parameter problems and have demonstrated its power through a wide range of such problems. It is quite surprising that this new method is much more powerful and versatile than expected, and we believe that it is a promising and effective tool for solving fixed-parameter problems. For further development, we list a few open problems for the reader to ponder.

1. We feel that the power of random separation has not been fully explored for graphs of bounded degeneracy. What kind of fixed-cardinality optimization problems can be solved for such graphs? Is it possible to obtain some general results, such as those for degree-bounded graphs, for planar graphs?

2. For degree-bounded graphs G, is random separation useful for problems that do not have the properties in Theorem 4? For instance, the problem of deleting k vertices from G to create as many components as possible.
3. For the complexity of our deterministic algorithms, is it possible to remove the $\log n$ factor? For the combined method, is there a direct way to derandomize the algorithms to reduce $\log^2 n$ to $\log n$?
4. Unless $P = NP$, it is unavoidable that functions $f(k, d)$ in the running times of most of our uniformly polynomial-time algorithms are exponential in both k and d. However, is it possible to reduce $f(k, d)$ to $c_1^k + c_2^d$ for some constants c_1 and c_2 independent of k and d?
5. Is there an $O(n)$ expected time algorithm for finding an induced k-cycle in a d-degenerate graph?
6. Finally, it will be interesting to see how fast our algorithms can run in practice. For implementation purpose, it is useful to fine tune the probability that a vertex is coloured green, and a preliminary test of our randomized algorithm for the exact vertex cover problem shows very encouraging results.

References

1. N. Alon, R. Yuster, and U. Zwick, Color-coding, *J. ACM* 42(4): 844-856, 1995.
2. S. Arnborg, J. Lagergren and D. Seese, Easy problems for tree-decomposable graphs, *J. Algorithms* 12: 308-340, 1991.
3. H.L. Bodlaender, A linear time algorithm for finding tree-decompositions of small treewidth, *SIAM J. Comput.* 25: 1305-1317, 1996.
4. L. Cai, Parameterized complexity of cardinality constrained optimization problems, submitted to a special issue of *The Computer Journal* on parameterized complexity, 2006.
5. L. Cai, S.M. Chan and S.O. Chan, research notes, 2006.
6. J. Chen, I. Kanj, and W. Jia, Vertex cover: further observations and further improvements, *J. Algorithms* 41: 280-301, 2001.
7. B. Courcelle, The monadic second-order logic of graphs I: recognisable sets of finite graphs, *Inform. Comput.* 85(1): 12-75, 1990.
8. R.G. Downey and M.R. Fellows, *Parameterized Complexity*, Springer-Verlag, 1999.
9. M. Frick and M. Grohe, Deciding first-order properties of locally tree-decomposable structures, *J. of the ACM* 48(6): 1184-1206, 2001.
10. J. Kleinberg and E. Tardos, *Algorithm Design*, Pearson, 2005.
11. M. Naor, L.J. Schulman, and A. Srinivasan, Splitters and near-optimal derandomization, *Proc. 36th Annual Symp Foundations of Computer Science*, pp. 182-191, 1995.
12. N. Robertson and P.D. Seymour, Graph minors XIII: the disjoint paths problem, *J. Comb. Ther.(B)* 63(1): 65-110, 1995.
13. J.P. Schmidt and A. Siegel, The spatial complexity of oblivious k-probe hash functions, *SIAM J. Comp.* 19: 775-786, 1990.
14. D. Seese, Linear time computable problems and first-order descriptions, *Mathematical Structures in Computer Science* 6(6): 505-526, 1996.

Towards a Taxonomy of Techniques for Designing Parameterized Algorithms

Christian Sloper and Jan Arne Telle

Department of Informatics, University of Bergen, Norway
{sloper, telle}@ii.uib.no

Abstract. A survey is given of the main techniques in parameterized algorithm design, with a focus on formal descriptions of the less familiar techniques. A taxonomy of techniques is proposed, under the four main headings of Branching, Kernelization, Induction and Win/Win. In this classification the Extremal Method is viewed as the natural maximization counterpart of Iterative Compression, under the heading of Induction. The formal description given of Greedy Localization generalizes the application of this technique to a larger class of problems.

1 Introduction

The field of parameterized algorithms continues to grow. It is a sign of its success that in the literature today there exists over twenty differently named techniques for designing parameterized algorithms[1]. Many parameterized algorithms build on the same ideas, and as a problem solver it is important to be familiar with these general themes and ideas. Several survey articles [F03, DFRS04] and books [N02, DF99] have identified common themes like Bounded Search Trees, Kernelization and Win/Win. In this paper we make an attempt at a full taxonomy encompassing all known techniques. In addition to its pedagogic value, we believe that such a taxonomy could help in developing both new techniques and new combinations of known techniques.

We classify the known techniques under the four main themes of Branching, Kernelization, Induction and Win/Win, see Figure 1. The four main themes are described in separate sections. Each technique is introduced, possibly with an example, and its placement in the taxonomy is argued for. In this extended abstract we focus on the most novel aspects, *e.g.* the grouping of Greedy Localization and Color Coding together with Bounded Search Trees under the heading of Branching algorithms, and the placement of the Extremal Method

---

[1] This includes Bounded Search Trees [DF99], Data Reduction [N02], Kernelization [DF99], The Extremal Method [FMRS00], The Algorithmic Method [P05], Catalytic Vertices [FMRS00], Crown Reductions [CFJ04], Modelled Crown Reductions [DFRS04], Either/Or [PS03], Reduction to Independent Set Structure [PS05], Greedy Localization [DFRS04], Win/Win [F03], Iterative Compression [DFRS04], Well-Quasi-Ordering [DF99], FPT through Treewidth [DF99], Search Trees [N02], Bounded Integer Linear Programming [N02], Color Coding [AYZ95], Method of Testsets [DF99], Interleaving [NR00].

H.L. Bodlaender and M.A. Langston (Eds.): IWPEC 2006, LNCS 4169, pp. 251–263, 2006.
© Springer-Verlag Berlin Heidelberg 2006

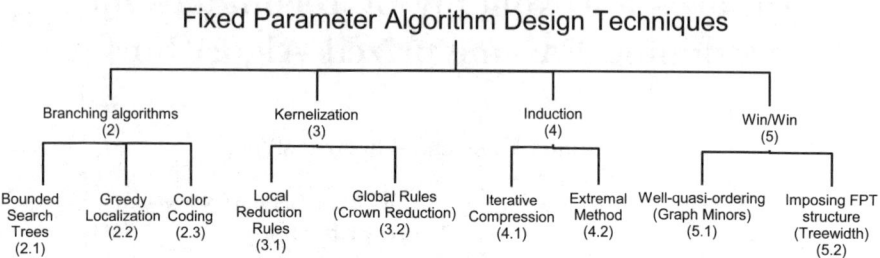

Fig. 1. The first two levels of the taxonomy, labelled by section number

as the maximization counterpart of Iterative Compression under the heading of Induction. We give a generic and formal description of Greedy Localization that encompasses problems hitherto not known to be FPT solvable by this technique.

Sometimes a technique is known under different names[2] or it is a variant of a more general technique[3]. Clearly, the fastest FPT algorithm for a problem will usually combine several techniques. In some cases such a combination has itself been given a separate name[4]. For simplicity and lack of space this extended abstract does not consider all these variations of the most general techniques, and our discussion is restricted to the first two levels of the hierarchy in Figure 1. A note on notation: the chosen parameter for a problem is usually not stated explicitly, instead we use the convention that the variable name k always denotes the parameter.

2   Branching Algorithms

We start by considering algorithm techniques that use a branching strategy to create a set of subproblems such that at least one of the subproblems is a yes-instance if and only if the original problem is a yes-instance. Techniques that create branching algorithms are Bounded Search Trees, Greedy Localization and Color Coding. For a branching algorithm to have FPT running time it suffices to require that (see [S06] for a proof):

---

[2] This is the case with Data Reduction=Local Reduction Rules, Either/Or=Win/Win, Search Trees=Bounded Search Trees and other names like Hashing [DF99]=Color Coding which do not seem to be in use anymore.

[3] This is the case with Catalytic Vertices ⊆ Local Reduction Rules, The Algorithmic Method ⊆ Extremal Method, Modelled Crown Reductions ⊆ Crown reductions, FPT by Treewidth ⊆ Imposing FPT structure, Reduction to Independent set structure ⊆ Imposing FPT structure, Method of Testsets ⊆ Imposing FPT structure.

[4] This is the case with Interleaving which combines Bounded Search Trees and Local Reduction Rules. The technique known as Bounded Integer Linear Programming can in a similar way be viewed as a combination of a Branching algorithm with Local Reduction Rules.

- each node in the execution tree has polynomial running time
- makes $\mathcal{O}((\log n)^{g(k)} f(k))$ branches at each step
- reaches a polynomial time base case in at most $h(k)$ nested calls

2.1 Bounded Search Trees

The method of Bounded Search Trees is arguably the most famous parameterized algorithm design technique and good general descriptions of it are plentiful, see for example the books [DF99, N02]. For lack of space we describe it very simply by saying that it is a branching algorithm which is strictly recursive.

2.2 Greedy Localization

Greedy Localization is a technique that uses a clever first branching to start off the recursive search for the solution. It was introduced in a paper that has since appeared in Journal of Algorithms [JZC04] and popularized in an IWPEC'04 paper [DFRS04]. Our aim is to show that if a parameterized problem satisfies the following conditions 1 and 2 then Greedy Localization will give an FPT algorithm.

1. The problem can be formulated as that of finding k pairwise non-overlapping 'objects' in an input instance G, with objects being special subsets of size depending only on k of a ground set W of G.
2. For any $R \subseteq W$ and $X \subseteq W$ we can decide in FPT time if there exists $S \subseteq W \setminus X$ such that $S \cup R$ is an object.

Not all bounded-size subsets of W are objects, and an obvious requirement for the problem to have an FPT algorithm is that for any $R \subseteq W$ we must be able to decide in FPT time if R is an object or not. Condition 2 can be seen as a strengthening of this obvious requirement and we will refer to S as an 'extension' of the 'partial object' R to a 'full object' $R \cup S$ avoiding X.

The figure on the next page gives the Greedy Localization algorithm, in non-deterministic style, for a problem satisfying these two conditions. It uses the notation that for a set of partial objects $B = \{B_1, B_2, \ldots, B_k\}$ the ground elements contained in B are denoted by $W_B = \bigcup_{B_i \in B} B_i$.

Theorem 1. *If a parameterized problem satisfies conditions 1 and 2 above then the algorithm 'GREEDY LOCALIZATION' is an FPT algorithm for this problem.*

Proof. The algorithm starts by computing an inclusion maximal non-overlapping set of objects A. By condition 2 this first step can be done in FPT time as follows: repeatedly extend the emptyset to a full object while avoiding a set X, by calling subroutine EXTEND$(\emptyset, X)$ with X initially empty, and adding the extension to X before the next iteration. When no extension exists we are assured that the sequence of extensions found must be an inclusion maximal non-overlapping set of objects A.

The crucial aspect that makes the algorithm correct is that if A and B are two inclusion maximal non-overlapping sets of objects then for any object B_i in

Algorithm. GREEDY LOCALIZATION /* non-deterministic */
Input: Instance G with ground set W, and an integer k
Output: Yes if G has k non-overlapping objects, otherwise No
compute an inclusion maximal non-overlapping set of objects A
if A contains at least k objects then **halt** and **answer** 'Yes'
else if $|W_A| < k$ then **halt** and **answer** 'No'
else guess $\{v_1, v_2, \ldots, v_k\} \subseteq W_A$ and let B_i be partial object containing v_i $(1 \le i \le k)$
BRANCHING$(B = \{B_1, B_2, \ldots, B_k\})$

Subroutine BRANCHING
Input: Set of partial objects $B = \{B_1, B_2, \ldots, B_k\}$
Output: Yes if B can be extended to a set of full objects, otherwise No
$F = S = \emptyset$
for $j = 1$ to k
   if B_j not full then {
      $S =$ EXTEND$(B_j, W_B \cup F)$
      if $S == \emptyset$ then **break**
      else $F = F \cup S$
      }
if all objects could be extended then **halt** and **answer** 'Yes'
else if $F == \emptyset$ then **halt** and **answer** 'No'
else guess $v \in F$ and add v to B_j /* j is value of parameter at break */
BRANCHING$(B = \{B_1, B_2, \ldots, B_k\})$

Subroutine EXTEND
Input: Partial object B_i, unavailable ground elements X
Output: Ground elements $S \subseteq W \setminus X$ whose addition to B_i gives a full object or
        $S = \emptyset$ if this is not possible

B there is an object A_j in A such that A_j and B_i overlap, since otherwise A is not maximal. Thus, if the instance contains such a set B of at least k objects, then we can guess k ground elements appearing in A, with A constructed in the first step of the algorithm, such that these ground elements belong to k separate objects of B (and if $|W_A| < k$ we can answer 'No'.) The Branching subroutine is called on these k one-element partial objects and we greedily try to extend them to full objects. If this fails for some object B_j, after having added extension elements F to objects $B_1, B_2, ..., B_{j-1}$, then there must exist an element v from F that should have instead been used to extend B_j. We then simply guess the element v and try again.

For a deterministic algorithm the guesses are replaced by branchings, and we give a No answer iff all branches answer No. The first branching is of size $\binom{|W_A|}{k}$, the remainder of the branches are of size $|F|$, and the total height of the tree is bounded by k times the maximum size of an object since at each level one new ground element is added to W_B. All these are bounded by a function depending on k as we assumed in condition 1 that each object had size depending on k only. The calls to the Extend subroutine are FPT by condition 2. Hence the algorithm is FPT. $\qquad\square$

For example, this implies that deciding if a graph G contains k vertex-disjoint cycles on k vertices is FPT by Greedy Localization. The ground set W will be the vertex set of G and the objects will be subsets of k vertices inducing a subgraph containing a k-cycle, to satisfy condition 1. Given $R, X \subseteq W$ we can by an FPT algorithm designed using the Color Coding technique, see the next section, decide if there exists $S \subseteq W \setminus X$ such that $R \cup S$ induces a subgraph containing a k-cycle, to satisfy condition 2. By Theorem 1 the Greedy Localization meta-algorithm therefore solves the problem in FPT time. For packing of edge-disjoint cycles a similar argument holds, with W being the edge set of the graph.

2.3   Color Coding

Color Coding is a technique that was introduced by Alon, Yuster, and Zwick in their paper 'Color Coding' [AYZ95] and is characterized by a powerful first branching step. Given an input to a parameterized graph problem we color the vertices with k colors such that the structure we are looking for will interact with the color classes in a specific way. To do this we create many branches of colored graphs, using a family of perfect hash functions for the coloring.

Definition 1. *A k-perfect family of hash functions is a family $\mathcal{H}$ of functions from $\{1, \ldots, n\}$ onto $\{1, \ldots, k\}$ such that for each $S \subset \{1, \ldots, n\}$ with $|S| = k$ there exists an $h \in \mathcal{H}$ that is bijective when restricted to S.*

Schmidt and Siegal [SS90] describe a construction of a k-perfect family of hash functions of size $2^{\mathcal{O}(k)} \log^2 n$, and [AYZ95] describes how to obtain an even smaller one of size $2^{\mathcal{O}(k)} \log n$.

The technique applies a family of perfect hash functions to partition vertices of the input graph into k color classes. By the property of perfect hash families we know that for any k-sized subset S of the vertices, one of the hash functions in the family will color each vertex in S with a different color. Thus, if we seek a k-set C with a specific property (e.g., a k-cycle), we know that if there is such a set C in the graph then its vertices will, for at least one function in the hash family, be colored with each of the k colors. The color coding technique gives an FPT algorithm whenever this colored subproblem can be solved in FPT time. Perhaps the strongest results using Color Coding are obtained in [FKNRSTW04] where it is combined with kernelization to give FPT algorithms for a large variety of problems.

A major drawback of these algorithms is that while the hash family has an asymptotically good size, the $\mathcal{O}$-notation hides a large constant. Thus, from a practical viewpoint the color coding algorithms could be slower than, for example, a $2^{k \log k}$ algorithm obtained through other techniques.

3   Kernelization

Under the heading of kernelization we combine techniques that reduce a general instance into an equivalent *kernel*, i.e. , an instance whose total size is bounded by a function depending only on the parameter. We distinguish between *local reductions* and *global reductions*.

3.1   Local Reductions

Local reduction is a well known technique. We say that a local *reduction rule* is a rule that identifies a certain constant-size structure LHS in the instance (the left-hand-side) and modifies it to RHS (the right-hand-side). This must be done in such a way that the original instance (G, k) has a positive solution iff the reduced instance (G', k') has one. The goal is to find a set of rules such that repeatedly applying the rules to an instance will either determine the answer directly or give a kernel.

3.2   Global Reduction Rules - Crown Reduction

Lately there has been a focus on reduction rules that do not follow the pattern of finding a local structure of constant size. In this section we describe reduction rules based on finding *crown decompositions* in graphs.

Definition 2. *A* crown decomposition *of a graph* $G = (V, E)$ *is a partitioning of* V *into sets* C, H, R, *where* C *and* H *are both nonempty, such that:*

1. C *is an independent set.*
2. *There is no edge between a vertex in* C *and a vertex in* R.
3. *There exists an injective map* $m : H \rightarrow C$, *such that* $m(a) = b$ *implies that* ab *is an edge. We call* ab *a matched edge if* $m(a) = b$.

When using a crown decomposition (C, H, R) in a reduction rule for a graph G we must show that we can can remove or modify $(C \cup H)$ to obtain a reduced instance (G', k') which is a Yes-instance if and only if (G, k) is a Yes-instance. For example, it is easy to see that G has a Vertex Cover of size k iff the graph G' resulting from removing $C \cup H$ has a Vertex Cover of size $k - |H|$. Usually more complicated reduced instances and arguments are necessary. For example, an FPT algorithm for k-Internal Spanning Tree [PS05] uses crown reduction rules that remove only the vertices of C not incident to a matched edge.

Although it is possible to determine if a graph has a crown decomposition in polynomial time [ACFL04], this technique is often combined with the following lemma by Chor, Fellows, and Juedes [CFJ04].

Lemma 1. *If a graph* $G = (V, E)$ *has an independent set* I *such that* $|N(I)| < |I|$, *then a crown decomposition* (C, H, R) *for* G *such that* $C \subseteq I$ *can be found in time* $\mathcal{O}(|V| + |E|)$.

The notation $N(I)$ denotes vertices in $V \setminus I$ that are adjacent to a vertex in I. Since it is W[1]-hard to find a large independent set we cannot directly apply Lemma 1. To see how the Lemma can be used, consider k-Vertex Cover on a graph G. We first compute a maximal matching M in G, and let V_M be the vertices incident to the edges in M. If $|V_M| > 2k$ then there does not exist a

Vertex Cover for G of size k. If $|V_M| \leq 2k$ and $|V \setminus V_M| \leq 2k$ then G is a kernel. Otherwise, $|V_M| \leq 2k$ and $|V \setminus V_M| > 2k$ and since $V \setminus V_M$ is an independent set with $|N(V \setminus V_M)| \leq |V_M| \leq 2k < |V \setminus V_M|$ we have, by Lemma 1, a crown decomposition of G, that can be used to reduce the graph as described above. Repeating this process gives an FPT algorithm for k-Vertex Cover.

Although crown reduction rules were independently discovered by Chor, Fellows, and Juedes [CFJ04] one should note that a similar type of structure has been studied in the field of boolean satisfiability problems (SAT). An *autarky* is a partial truth assignment (assigning true/false to only a subset of the variables) such that each clause that contains a variable determined by the partial truth assignment is satisfied. In a *matching autarky* we require in addition that the clauses satisfied and the satisfying variables form a matching cover in the natural bipartite graph description of the satisfiability problem. It is easy to see that the matching autarky is a crown decomposition in this bipartite graph. The main protagonist in this field is Oliver Kullmann [K00, K03], who has developed an extensive theory on different types of autarkies.

4   FPT by Induction

We discuss techniques closely related to mathematical induction. If we are provided a solution for a smaller instance (G, k) we can for some problems use the information to determine the solution for one of the larger instances $(G + v, k)$ or $(G, k + 1)$. We will argue that the two techniques Iterative Compression and Extremal Method are actually two facets of this inductive technique, depending on wether the problem is a minimization problem or a maximization problem.

In his book *Introduction to Algorithms* [M89] Udi Manber shows how induction can be used as a design technique to create remarkably simple algorithms for a range of problems. He suggests that one should always try to construct a solution based on the inductive assumption that we have a solution to smaller problems. For example, this leads to the well known INSERTION SORT algorithm by noting that we can sort sequences of n elements by first sorting $n-1$ elements and then inserting the last element at its correct place.

This inductive technique may also be applied to the design of FPT algorithms but more care must be taken on two accounts: (a) we have one or more parameters and (b) we are dealing with decision problems. The core idea of the technique is based on using the information provided by a solution for a smaller instance. When an instance contains both a main input and a parameter input, we must be clear about what we mean by 'smaller' instances. Let (G, k) be an instance, where G is the main input and k the parameter. We can now construct three distinctly different 'smaller' instances $(G - v, k)$, $(G, k - 1)$ and $(G - v, k - 1)$. Which one of these to use?

We first show that using smaller instances of the type $(G - v, k)$ is very suitable for minimization problems and leads to a technique known as Iterative Compression. Then we show that using smaller instances of the type $(G, k - 1)$

can be used to construct algorithms for maximization problems and is in fact the technique known as the Extremal Method.

4.1   For Minimization - Iterative Compression

In this section we present Iterative Compression which works well on certain parameterized minimization problems. Let us assume that we can inductively (recursively) compute the solution for the smaller instance $(G - v, k)$. Since our problems are decision problems, we get either a 'Yes'-answer or a 'No'-answer. In both cases we must use the information provided by the answer to compute the solution for (G, k). We must assume that for a 'Yes'-instance we also have a certificate that verifies that the instance is a 'Yes'-instance and it is this certificate that must be used to compute the solution for (G, k). However, for a 'No'-answer we may receive no extra information. A class of problems where 'No'-answers carry sufficient information is the class of *monotone* problems in which the 'No'-instances are closed under element addition. Thus if a problem is monotone we can immediately answer 'No' for (G, k) whenever $(G - v, k)$ is a 'No'-instance.

Two papers that use this type of induction on monotone graph minimization problems are [RSV03] which shows that k-Odd Cycle Cover (is it possible to delete k vertices from G to obtain a bipartite graph) is FPT, and [DFRS04] where a $2k$ kernel is given for k-Vertex Cover without using the complicated Nemhauser-Trotter results [NT75].

Note that many minimization problems are not monotone, like DOMINATING SET where the addition of a universal vertex always changes a 'No' answer to 'Yes' (unless $k \geq n$). For such problems we believe that Iterative Compression is ill suited.

4.2   For Maximization - The Extremal Method

For maximization problems we consider smaller instances of the type $(G, k - 1)$, and induct on k instead of n. We say that a problem is *parameter monotone* if the 'No'-instances are closed under parameter increment, *i.e.* if instance (G, k) is a 'No'-instance then (G, k') is also a 'No'-instance for all $k' > k$.

The *Method of Extremal Structure*[5] is a design technique that works well for parameter monotone maximization problems. In this technique we do not focus on any particular instance (G, k), but instead investigate the structure of graphs that are 'Yes'-instances for k, but 'No'-instances for $k + 1$. Let $\mathcal{G}(k)$ be the class of such graphs, i.e., $\mathcal{G}(k) = \{G \mid (G, k)$ is a 'Yes'-instance, and $(G, k + 1)$ is a 'No'-instance $\}$.

Our ultimate goal is to prove that there exists a function $f(k)$ such that $h \max\{|V(G)| \mid G \in \mathcal{G}(k)\} \leq f(k)$. This is usually not possible without some refinement of $\mathcal{G}(k)$, to do this we make a set of observations E of the following type:

---

[5] An exposition of this design technique can be found in E. Prieto's PhD thesis [P05].

Since (G, k) is a 'Yes'-instance, but $(G, k + 1)$ is a 'No'-instance, G has
property p. (1)

Given a set of such observations E and consequently a set of properties P we
try to devise a set of reduction rules R that apply specifically to large graphs
having the properties P. We call our refined class $\mathcal{G}_R(k) = \{G \mid$ no reduction
rule in R applies to (G, k), and (G, k) is a 'Yes'-instance, and $(G, k+1)$ is a 'No'-
instance $\}$. If we can add enough observations to E and reductions rules to R to
prove that there is a function $f(k)$ such that $\max\{|V(G)| \mid G \in \mathcal{G}_R(k)\} \le f(k)$
we have proven that:

If i) no rule in R applies to (G, k) and ii) (G, k) is a 'Yes'-instance and
iii) $(G, k + 1)$ is a 'No'-instance, then $|V(G)| \le f(k)$

Given such a boundary lemma and the fact that the problem is a *parameter
monotone* maximization problem a *kernelization* lemma follows, saying that 'If
no rule in R applies to (G, k) and $|V(G)| > f(k)$, then (G, k) is a 'Yes'-instance.

It is not immediately obvious that this can be viewed as an inductive process,
but we will now make this clear by presenting the *Algorithmic Method*, a version of the 'Extremal Method'. Here the 'Extremal Method' can be used as the
inductive step, going from k to $k + 1$, in an inductive algorithm.

As its base case, the algorithm decides $(G, 0)$, which is usually a trivial 'Yes'-
instance for a maximization problem. Our induction hypothesis is that we can
decide (G, k'). Then as long as $k' + 1 \le k$ we try to compute $(G, k' + 1)$. If (G, k')
is a 'No'-instance we can immediately answer 'No' for $(G, k' + 1)$ as the problem
is parameter monotone. Otherwise we can now make an algorithmic use of observations of the type defined for extremal method ((1) above). For each of the properties $p \in P$ we check if G has the property p. If G does not have property p then
since (G, k') is a 'Yes'-instance it follows that $(G, k' + 1)$ is also a 'Yes'-instance.
By the same reductions and observations (although the reader should observe that
we here require properties to be FPT time verifiable), we obtain that

If no observation in E or reduction rule R applies to $(G, k' + 1)$ then
$|V(G)| < f(k)$.

At that point we can invoke a brute force algorithm to obtain either a solution
S or a 'No'-answer for $(G, k' + 1)$. This answer for $(G, k' + 1)$ can then be used
in the next step, $k' + 2$, of our inductive algorithm.

This technique, either the 'Extremal Method' or its variant the 'Algorithmic
Method', can be applied successfully to a range of problems, such as: k-Max
Cut [P04], k-Leaf Spanning Tree [P05], k-Non-Blocker [P05], k-Edge-disjoint
Triangle-Packing [MPS04], k-$K_{1,s}$-packing [PS04], k-K_3-packing [FHRST04], k-
Set Splitting [DFR03], and k-Internal Spanning Tree [PS05].

5   Win/Win

Imagine that we solve our problem by first calling an FPT algorithm for another
problem and use both its 'Yes' and 'No' answer to decide in FPT time the

answer to our problem. Since we then 'win' if its answer is 'Yes' and we also 'win' if its answer is 'No', this is called a 'Win/Win' situation. In this section we focus on techniques exploting this behavior. According to [DH05] the only known algorithms with sub-exponential running time $\mathcal{O}^*(c^{\sqrt{k}})$ are algorithms based on imposing treewidth and branchwidth structure on the complicated cases, and these fall into the Win/Win category.

5.1   Well-Quasi-ordering and Graph Minors

Robertson and Seymour have shown that i) the set of finite graphs are well-quasi-ordered under minors and ii) the H-Minor problem that checks if H is a minor of some input graph, with $k = |V(H)|$, is FPT. These two facts suffice to prove that any parameterized graph problem whose Yes-instances (or No-instances) are closed under minors is FPT. If analogous structural results could be shown for some other relation, besides minors, then for problems closed under this other relation we would also get FPT algorithms. Thus the general technique is called 'well-quasi-ordering'. We consider this a Win/Win algorithm as we relate the problem we wish to solve to the FPT problem of checking if one of the forbidden minors (or whatever other relation is involved) appear in our problem instance.

Let us briefly explain the main ideas. A well-quasi-ordering is a reflexive and transitive ordering which has no infinite antichain, meaning that any set of elements no two of which are comparable in the ordering must be finite. A graph H is a *minor* of a graph G, denoted $H \preceq_m G$, if a graph isomorphic to H can be obtained from contracting edges of a subgraph of G. The Graph Minors Theorem [RS99] states that 'The set of graphs are well-quasi-ordered by the minor relation'. Combined with H-minor testing this can be used to prove existence of an FPT algorithm for any problem A with the property that for any k the 'Yes'-instances are closed under minors. In other words, let us assume that if A_k is the class of graphs G such that (G, k) is a 'Yes'-instance to problem A and $H \preceq_m G$ for some $G \in A_k$ then $H \in A_k$ as well. Consider the Minimal Forbidden Minors of A_k, denoted $MFM(A_k)$, defined as follows: $MFM(A_k) = \{G \mid G \notin A_k \text{ and } (\forall H \preceq_m G, H \in A_k \vee H = G)\}$.

By definition, $MFM(A_k)$ is an antichain of the $\preceq_m$ ordering of graphs so by the Graph Minors Theorem it is finite. Beware that the non-constructive nature of the proof of the Graph Minors Theorem implies that we can in general not construct the set $MFM(A_k)$ and thus we can only argue for the existence of an FPT algorithm. We do this by noting that (G, k) is a Yes-instance of problem A iff there is no $H \in MFM(A_k)$ such that $H \preceq_m G$. Since $MFM(A_k)$ is independent of $|G|$ we can therefore decide if (G, k) is a Yes-instance in FPT time by $|MFM(A_k)|$ calls of H-Minor. Armed with this powerful tool, all we have to do to prove that a parameterized graph problem is FPT is to show that the Yes-instances are closed under the operations of edge contraction, edge deletion and vertex deletion.

5.2   Imposing FPT Structure and Bounded Treewidth

In the literature on parameterized graph algorithms there are several notable occurrences of a Win/Win strategy that imposes a tree-like structure on the class of problematic graphs, in particular by showing that they must have treewidth bounded by a function of the parameter. This is then combined with the fact that many NP-hard problems are solvable in FPT time if the parameter is the treewidth of the input graph.

Let us briefly explain this technique in the case of finding k-dominating sets in planar graphs, where a very low treewidth bound on Yes-instances gives very fast FPT algorithms. In [ABFKN02] it is shown that a planar graph that has a k-dominating set has treewidth at most $c\sqrt{k}$ [ST94], for an appropriate constant c. Thus we have a win/win relationship, since we can check in polynomial time if a planar graph has treewidth at most $c'\sqrt{k}$, for some slightly larger constant c', and if so find a tree-decomposition of this width. If the treewidth is higher we can safely reject the instance, and otherwise we can run a dynamic programming algorithm on its tree-decomposition, parameterized by $c'\sqrt{k}$, to find in FPT time the optimal solution. In total this gives a $\mathcal{O}^*(c''^{\sqrt{k}})$ algorithm for deciding if a planar graph has a dominating set of size k.

A series of papers have lowered the constant c'' of this algorithm, by several techniques, like moving to branchwidth instead of treewidth, by improving the constant c, and by improving the FPT runtime of the dynamic programming stage. Yet another series of papers have generalized these 'subexponential in k' FPT algorithms from dominating set to all so-called bidimensional parameters and also from planar graphs to all graphs not having a fixed graph H as minor [DH05].

6   Conclusion

We believe that a taxonomy of techniques for designing parameterized algorithms will invariably develop over time and in this paper we made a comprehensive attempt. Many such classification schemes are possible, and the one proposed in this paper is the result of many discussions, primarily between the two authors, but also with other people in the field, see [S06]. Given the nature of such classifications and the continual development of the field we expect that this proposal will be criticized and altered over time, but it is our hope that it will lay the ground for fruitful discussions.

References

[ACFL04]   F. Abu-Khzam, R. Collins, M. Fellows and M. Langston. Kernelization Algorithms for the Vertex Cover Problem: Theory and Experiments. *Proceedings ALENEX 2004*, Springer-Verlag, *Lecture Notes in Computer Science* 3353, p 235-244(2004).

[ABFKN02]   J. Alber, H. L. Bodlaender, H. Fernau, T. Kloks, and R. Niedermeier. Fixed parameter algorithms for dominating set and related problems on planar graphs, *Algorithmica*, vol. 33, pages 461–493, 2002.

[AYZ95]   N. Alon, R. Yuster, and U. Zwick. Color-Coding, *Journal of the ACM*, Volume 42(4), pages 844–856, 1995.

[CFJ04]   B. Chor, M. Fellows, and D. Juedes. Linear Kernels in Linear Time, or How to Save k Colors in $\mathcal{O}(n^2)$ steps. *Proceedings of WG2004, LNCS*, 2004.

[DF99]   R. Downey and M. Fellows. Parameterized Complexity, *Springer-Verlag*, 1999.

[DFR03]   F. Dehne, M. Fellows, and F. Rosamond. An FPT Algorithm for Set Splitting, in *Proceedings WG2004 - 30th Workshop on Graph Theoretic Concepts in Computer science*, LNCS. Springer Verlag, 2004.

[DFRS04]   F. Dehne, M. Fellows, F. Rosamond, P.Shaw. Greedy Localization, Iterative Compression and Modeled Crown Reductions: New FPT Techniques and Improved Algorithms for Max Set Splitting and Vertex Cover, *Proceedings of IWPEC04*, LNCS 3162, pages 271–281, 2004

[DH05]   E. Demaine and M. Hajiaghayi. Bidimensionality: New Connections between FPT Algorithms and PTASs, *Proceedings of the 16th Annual ACM-SIAM Symposium on Discrete Algorithms (SODA 2005)*, January 23–25, pages 590–601, 2005.

[F03]   M.Fellows. Blow-ups, Win/Wins and Crown Rules: Some New Directions in FPT, *Proceedings WG 2003*, Springer Verlag LNCS 2880, pages 1–12, 2003.

[FKNRSTW04]   M. Fellows, C. Knauer, N. Nishimura, P. Ragde, F. Rosamond, U. Stege, D. Thilikos, and S. Whitesides. Faster fixed-parameter tractable algorithms for matching and packing problems, *Proceedings of the 12th Annual European Symposium on Algorithms (ESA 2004)*, 2004.

[FHRST04]   M.Fellows, P.Heggernes, F.Rosamond, C. Sloper, J.A.Telle, Finding k disjoint triangles in an arbitrary graph. To appear in proceedings *30th Workshop on Graph Theoretic Concepts in Computer Science (WG '04)*, Springer Lecture Notes in Computer Science, (2004).

[FMRS00]   M.R. Fellows, C. McCartin, F. Rosamond, and U.Stege. Coordinatized Kernels and Catalytic Reductions: An Improved FPT Algorithm for Max Leaf Spanning Tree and Other Problems, *Foundations of Software Technology and Theoretical Computer Science*, 2000.

[JZC04]   W. Jia, C. Zhang and J. Chen. An efficient parameterized algorithm for m-set packing, *Journal of Algorithms*, 2004, vol. 50(1), pages 106–117

[K00]   O. Kullmann. Investigations on autark assignments, *Discrete Applied Mathematics*, vol. 107, pages 99–138, 2000.

[K03]   O. Kullmann. Lean clause-sets: Generalizations of minimally unsatisfiable clause-sets, *Discrete Applied Mathematics*, vol 130, pages 209–249, 2003.

[MPS04]   L. Mathieson, E. Prieto, P. Shaw. Packing Edge Disjoint Triangles: A Parameterized View. *Proceedings of IWPEC 04*, 2004, LNCS 3162,pages 127-137.

[M89]   U. Manber. Introduction to algorithms, a creative approach, *Addison Wesley Publishing*, 1989.

[MR99]   M. Mahajan, V. Raman. Parameterizing above guaranteed values: MaxSat and MaxCut, *Journal of Algorithms*, vol. 31, issue 2, pages 335-354, 1999.

[N02]   R. Niedermeier. Invitation to Fixed-Parameter Algorithms. Manuscript, 2002, a book version was recently announced by Oxford University Press.

[NR00]   R. Niedermeier and P. Rossmanith. A general method to speed up fixed parameter algorithms, *Information Processing Letters*, 73, pages 125–129, 2000.

[NT75]   G. Nemhauser and L. Trotter Jr. Vertex Packings: Structural properties and algorithms, *Mathematical Programming*, 8, pages 232–248, 1975.

[P04]   E. Prieto. The Method of Extremal Structure on the k-Maximum Cut Problem. *Proc. Computing: The Australasian Theory Symposium (CATS 2005). Conferences in Research and Practice in Information Technology*, 2005, vol. 41, pages 119-126.

[P05]   E. Prieto. Systematic kernelization in FPT algorithm design. *PhD thesis, University of Newcastle, Australia* 2005.

[PS03]   E. Prieto, C. Sloper. Either/Or: Using Vertex Cover Structure in designing FPT-algorithms - the case of k-Internal Spanning Tree, *Proceedings of WADS 2003*, LNCS vol 2748, pages 465–483.

[PS04]   E. Prieto, C. Sloper. Looking at the Stars,*Proceedings of International Workshop on Parameterized and Exact Computation (IWPEC 04)*, LNCS vol. 3162, pages 138-149, 2004.

[PS05]   E. Prieto, C. Sloper. Reducing to Independent Set Structure — the Case of k-INTERNAL SPANNING TREE', *Nordic Journal of Computing*, 2005, vol 12, nr 3, pp. 308-318.

[RS99]   N. Robertson, PD. Seymour. Graph Minors XX Wagner's conjecture. *To appear.*

[RSV03]   B.Reed, K.Smith, and A. Vetta. Finding Odd Cycle Transversals, *Operations Research Letters*, 32, pages 299-301, 2003.

[S06]   C.Sloper. Techniques in Parameterized Algorithm Design. *PhD thesis, University of Bergen*, 2006, (http://www.ii.uib.no/ sloper).

[SS90]   J.P. Schmidt and A. Siegel. The spatial complexity of oblivious k-probe hash functions. *SIAM Journal of Computing*, 19(5), pages 775-786, 1990.

[ST94]   P. D. Seymour and R. Thomas. Call routing and the ratcatcher. *Combinatorica*, vol 14(2), pages 217 – 241, 1994.

Kernels: Annotated, Proper and Induced

Faisal N. Abu-Khzam[1] and Henning Fernau[2]

[1] Lebanese American University, Division of Comp.Sci. & Math., Beirut, Lebanon
[2] Universität Trier, FB 4—Abteilung Informatik, 54286 Trier, Germany

Abstract. The notion of a "problem kernel" plays a central role in the design of fixed-parameter algorithms. The $\mathcal{FPT}$ literature is rich in kernelization algorithms that exhibit fundamentally different approaches. We highlight these differences and discuss several generalizations and restrictions of the standard notion.

1 Introduction

A *parameterized problem* $\mathcal{P}$ is a usual decision problem together with a special entity called *parameter*. Formally, this means that the language of YES-*instances* of $\mathcal{P}$, written $L(\mathcal{P})$, is a subset of $\Sigma^* \times \mathbb{N}$. An *instance* of a parameterized problem $\mathcal{P}$ is therefore a pair $(I, k) \in \Sigma^* \times \mathbb{N}$. $\mathcal{P}$ is called *fixed-parameter tractable* if there exists an algorithm for $\mathcal{P}$ running in time $\mathcal{O}(f(k)p(\text{size}(I)))$ on instance (I, k) for some function f and some polynomial p, i.e., the question if $(I, k) \in L(\mathcal{P})$ or not can be decided in time $\mathcal{O}(f(k)p(\text{size}(I)))$. The class of all fixed-parameter tractable parameterized problems is called $\mathcal{FPT}$. Some other notions of fixed-parameter tractability are also used in the literature (e.g., imposing additional computability constraints), but the above definition is particularly fit for our discussions (especially because of the validity of Theorem 1 below).

The notion of *problem reduction* is at the very core of parameterized complexity. Generally speaking, it can be seen as a particular form of "self-reduction." However, there seem to be various definitions around that formalize the idea of a self-reduction. A quite general formulation (as can be distilled from [7, p. 39]) seems to be the following one:

Definition 1 (Kernelization). *Let $\mathcal{P}$ be a parameterized problem. A kernelization (reduction) is a function K that is computable in polynomial time and maps an instance (I, k) of $\mathcal{P}$ onto an instance (I', k') of $\mathcal{P}$ such that*

- *(I, k) is a YES-instance of $\mathcal{P}$ if and only if (I', k') is a YES-instance of $\mathcal{P}$*
- *$\text{size}(I') \leq f(k)$, and*
- *$k' \leq g(k)$ for some arbitrary functions f and g.*

To underpin the algorithmic nature of a kernelization, K may be referred to as a *kernelization reduction*. (I', k') is also called the *(problem) kernel* (of I), and $\text{size}(I')$ the *kernel size*. Of special interest are *polynomial-size kernels* and *linear-size kernels*, where f is a polynomial or a linear function, respectively. A parameterized problem that admits a kernelization is also called *kernelizable*.

H.L. Bodlaender and M.A. Langston (Eds.): IWPEC 2006, LNCS 4169, pp. 264–275, 2006.

Here, size(I) measures the size of instance I in some "reasonable" manner. In general, it is only the number of bits required to encode I (again, in some "reasonable" manner). More specific size measures are used for example with graphs, where the number of vertices is a further possibility. Observe that the number of bits n' required to store say an adjacency matrix of an n-vertex graph is $n' = \Theta(n^2)$, i.e., there is a polynomial inter-relation between both size measures.

Often, a kernelization function is provided by a set of *kernelization rules* that are to be exhaustively applied to a given instance; we then also speak about a *rule-induced kernelization*.

The notion of a problem kernel is so central to parameterized complexity because of the following well-known result [8]:

Theorem 1. *A parameterized problem is in $\mathcal{FPT}$ iff it is kernelizable.*

The proof of Theorem 1 is pretty simple; we indicate the non-trivial direction $\Rightarrow$: if each problem instance (I, k) is solved in time $f(k)(\text{size}(I))^c$, a reduction running in time $(\text{size}(I))^{c+1}$ can already solve instances with $f(k) \leq \text{size}(I)$. Otherwise, $\text{size}(I) < f(k)$, and then (I, k) can be seen as a "reduced instance."

However, the given definition of kernelization is not appropriate for several applications. We will discuss the following problems and according modifications of the definition in this paper:

(a) When proving lower bounds, the notion presented so far seems to be too strong. A slightly weaker version is more convenient to establish a simple relation between hardness of approximation results and lower bounds on problem kernels.
(b) Some "kernels" as presented in the literature do not fit into our definition, since rather a reduction to a more general problem instance is provided.
(c) Finally, some problems can be easily solved by search tree techniques through a reduction to a "master problem" like HITTING SET. However, in general the kernel result known for the master problem does not provide an immediate kernel result for the related problem one is interested in, since some ingredients of the master problem kernelization may not have counterparts in the related problem. Thus, it is interesting to investigate *induced kernels* that allow for transferring kernelization rules to a new problem.

2 Proper Kernelizations

A kernelization is a *proper kernelization* if $g(k) \leq k$ in Def. 1, i.e., we have $k' \leq k$. This additional requirement is backed by the intuition that kernelizations are meant to provide small instances (smaller than the given one), and a blow-up (even if only in the parameter) would counter this intuition. Secondly, as mentioned above, kernelizations are often described purely in terms of reduction rules. As long as $k' \leq k$ and $\text{size}(I') \leq \text{size}(I)$ (and not both inequalities may turn into equalities at the same time) is valid for each kernelization rule, rule-induced kernelizations can be rather easily seen to work in polynomial time; this can be no longer guaranteed if some rules that constitute the kernelization are not proper. Thirdly, most kernelizations that can be found in the literature are rule-induced and therefore proper. For example, there is only one kernelization among dozens presented in [11] that is not proper, namely, the one for NONBLOCKER SET, also described in [4]. Also observe that rule-induced kernelizations are a neat formalization of the heuristic idea of data reduction that is very successful in practice. Fourthly, (rule-induced) kernelizations have been quite successfully used as a means to speed up search tree algorithms, see [11] for a couple of examples. To this end, kernelizations and branchings are interleaved. If non-proper kernelizations would be used, this would at least complicate the run-time analysis of the obtained algorithm; properness is in fact at least implicitly used to show that applying the reduction rules never worsens the overall running time.

Browsing through the literature, we can often find problem kernels to be defined via proper kernelizations. A quick analysis of the proof sketch of Theorem 1 reveals that this is no loss of generality:

Corollary 1. *A parameterized problem is in $\mathcal{FPT}$ iff it is properly kernelizable.*

It would be interesting to see if also the quality of the kernelization, measured in terms of kernel size, is never worse when insisting on proper kernels. For example, in the case of NONBLOCKER SET, see [4], the best proper kernel we know of is of size $2k$, while the best non-proper (and also the best annotated kernel, see the discussion below) is of size $5/3k$.

3   Parameterized Kernelizations

Conversely, we now discuss the possibility to further generalize the notion of kernelization we had so far. Notice that in terms of a general philosophy, kernelizations can be seen as a sort of "self-reduction" of a problem instance to another instance of the same kind. But is this intuition really reflected in Definition 1? Possibly not, since the proper notion of a reduction in our case would be that of a parameterized reduction. This would mean that in Definition 1, we would weaken the requirement of a kernelization being computable in polynomial time to being computable in $\mathcal{FPT}$ time. This would render Theorem 1 completely trivial, since all work could be done by the reduction.

However, such a notion could make perfect sense from the practical point of view of data reduction: reductions "cheaper" than the final solving methodology should be always used as good preprocessing. In a sense, the customary branching rules that are employed in search tree processing in order to prefer good branches could be likewise seen as intercalated parameterized kernelizations.

When viewing branching rules as kernelizations, we are automatically lead to two further generalizations of the notion of kernelization as discussed so far: Firstly, it is quite natural to consider branchings as sorts of Turing-type reductions instead of many-one reductions. We will not discuss Turing reductions further here, since we did not find any other use of them in parameterized algorithms so far. Secondly, branching might change the nature of the problem by providing what we will call annotations in the next section. For example, when considering a branching algorithm for DOMINATING SET, putting a vertex x into the dominating set to be constructed causes that the neighborhood $N(x)$ need not be dominated, although $N(x)$ cannot be deleted, since it still might be a good idea to put some $y \in N(x)$ into the dominating set to dominate $N(y)$.

4   Annotated Kernelizations

In the literature, there are cases where kernelization results are claimed that do not fall under the notions discussed so far. For example, in [1], a "kernelization" algorithm for FACE COVER (FC) is presented. However, that reduction takes an instance of FC and produces a kernel instance of ANNOTATED FACE COVER (FCANN). While an instance of FC is just a usual plane graph (plus the parameter bounding the size of the face cover), an instance of FCANN allows annotating (marking) the vertices and faces. Furthermore, an instance of FCANN may contain loops and multiple edges. Observe: any instance of FC can be trivially seen as an instance of FCANN by considering all vertices and faces as unmarked.

We therefore propose the following definition:

Definition 2. *Let $\mathcal{P}$ and $\mathcal{P}'$ be parameterized problems. $\mathcal{P}'$ is called an* annotation *of $\mathcal{P}$ iff there is a mapping r that maps I onto I' such that for all parameter values k: (I, k) is a* YES-*instance of $\mathcal{P}$ iff (I', k) is a* YES-*instance of $\mathcal{P}'$ and r can be implemented by a deterministic finite transducer.*

Recall that a deterministic finite transducer is a deterministic finite automaton with output that may output longer strings than seen in the input. In our example, the transducer may output "unmarked" (besides the mere copying of the input) upon seeing any occurrence of a vertex or face in the input (and this should be viewed as a possible encoding of an annotated instance). Composing two deterministic finite transducers gives a deterministic finite transducer.

Proposition 1. *Let $\mathcal{P}$, $\mathcal{P}'$ and $\mathcal{P}''$ be parameterized problems. If $\mathcal{P}'$ is an annotation of $\mathcal{P}$ and if $\mathcal{P}''$ is an annotation of $\mathcal{P}'$, then $\mathcal{P}''$ is an annotation of $\mathcal{P}$.*

For example, FCANN is an annotation of FC. We could also consider the variant of FCANN that has, in addition, real-number weights $w(f) \geq 1$ associated to each face f. Then, WEIGHTED FCANN is an annotation of FCANN and hence WEIGHTED FCANN is an annotation of FC.

Not all "closely related" problems are annotations. For example, consider the problem TRIANGLE VERTEX DELETION (TVD), first investigated in [14].

Given: A graph $G = (V, E)$
Parameter: a positive integer k
Question: Is there an vertex set $C \subseteq V$ with $|C| \leq k$ whose removal produces a graph without triangles as vertex-induced subgraphs?
We can abuse earlier results [10] of ours to see:

Corollary 2. TVD *can be solved in time $\mathcal{O}((\text{size}(V(G)))^3 + 2.1788^k)$, given G.*

Namely, we can translate every TVD instance (G, k) into a 3-HITTING SET (3-HS) instance by mapping every triangle of G onto a hyperedge of the constructed hypergraph. However, doing this could take cubic time (for example when dealing with a complete graph K_n). This is surely not an annotation (with respect to the usual representation(s) of a graph). However, we could also consider graph representations that explicitly list all triangles in a graph. This shows that the notion of annotation is dependent on the chosen representation of the instance. However, on the very low level, a problem would also fix the coding of the instances (although usually not explicitly given), since only then problems would become formal languages and complexity classes would correspond to formal language classes. In this sense, Def. 2 is sound.

How do we get the running time claimed in Cor. 2? After having translated the TVD instance to the corresponding 3-HS instance, we can run the known kernelization procedure for 3-HS (again in cubic time when measured against the number of vertices, see [16]) and then the search tree algorithm exhibited in [10]. Observe that we cannot claim to have yet a kernel for TRIANGLE VERTEX DELETION, but still we get the "additive parameterized complexity" seen to be typical for $\mathcal{FPT}$ algorithms involving kernelization. We will see in the following section that with additional assumptions it is even possible to inherit a kernelization algorithm for TVD from 3-HS.

We can say that the parameterized problem $\mathcal{P}$ possesses a (proper) annotated kernel if there is an annotation of $\mathcal{P}'$ that possesses a (proper) kernel. Due to the very restricted character of operation we allow for annotation, also the size measures and the running times immediately translate from the known kernelization result for $\mathcal{P}'$ to annotated kernelization results for $\mathcal{P}$. We will study an example (namely again TVD) in more detail in the following section.

Let us turn to another example, namely nonblocker set (NB): Given a graph $G = (V, E)$ and a positive integer k, is there a *nonblocker set* $N \subseteq V$ with $|N| \geq k$? In [4], it was recently shown how to produce a kernel of size $5/3k + 3$. Without giving the reduction rules here, we like to mention that in fact we provided a kernel of size $5/3k$ for the annotation where, in addition, a vertex d is specified and we require that $d \notin N$.

In that particular example, we could produce a reduction rule that gets rid of the annotation (called catalyzation in [4]) at the expense of introducing three more vertices to the kernel. So, this gives us an example where a kernel for an annotated version could be used to produce a kernel for the original version of the problem. It would be interesting to see if there are more examples along this venue or if the notion of an annotated kernel should be standing as a notion on its own right.

5   Induced Kernels

The main motivation for this section is to provide a framework that allows to transfer kernelization results (in the classical sense) from one problem to another. A further motivation for studying linear-size induced kernels would be the possibility to use exponential space in order to improve on the running times. Let us continue our discussion of the relationship between TVD and 3-HS as a running example. Let us first repeat the reduction rules for 3-HS from [16] in a slightly modified (and corrected) form:

1. (hyper)edge domination: A hyperedge e is *dominated* by another hyperedge f if $f \subset e$. In that case, delete e.
2. tiny edges: Delete all hyperedges of degree one and place the corresponding vertices into the hitting set (reducing the parameter accordingly).
3. vertex domination: A vertex x is *dominated* by a vertex y if, whenever x belongs to some hyperedge e, then y also belongs to e. Then, we can simply delete x from the vertex set and from all edges it belongs to.
4. too many edges containing a vertex pair: If $\{x, y\}$ is a pair of vertices such that the number of edges that contain both x and y exceeds k, then add an edge $\{x, y\}$.
5. too many edges containing a vertex: If x is a vertex such that the number of edges that contain x exceeds k^2, then add an edge $\{x\}$.

Notice that the last two rules only seemingly increase the size of the instance, since immediately one of the first two rules would be triggered. Hence, we could reformulate them as follows:

4′ too many edges containing a vertex pair: If $\{x, y\}$ is a pair of of vertices such that the number of edges that contain both x and y exceeds k, then delete all edges containing both x and y and replace them by an edge $\{x, y\}$.
5′ too many edges containing a vertex: If x is a vertex such that the number of edges that contain x exceeds k^2, then delete all edges containing x and reduce the parameter by one.

Now, if none of the above rules applies to a 3-HS instance, the following cutting rule applies:

- If the instance has more than k^3 hyperedges, then NO.

The cutting rule implies that there is a $\mathcal{O}(k^3)$ rule-induced kernel for 3-HS. Annotated kernels. One immediate problem with the idea of translating these rules into rules for TVD is that only hyperedges with exactly three vertices have an interpretation as "triangles" in the TVD "world." This is still true when observing that rules 4′ and 5′ together with the cutting rule are sufficient to provide the known kernelization result for 3-HS, since rule 4′ might introduce hyperedges of size two even if the original instance only contains hyperedges of size three (this would correspond to a 3-HS instance obtained from a given TVD instance). One way out of this sort of dilemma might be a specific form of annotation of graphs: in this sense, an ANNOTATED TVD instance would consist of specifying a graph $G = (V, E)$, a parameter k and another graph $G' = (V, E')$ specified by E', and the question would be to find a vertex set C, $|C| \leq k$ that is a vertex cover of G' and whose removal destroys all triangles of G. What would be the kernelization rules for ANNOTATED TVD?

1. Delete vertices of degree zero in G that are not contained in edges from E'.
2. If x is a vertex of degree one in G, then delete the incident edge from E.
3. If $\{x, y\}$ is a pair of of vertices such that the number of triangles in G that contain both x and y exceeds k, then move the edge $\{x, y\}$ from E into E'. (Hence, we "remove" the triangles containing x and y from E.)
4. If x is a vertex such that the number of triangles that contain x exceeds k^2, then delete all edges containing x from E and from E' and reduce the parameter by one.

Now, if none of the above rules applies to the ANNOTATED TVD instance, the following <u>cutting rule</u> applies:

- If the reduced instance has more than k^3 triangles, then NO.

Recently, a very similar kernelization for TRIANGLE VERTEX DELETION was proposed in [6], including suggesting a proper de-annotation rule. However, the kernel obtained that way is not a subgraph of the original graph, in contrast to our construction given below.

Another case where an annotated kernel could be smaller than a non-annotated kernel is EDGE DOMINATING SET, as presented in another paper of second author in these proceedings.

<u>Induced kernels</u>. A subtle drawback of the given kernelization for 3-HS is the fact that a reduced instance is obtained by changing some edges without deleting their vertices. In other words, the kernel is a subgraph that is not vertex-induced. This complication may seem harmless in general, but could be serious when the HS instance is used to model problems in such a way that certain forbidden relations between the objects (vertices) translate to edges in the input to HS.

We offer a new kernelization algorithm to remedy this situation at the cost of more involved reduction rules. We will also use the tiny edge rule and isolated vertex rule to obtain the kernel. However, we cannot use the hyperedge domination rule, since this rule destroys inducedness. Whenever we face an instance of 3-HITTING SET, we define: Let $F = \{x \in V \mid x$ shares more than k edges with some y in $V\}$ and let S be the complement of F in V. Elements of F and S will be referred to in the sequel as *fat* and *slim* vertices, respectively. The set of (hyper-)edges E can be partitioned as follows:

$$E_{j*s} = \{e \in E \mid |e| = 3 \wedge |e \cap S| = j\} \quad \text{for } j = 0, 1, 2, 3; \text{ and}$$

$$E_p = \{e \in E \mid |e| = 2\}.$$

Observe that edges that only contain one vertex can be dealt with by using the tiny edge rule, so that in fact we can assume that $E = \bigcup_{j=0}^3 E_{j*s} \cup E_p$.

Let us define the *co-occurrence* of a pair $\{x, y\}$ of vertices to be the number of edges that "contain" the two vertices simultaneously. Denote by $co(x, y)$ the co-occurrence of $\{x, y\}$. We say that x, y co-occur iff $co(x, y) > 0$. In the rules listed below, whenever we say that we put x into the hitting set, then this means that we reduce the parameter by one and delete x and all edges containing x from the instance. Denote by H the target hitting set. In order to justify the actions taken by the following reduction rules, we shall assume the input is a YES-instance (so, $|H| \leq k$).

1. If x is a vertex that occurs more than k times in edges from E_p, then put x into the hitting set.

   The soundness of this first rule is obvious. If x is not in H, then its (more than k) neighbors in E_p are all needed to cover the edges of E_p.

2. If x is a slim vertex of degree larger than k^2, then put x into the hitting set.

   If x does not belong to H, then the degree of x must be bounded above by k^2. Otherwise, x appears more than k times with an element of H (which violates the definition of S). Therefore this second rule is sound.

3. If x is a fat vertex that appears more than k times with more than k different other fat vertices, then put x into the hitting set.

   To see this, note that if $co(x, y) > k$, then at least one element of $\{x, y\}$ must be in H. It follows that if x is excluded from H, then more than k elements (co-occurring more than k times with x) would be needed in H, a contradiction to the assumption that the input is a YES-instance.

4. If x is a fat vertex that belongs to more than k^2 edges of E_{2*s}, then put x into the hitting set.

   If x were not in H, then its edges in E_{2*s} must be hit by slim vertices only. Since it appears in more than k^2 such edges, we conclude that a slim neighbor of s (in E_{2*s}) must have more that k common edges with x, contradicting the definition of slim vertices.

Consider the simple graph G_F, constructed as follows: (i) $V(G_F) = F$, and (ii) $E(G_F) = \{(u, v) : co(u, v) > k\}$. Then G_F must have a vertex cover of size k or less. It follows that, after applying rule 3 above, F contains at most $k^2 + k$ vertices. A similar argument shows that the total number of vertices that appear in E_p is bounded above by $k^2 + k$. Moreover, every vertex appears in at most k^2 edges of $E_{2*s} \cup E_{3*s}$. Therefore, unless we have a NO instance, the number of edges in $E_{2*s} \cup E_{3*s}$ is bounded above by k^3 (H has at most k vertices, each appearing in at most k^2 edges). So, it remains to find an upper bound on the number of edges (or slim vertices) appearing in E_{1*s}. To do this, we partition E_{1*s} further. Let $E_{1*s,<} = \{\{x, y, z\} \in E_{1*s} \mid x, y \in F$ and $co(x, y) \leq k$ in $E\}$. Moreover, let $E_{1*s,>} = E_{1*s} \setminus E_{1*s,<}$.

- If x is a vertex that occurs more than k^2 times in edges of $E_{1*s,<}$, then put x into the hitting set.

   Let x be any vertex that belongs to more than k^2 edges in $E_{1*s,<}$. If $x \notin H$, then x would share more than k edges (in $E_{1*s,<}$) with another vertex, violating the definitions of $E_{1*s,<}$ or S (depending on whether x is fat vertex appearing more than k times with another fat vertex in $E_{1*s,<}$ or x is slim one appearing more than k times with another vertex). Therefore, $E_{1*s,<}$ has at most k^3 edges.

- If x is a slim vertex that occurs in edges of $E_{1*s,>}$ and does not appear elsewhere in E, then delete x.

   Any such slim vertex would be dominated. Every edge that contains such a vertex is guaranteed to be covered by some fat vertex.

Finally, we have showed that, after applying our reduction rules, the number of edges containing slim vertices is bounded above by $2k^3$ (k^3 in $E_{1*s,<}$ and k^3 in $E_{2*s} \cup E_{3*s}$). Therefore. the number of slim vertices is in $O(k^3)$ (due to the vertex domination rule, the upper bound is $3k^3$). Since the number of fat vertices is quadratic in k, we can conclude:

Theorem 2. 3-HITTING SET *admits a vertex-induced problem kernel of size* $O(k^3)$, *measured both in the number of vertices and in the number of edges.*

It is tedious but possible to generalize our approach to obtain induced kernels for d-HS for any $d > 2$.

Corollary 3. TVD *admits a problem kernel of size* $O(k^3)$.

Proof. (Sketch) Since all rules presented above only delete vertices or conclude that we face a NO-instance, the rules can be immediately interpreted as TVD reduction rules.

We have used a similar translation to provide the first ever published small-size kernel for ONE-LAYER PLANARIZATION, a problem arising in the area of graph drawing, see [13].

The reader might have noticed that we have not yet given a proper general definition of what we might mean by "induced kernels." This notion is quite clear for (hyper)graphs (in particular for vertex set minimization problems, vertex-induced would be the appropriate formalization), but less clear otherwise.

To provide our general notion of induced kernels, we consider only $\mathcal{FPT}$ problems $\mathcal{P}$ that are *naturally-parameterized*, meaning the following:

- $\mathcal{P}$ is parameterized only by its target solution size k, and
- every instance I of $\mathcal{P}$ contains a set of distinguished elements that qualify for membership in the solution set. Such elements form what we call the *candidate set* and what we denote $\kappa(I)$.
- Finally, we require that $k \leq |\kappa(I)| \leq \text{size}(I)$.

In the case of HITTING SET, the vertex set would be the candidate set. Let $\mathcal{P}$ be a naturally-parameterized $\mathcal{FPT}$ problem and (I, k) be an instance of $\mathcal{P}$. A kernelization K for $\mathcal{P}$ is said to *produce induced kernels* if, on (I, k), K outputs:

1. the reduced instance (I', k') with $k' \leq k$;
2. a *partial solution* set $S \subseteq \kappa(I)$ with $\kappa(I') \cap S = \emptyset$ of size $k - k'$ consisting of elements of the candidate set that are selected into the solution;
3. a *trash* set $T \subseteq \kappa(I)$ with $(\kappa(I') \cup S) \cap T = \emptyset$.

In the 3-HS example, the problem is naturally-parameterized and the candidate set is the set of vertices. Also note that our vertex-induced kernel of 3-HS does not entail any direct deletion of edges. In other words, constraints that relate a set of candidates will be deleted (automatically) only when the candidates are removed. In the case of VC, the Nemhauser-Trotter reduction explicitly provides the sets S and T, see [11]. This generalized concept of induced kernels yields an important relationship between hardness of approximation and *hardness of kernelization.*

Theorem 3. *Let $\mathcal{P}$ be a minimization problem. If the naturally-parameterized version of $\mathcal{P}$ has a kernelization algorithm that produces, for any instance (I, k), an induced kernel (I', k') such that $\text{size}(I') \leq \alpha \cdot k'$ for some constant $\alpha > 1$, then $\mathcal{P}$ has an approximation algorithm with ratio α.*

Proof. Let I be an instance of $\mathcal{P}$ for which we seek an optimum solution. Let K be a kernelization algorithm that satisfies the statement of this theorem. The corresponding approximation algorithm $A_K(k)$ works as follows: (1) produce $I'(k) = K(I, k)$; let $S(k)$ be the partial solution and $T(k)$ the trash set obtained as a by-product of applying K; and (2) output $S(k) \cup \kappa(I'(k))$. Notice that the output of $A_K(k)$ is a feasible solution (by the definition of an induced kernel, since elements of T are excluded from some solution that contains S).

Let $S_{opt} \subseteq \kappa(I)$ be an optimum solution of the instance I. Then,

$$\frac{|S(k) \cup \kappa(I'(k))|}{|S_{opt}|} \leq \frac{k - k' + \alpha \cdot k'}{|S_{opt}|} \leq \frac{\alpha \cdot k}{|S_{opt}|}$$

This is not yet quite our result, but we can iterate $A_K(k)$ for all $k = 0, \ldots, |\kappa(I)|$. The parameter k that yields the smallest value of $|S(k) \cup \kappa(I'(k))|$ surely satisfies $|S(k) \cup \kappa(I'(k))| \leq |S(k_{opt}) \cup \kappa(I'(k_{opt}))|$, where $k_{opt} = |S_{opt}|$. Hence, an algorithm A_K that finally outputs the smallest solution found by any of the $A_K(k)$ is an α-approximation. Notice that the overall algorithm runs in polynomial time, since K does so and $|\kappa(I)| \leq \text{size}(I)$.

It was shown in [5] that VERTEX COVER is hard to approximate with a ratio bound of ≈ 1.36. Combining this result with Theorem 3 yields the following.

Corollary 4. *Unless $\mathcal{P} = \mathcal{NP}$, we can claim: for any $\epsilon > 0$, VERTEX COVER does not have a $(1.36 - \epsilon)k$ induced kernel.*

6 Kernelization Schemes

In the design of approximation algorithms, it has been quite popular to design *algorithm schemes*, i.e., families of algorithms that, in the case of approximation, provide better and better approximation guarantees, at the expense of higher and higher running times. In the case of approximation algorithms, such schemes were called *approximation schemes*. Is it possible to translate this idea into the realm of parameterized algorithmics?

To this end, let us turn to LINEAR ARRANGEMENT (LA), see [12,15,18] for a treatment from a parameterized perspective. Given a graph $G = (V, E)$ and a positive integer k, is there a one-to-one mapping $\sigma : V \rightarrow \{1, \ldots, |V|\}$ such that $\sum_{\{u,v\} \in E} |\sigma(u) - \sigma(v)| \leq k$? It is easily seen that LA is fixed parameter tractable, based on the following observation cast in the form of reduction rules that immediately give a problem kernel:

Rule 1. *Let (G, k) be an instance of LA. (a) If v is an isolated vertex, then reduce to $(G - v, k)$. (b) If e is an isolated edge, then reduce to $(G - e, k)$.*

Namely, the endpoints of an isolated edge can be arbitrarily arranged.

Rule 2. *If (G, k) is an instance of* LA *and if $G = (V, E)$ is reduced with respect to Rule 1, then return* NO *if $|V| > 1.5k$.*

Proposition 2. *A reduced instance of* LA *has at most $1.5k$ vertices.*

Our reduction rules can be generalized: optimal arrangements for all graphs up to $q - 1$ vertices may be precomputed (taking exponential time, measured against q) and stored in a huge table; this shows that each component then has at least q vertices. With an appropriate reduction rule, we may deduce:

Proposition 3. *For fixed q, a reduced instance of* LA *has at most $\frac{q}{q-1}k$ vertices.*

This certainly has the flavor of an algorithmic scheme: at the expense of larger and larger running times, we may get smaller and smaller kernels (let q grow). Notice however that it is not at all clear whether the bound k is sharp is some sense. This (still) contrasts the analogous idea of approximation schemes. We have encountered a similar scheme for other problems, as well, e.g., for POSITIVE WEIGHTED COMPLETION OF AN ORDERING, see [9]. Finally, in [2] generalizations of reduction rules for PLANAR DOMINATING SET have been presented; unfortunately, it is unknown if their so-called data reduction scheme is really providing smaller kernels. It would be interesting to see examples of problems that admit kernelization schemes that actually "converge" to a provable lower bound.

7   Conclusions

We have tried to raise some questions concerning one of the very basic notions of parameterized complexity theory: namely, kernels. This is rather a conceptual paper that classifies kernelization strategies found in practice than a paper presenting many mathematical results. However, we hope to stir discussions of the notions suggested in this paper. We also raised many open questions regarding the notions. The most important conceptual ones are:

Is "properness" a real restriction for kernel sizes?
Does "annotation" really give more flexibility?
What are good "master problems" to work on induced kernels?
Can we get lower bound results on kernel sizes stronger than what was presented in [3] or in this paper, or possibly with fewer assumptions?
Are there good examples of kernelization schemes (with sharp lower bounds)?

More generally speaking, the relationships between parameterized complexity and approximability need further research.

Acknowledgment. We are grateful for discussions with Peter Shaw.

References

1. F. Abu-Khzam and M. Langston. A direct algorithm for the parameterized face cover problem. In *International Workshop on Parameterized and Exact Computation IWPEC 2004*, volume 3162 of *LNCS*, pages 213–222. Springer, 2004.
2. J. Alber, B. Dorn, and R. Niedermeier. A general data reduction scheme for domination in graphs. In *Software Seminar SOFSEM*, volume 3831 of *LNCS*, pages 137–147. Springer, 2006.
3. J. Chen, H. Fernau, I. A. Kanj, and Ge Xia. Parametric duality and kernelization: Lower bounds and upper bounds on kernel size. In *Symposium on Theoretical Aspects of Computer Science STACS 2005*, volume 3404 of *LNCS*, pages 269–280. Springer, 2005.
4. F. Dehne, M. Fellows, H. Fernau, E. Prieto, and F. Rosamond. NONBLOCKER: parameterized algorithmics for MINIMUM DOMINATING SET. In *Software Seminar SOFSEM*, volume 3831 of *LNCS*, pages 237–245. Springer, 2006.
5. I. Dinur and S. Safra. On the hardness of approximating minimum vertex cover. *Annals of Mathematics*, 162:439–485, 2005.
6. M. Dom, J. Guo, F. Hüffner, R. Niedermeier, and A. Truß. Fixed-parameter tractability results for feedback set problems in tournaments In *Conference on Algorithms and Complexity CIAC*, volume 3998 of *LNCS*, pages 320–331. Springer, 2006.
7. R. G. Downey and M. R. Fellows. *Parameterized Complexity*. Springer, 1999.
8. R. G. Downey, M. R. Fellows, and U. Stege. Parameterized complexity: A framework for systematically confronting computational intractability. In *Contemporary Trends in Discrete Mathematics: From DIMACS and DIMATIA to the Future*, pages 49–99. 1999.
9. V. Dujmović, H. Fernau, and M. Kaufmann. Fixed parameter algorithms for one-sided crossing minimization revisited. In *Graph Drawing, 11th International Symposium GD 2003*, volume 2912 of *LNCS*, pages 332–344. Springer, 2004.
10. H. Fernau. A top-down approach to search-trees: Improved algorithmics for 3-Hitting Set. Technical Report TR04-073, Electronic Colloquium on Computational Complexity ECCC, 2004.
11. H. Fernau. *Parameterized Algorithmics: A Graph-Theoretic Approach*. Habilitationsschrift, Universität Tübingen, Germany, 2005.
12. H. Fernau. Parameterized algorithmics for linear arrangement problems. In *CTW 2005: Workshop on Graphs and Combinatorial Optimization*, pages 27–31. University of Cologne, Germany, 2005. Long version submitted.
13. H. Fernau. Two-layer planarization: improving on parameterized algorithmics. *Journal of Graph Algorithms and Applications*, 9:205–238, 2005.
14. J. Gramm, J. Guo, F. Hüffner, and R. Niedermeier. Automated generation of search tree algorithms for hard graph modification problems. *Algorithmica*, 39:321–347, 2004.
15. G. Gutin, A. Rafiey, S. Szeider, and A. Yeo. The linear arrangement problem parameterized above guaranteed value. In *Conference on Algorithms and Complexity CIAC*, volume 3998 of *LNCS*, pages 356–367. Springer, 2006.
16. R. Niedermeier and P. Rossmanith. An efficient fixed-parameter algorithm for 3-Hitting Set. *Journal of Discrete Algorithms*, 1:89–102, 2003.
17. R. Niedermeier and P. Rossmanith. On efficient fixed parameter algorithms for weighted vertex cover. *Journal of Algorithms*, 47:63–77, 2003.
18. M. Serna and D. M. Thilikos. Parameterized complexity for graph layout problems. *EATCS Bulletin*, 86:41–65, 2005.

The Lost Continent of Polynomial Time: Preprocessing and Kernelization

Michael R. Fellows

School of Electrical Engineering and Computer Science
University of Newcastle, University Drive, Callaghan NSW 2308, Australia
mfellows@cs.newcastle.edu.au

Abstract. One of the main objectives of the talk is to survey the history of the practical algorithmic strategy of *preprocessing* (also called *data-reduction* and *kernelization*) since the beginnings of computer science, and to overview what theoretical computer science has been able to say about it.

Parameterized complexity affords the subject of preprocessing (kernelization) a central place *via* the (trivial) lemma that states:

Lemma. *A parameterized problem Π is* fixed-parameter tractable *if and only if there is a transformation τ from Π to itself, that takes an instance (x, k) to an instance (x', k') where:*

(1) (x', k') is a yes-instance if and only if (x, k) is a yes-instance,
(2) $|x'| \leq g(k)$ for some function g associated to τ,
(3) $k' \leq k$,
(4) τ runs in polynomial time, that is, in time polynomial in $|(x, k)|$.

In the situation described by the lemma, we say that Π can be *kernelized* to instances of size $g(k)$. The lemma is built around a transformation τ that is many:1. This can be generalized to a notion of P-time Turing kernelization. If the parameterized problem Π is solvable in time $O(f(k)n^c)$, then the lemma provides only a P-time kernelization bound of $g(k) = f(k)$. Hence, membership in FPT generally insures only an exponential kernelization.

Many parameterized problems admit P-time kernelization bounds $g(k)$ where g is a polynomial, or even linear function of k. Sometimes, the bounds are stated in terms of other instance measures than total size, for example, a VERTEX COVER instance (G, k) can be kernelized to an instance (G', k') where G' has at most $2k$ vertices. Another avenue for generalization is therefore to consider kernelization as a P-time transformation that bounds one parameter in terms of another (the overall input size, the number of vertices or edges, the treewidth, etc.).

Pre-processing is a humble strategy for coping with hard problems, almost universally employed. It has become clear, however, that far from being trivial and uninteresting, that pre-processing has unexpected practical power for real-world input distributions, and is mathematically a much deeper subject than has generally been understood. It is almost impossible to talk about pre-processing

H.L. Bodlaender and M.A. Langston (Eds.): IWPEC 2006, LNCS 4169, pp. 276–277, 2006.

in the classical complexity framework in any sensible and interesting way, and the historical relative neglect of this vital subject by theoretical computer science may be related to this fact.

Here is the difficulty. If my problem Π is NP-hard, then probably there is no P-time algorithm to *solve* the problem, that is, to completely dispose of the input. If you suggested that perhaps I should settle for a P-time algorithm that instead of completely disposing of the input, at least simplifies it by getting rid of, or reducing away, the easy parts — then this would seem a highly compelling suggestion. But how can this be formalized? The obvious first shot is to ask for a P-time algorithm that reduces the input I to an input I' where $|I'| < |I|$ in a way that loses no essential information (i.e., trades the original input for smaller input, which can be called *data reduction*). The difficulty with this "obvious" formalization of the compelling suggestion is that if you had such a P-time data reduction algorithm, then by repeatedly applying it, you could dispose of the entire input in polynomial time, and this is impossible, since Π is NP-hard. Thus, in the classical framework, an effort to formulate a mathematically interesting program to explore polynomial-time preprocessing immediately crashes.

In the parameterized complexity framework, however, such a program can be formulated in an absolutely interesting and productive way. The effectiveness of P-time preprocessing is measured against the structure represented by the parameter. You might reasonably call the subject of FPT kernelization the *Lost Continent of Polynomial Time* (a *lost continent* being something that is large and interesting, that "should have been" explored long ago, and that was somehow overlooked).

In the last few years, a number of researchers have made important, pioneering investigations of subclasses of FPT that have stronger definitional claims on capturing *practical* fixed-parameter tractability. The effectiveness of kernelization offers another approach to such exploration of the internal structure of FPT, for example, the subclasses

$$lin(k) \subseteq poly(k) \subseteq FPT$$

of parameterized problems that admit problem kernels of size linear (or polynomial) in k. The talk will survey some of the important open problems that attend this perspective on the structure of FPT.

The challenges of finding effective P-time kernelization algorithms for parameterized problems in FPT seems to offer a rich combinatorial landscape for novel strategies with strong payoffs for practical computing. Beyond such concrete challenges, there also seem to be opportunities for developing systematic methodologies. Essentially, the issue seems to be the development of *P-time extremal structure theory* relating a source parameter to a target parameter (which might be, as in the lemma, the overall instance size), modulo polynomial-time processing.

The talk will survey some of the intriguing concrete open problems and new approaches in this area.

FPT at Work:
Using Fixed Parameter Tractability to Solve Larger Instances of Hard Problems*

Frank Dehne

School of Computer Science, Carleton University, Ottawa, Canada
http://www.dehne.net

When dealing with hard computational problems (NP-complete or worse), Scientists, Engineers and other users of software provided by Computer Scientists care most about how large a problem instance a particular method is able to solve in reasonable time. In this talk, we discuss the contributions and deficiencies of current fixed parameter tractability methods within this context.

---

* This research has been supported in part by the Natural Sciences and Engineering Research Council of Canada.

H.L. Bodlaender and M.A. Langston (Eds.): IWPEC 2006, LNCS 4169, p. 278, 2006.

Author Index

Lecture Notes in Computer Science

For information about Vols. 1–4071

please contact your bookseller or Springer